Lecture Notes in Geoinformation and Cartography

For further volumes:
http://www.springer.com/series/7418

Sisi Zlatanova · Rob Peters
Arta Dilo · Hans Scholten
Editors

Intelligent Systems for Crisis Management

Geo-information for Disaster Management (Gi4DM) 2012

Gi4DM

Springer

Editors
Sisi Zlatanova
GISt, OTB Research Institute
for the Built Environment
Delft University of Technology
Delft
The Netherlands

Rob Peters
Public Safety & Health Region
Kennemerland
Haarlem
The Netherlands

Arta Dilo
Pervasive Systems, Computer Science
University of Twente
Enschede
The Netherlands

Hans Scholten
Pervasive Systems, Computer Science
University of Twente
Enschede
The Netherlands

ISSN 1863-2246 ISSN 1863-2351 (electronic)
ISBN 978-3-642-33217-3 ISBN 978-3-642-33218-0 (eBook)
DOI 10.1007/978-3-642-33218-0
Springer Heidelberg New York Dordrecht London

Library of Congress Control Number: 2012954368

Printed on acid-free paper

Springer is part of Springer Science+Business Media (www.springer.com)

Foreword

Dear Reader,

It is my pleasure to present to you the results of the 8th Gi4DM Conference. To my knowledge it is the first time that scientists, R&D technicians, and emergency management officers have collaborated to produce joint books and learn from each other's expertise in this manner. I think there is great need for such interaction. With the rapidly improving digital means of our time, a number of questions arise concerning our job. We need technological support. The lack of relevant information is a reoccurring theme in almost every inspection report that I have seen. The effectiveness of my colleagues, both within the firebrigade and between the first responder agencies, is greatly enhanced with timely data about all aspects of the incident. We spend a considerable amount of resources on the project Net Centric Crisis Response to enable such information exchange. But there are a number of challenges to these developments. What are the consequences to our ways of working given the improved information position? What amount of information processing can we handle in what phase of the incident management? How do we enforce more interoperability between all those systems on which we have become more and more dependent? And who keeps Murphy at bay with all that complexity and technology? These questions cannot be answered on paper alone. They require iteration, inquiry, and inspiration. They require the willingness to cross the boundaries of cultures, tribes, and the comfort zones of our own four walls.

GI4DM is about maps.

I believe that Geographic information systems could be the core discipline for further learning about possibilities and limitations of technologies that support emergency management. Maps are the key to the common operational picture. Maps help to envision future risks that turn out to be closer than what both citizens and councilors see as comfortable. Maps help to see a common cause in times of need. It is for that reason that the firebrigade invests in digital maps, both in vehicles, in command containers as in the crisis rooms all over the country. It is for that reason that my fellow commanders and I have embraced a new reference architecture (VERA) for information management this very year.

I am grateful for this opportunity to foster more research collaboration. I hope it will produce long-standing relationships and open up new communication channels. I hope you find some of the content of this book intriguing. Emergency management will always be about dealing with uncertainty, but we can do our best.

Thank you for your time.

Frans Schippers
National Chief Information Officer of the Dutch Firebrigade
Director of Regional Safety Authority Kennemerland

Contents

Part II Sensors and Data Integration

Introduction

Technology and systems for risk and disaster management have advanced greatly in the last few years. Systems for early warning, command and control, and decision making have been successfully implemented in many countries and regions all over the world. However, many aspects related to efficient collection and integration of geoinformation, applied semantics, and situation awareness are still open. Could sensor technology help to find victims? How can new paradigms such as crowd sourcing and granular computing help in intelligent and fast response? What is the role of iPhone/iPad apps in the frame of large information systems for disaster management? What kind of vocabularies and semantics are required to enrich the spatial information, and how can they be translated on the fly? What other tools are needed in command and control systems for efficient coordination in an emergency response scene? Which icons are most appropriate in case of cross border emergency situations? To advance the systems and make them intelligent, an extensive collaboration is required between emergency responders, disaster managers, system designers, and researchers.

Noting the great importance of geoinformation for disaster and risk management, a group of researchers, professionals, and vendors has been getting together for eight years at the Gi4DM Conference. Gi4DM is coordinated by the ad hoc committee on risk and disaster management by the Joint Board of the Geo-information Societies (JB of GIS). Seven editions of this series have taken place in Delft, The Netherlands, (March 2005), in Goa, India, (September 2006), Toronto, Canada, (May 2007), Harbin, China (August 2008), Prague, Czech Republic (January 2009), Toronto, Italy (February 2010), and Antalya, Turkey (May 2011).

Gi4DM 2012 is specifically important because it was initiated by the Public Safety Regions of The Netherlands. For the first time, the emergency responders contacted researchers proposing to share several years of experience, discuss problems, and brainstorm about possible solutions. The Safety Regions took a significant role in the preparation and organization of the conference, which took place at two locations: the University of Twente Holland Casino, Enschede for the scientific sessions, and the former Twente Military airport for the demonstrations and field tests. The objective was twofold: to give researchers possibilities to talk with officers and validate their direction of research and to give technicians the chance to work with their prototype systems in real field tests. Vendors were also largely encouraged to participate and show products that can pass interoperability tests. The emergency response officers were able to see tests and demos displaying the state-of-the art technologies. One of the most remarkable parallel events was the European urban search and rescue exercise held at Troned Safety Campus,[1] which provided a test site for an earthquake scenario at the former Twente airport.

The focus of the Gi4DM 2012 Conference has emerged from an intensive discussion between researchers and practitioners. A number of important conference topics were identified such as geospatial data modeling and visualization, sensors and processing of sensor data, requirements and analysis of systems for emergency response, and disaster management, as well as best practice examples from running systems. The topics covered by the Gi4DM 2012 papers were: Cross-border and cross-sector semantics, Semantics and situational awareness, Agent-based systems, Multiplatform and multisensor data collection and processing, Crowd sourcing and volunteered geographic information, Design requirements and design processes for information systems, Simulation, decision enhancement systems, and Evacuation and navigation systems.

This volume is inspired by the topics presented above and addresses the main goals of the Gi4DM 2012 Conference toward sharing research achievements and practice experiences. The volume consists of 29 peer-reviewed chapters, of which 20 are scientific papers and 9 are short best practice papers. These were selected on the basis of double-blind reviews from among the 65 papers submitted to the Gi4DM Conference. A new review approach was applied for all received papers. Each paper was reviewed by two scientific reviewers and one practitioner. The purpose of this approach was to evaluate not only the scientific contribution of the papers but also the practical relevance of the research. The authors of the papers were encouraged to revise, extend, and adapt their papers to fit the goal of this volume.

The selected papers are organized in four parts: *Data Modeling and Visualization*, *Sensors and Data Integration*, *System Requirements and Analysis*, and *Best Practices*.

The first part consists of seven papers which illustrate how simulation and navigation techniques can advance and facilitate decision making during emergencies. The papers in this group clearly state the importance of semantics in emergency response. Bakillah et al. present a generic data model for agent-based evacuation simulation, which takes into consideration relevant social parameters such as type and size of group under risk, socioeconomic status, previous experience with disasters, and so on. Wang and Zlatanova propose an extension of a routing algorithm to deal with moving obstacles. The approach aims at supporting emergency responders' navigation within dynamically changing environments. A two-level routing strategy within complex indoor environments avoiding static

[1] http://www.vrtwente.nl/troned/

obstacles is proposed by Liu and Zlatanova. This strategy uses indoor semantic models to reduce computational complexity and still provide several types of passable routes. Al-Salman et al. discuss an approach for decision making that allows disaster managers to specify qualitative queries based on semantic descriptions. The last three papers concentrate on simulation and visualization of hazard events. Zelle et al. discuss a new approach for smoke plume modeling. Ishida et al. present a simulation system for evacuation of people as tsunami precaution, considering information about the ground, underground, and within buildings. The goal is also to provide data about possible refuge buildings and evacuation guidance methods and systems. The last paper in this part presents an advanced method for flood simulation and visualization making use of high-resolution point cloud data. Despite the large volumes of data to be processed the system allows real-time interaction.

The second part *Sensors and Data Integration* consists of eight papers which demonstrate the integration of sensor and GIS data. The papers clearly demonstrate that GIS and SDI have become important instruments for data integration and decision making in risk and disaster management. Hassanzadeh and Nedovic-Budic discuss the use of crowdsourced to outline damaged areas after an earthquake. The study concludes that such data are better suited for the determination of hot and cold spot areas rather than to provide exact locations. Kerle argues that the traditional charter-type mapping needs to move away from one-directional mapping, and proposes collaborative mapping as an alternative for a better understanding of maps. Two papers present mathematical models. Youn et al. suggest that large amounts of sensor information require good sensor models and present their mathematical model. Khamespanah et al. illustrate how the Dempster-Shafer theory can be used to integrate and resolve the conflict among different experts' viewpoints to arrive at a decision regarding the measure of seismic vulnerability. The remaining papers in this part concentrate on different approaches for information extraction by integrating sensor information and GIS data. Borovelli et al. discuss the provision of satellite precipitation data via GeoServices; Vatseva et al. discuss the added value of GIS for seismic data assessment, which also allows for real-time support; Fernández et al. elaborate on the use of GIS and SDI for landslides and propose a methodology for determining the susceptibility in different return periods from landslide inventories. Finally, Dragos proposes a methodology to create a GIS-integrated complex tool, which allows flexible integration of data and procedures to assess the vulnerability of a transportation network.

The third part *System Requirements and Analysis* contains five papers, which reflect advances in command and control systems. The papers highlight that geo-information has become an important component of command and control systems. Steenbruggen et al. present an extended investigation of the net centric working, which is considered as one of the most promising approaches for information sharing in emergency situations. Mäkelä and Virrantaus concentrate on the evaluation of collaboration. The authors argue that verbal reports created after SAR exercises are not sufficient to estimate the level of collaboration and propose a formal model. Řezník et al. provide an extensive overview of command and control systems and present a system architecture for emergency support. Nushi and van Loenen concentrate on accessibility of SDI for disaster management and propose a method for evaluating SDI. In the same direction but for a specific hazard is the contribution of Abidin et al.

The fourth part which is the last, *Best Practices*, is a selection of short papers presenting systems and approaches that have been proved in real-world situations or are in development. Nico van de Weghe et al. present a crowd sourcing platform, which makes use of Bluetooth. Kinugasa et al. elaborate on evacuation system tested in Kyoto. Broer elaborates on the new Dutch program Virtual Police Korps and its attempt to provide real-time intelligence to the first responders with linked open data and a secure 'Appstore'. Van Persie et al. present a system, which allows integration of UAV videos in firefighting. De Bruin and Wijngaards present a system for secure sharing of information according to which participants obtain only those data relevant to their role in the emergency management process and their context. Genc et al present an approach for information flow management, which intends to overcome security problems. Van Aalst et al. discuss an option to link all information with generic keys using base registries, an approach used in the Dutch emergency response sector. Peters et al. elaborates on use of icons as semantic vehicle to deal with vocabulary challenges. Panneman et al. discuss network information management as an alternative to other approaches for adoption of innovative technologies.

The papers in this volume identify important tendencies of using geoinformation for emergency response. Clearly, geoinformation or all the information with location or spatial extent is of critical importance. The research and developments are now focused on making the system intelligent and interoperable. The visualization approaches have been significantly advanced. The geodata are not only visualized but also used for simulations and predictions. The importance of semantics and semantically rich models is appreciated not only by researchers but also by practitioners. Sensors information is becoming increasingly important since much value is given to crowd sourcing. This fact was also confirmed by the interest in the special session dedicated to crowdsourcing.

The editors of this volume acknowledge all members of the scientific committee for their time, careful review, and valuable comments: Andrea Ajmar (Italy), Orhan Altan (Turkey), Costas Armenakis (Canada), Robert Backhaus (Germany), Vera Banki (The Netherlands), Jakob Beetz (The Netherlands), Piero Boccardo (Italy), Alexander Bouwman (The Netherlands), Martin Breunig (Germany), Jeroen Broekhuijsen (The Netherlands), Eliseo Clementini (Italy), Joep Crompvoets (Belgium), Tom De Groeve (Italy), Marian de Vries (The Netherlands), Ioannis Delikostidis (Germany), Eduardo Diaz (The Netherlands), Arta Dilo (The Netherlands), Gilles Falquet (Switzerland), Yoshikazu Fukushima (Japan), Marcus Goetz (Germany), Ben Gorte (The Netherlands), Ihab Hidjazi (Palestine), Liu Hua (China), Bo Huang (China), Umit Isikdag (Turkey), Alik Ismail-Zadeh (USA), Ivana Ivanova (The Netherlands), Zhizhong Kang (China), Milan Konecný (Czech Republic), Petr Kubicek (Czech Republic), Monika Kuffer (The Netherlands), Werner Kuhn (Germany), Zentai Laszlo (Hungary), Hugo Ledoux (The

Netherlands), Ki-Joune Li (South Korea), Martijn Meijers (The Netherlands), Nirvana Meratnia (The Netherlands), Darka Mioc (Denmark), Aitor Moreno (Spain), Claus Nagel (Germany), Lei Niu (China), Masafumi Nakagawa (Japan), Rob Peters (The Netherlands), Hans Schevers (The Netherlands), Henk Scholten (The Netherlands), George Sithole (South Africa), Tuong Thuy Vu (Malaysia), Fabio Tonolo (Italy), John Trinder (Australia), Herman van Hijum (The Netherlands), Berend-Jan van der Zwaag (The Netherlands), Niels van Mannen (The Netherlands), Peter van Oosterom (The Netherlands), Luud Verheijen (The Netherlands), John Zhao (The Netherlands), Qing Zhu (China), and Sisi Zlatanova (The Netherlands).

The editors are especially thankful to our practitioners, who took the great challenge to advise on the value of scientific research for the emergency response sector Henk Djurrema (The Netherlands), Dave Fortune (UK), Arthur Haasbroek (The Netherlands), Robert Kieboom (The Netherlands), Marcos Sacristán Cepeda (Spain), Josien Oosterhoff (The Netherlands), Frederik Schütte (Germany), Cor Snijders (The Netherlands), Gerke Spaling (The Netherlands), Simon Stenneberg (The Netherlands), Jan Willem van Aalst (The Netherlands), Bart van Leeuwen (The Netherlands), Benedikt Weber (Germany), Frank Wilson (UK), and Guus Zijlstra (The Netherlands). We are also grateful to Frank Wilson and Dave Fortune for the thorough proofreading of most of the papers.

The editors express their gratitude to all the contributors who made this volume possible. Many thanks go to all the supporting organizations EuroSDR, ICA, IAG, ITHACA, IHO, ISPRS, GRSS, FIG, IUGG, UNOOSA, ISCGM, GSDI and JBGIS, companies Bentely Systems, ESRI, Geodana, and Cyclomedia. We are especially thankful to our sponsors IBridge, StudioVeiligheid, Disaster Project, and Sen Safety. The editors are grateful to our media sponsor Geoconnexion, Informed Infrastructure and Directions Magazine, Veiligheidsregio Twente, and Veiligheidsregio Kennemerland who endorsed this conference.

Part I
Data Modelling and Visualisation

Multi-agent Evacuation Simulation Data Model with Social Considerations for Disaster Management Context

Mohamed Bakillah, J. Andrés Domínguez, Alexander Zipf, Steve H. L. Liang and M. A. Mostafavi

Abstract Large scale disasters often create the need for evacuating affected regions to save lives. Disaster management authorities need evacuation simulation tools to assess the efficiency of various evacuation scenarios and the impact of a variety of environmental and social factors on the evacuation process. Therefore, sound simulation models should include the relevant factors influencing the evacuation process and allow for the representation of different levels of detail, in order to support large scale evacuation simulation while also offering the option of considering factors operating at a finer level of detail, such as at the single individual level. In particular, the impact of social factors, such as interaction between agents, should be integrated into the simulation model to reflect the reality of evacuation processes. In this paper, we present a generic data model for agent-based evacuation simulation that includes the relevant social parameters identified in the emergency literature. The model is composed of three sub-models that describe the agents, their context and behaviour, the dynamic environment in which the agents evolve and the parameters of the evacuation scenario. The objective of this model is to improve the simulation so that it can be better represent reality.

M. Bakillah (✉) · A. Zipf
Rupprecht-Karls-Universität Institute for GI-Science, Heidelberg, Germany
e-mail: mohamed.bakillah@geog.uni-heidelberg.de

J. Andrés Domínguez
Department of Sociology and Social Work, University of Huelva, Huelva, Spain

Steve H. L. Liang
Department of Geomatics Engineering, University of Calgary, Canada

M. A. Mostafavi
Geomatics Research Center, Laval University, Canada

S. Zlatanova et al. (eds.), *Intelligent Systems for Crisis Management*, Lecture Notes in Geoinformation and Cartography, DOI: 10.1007/978-3-642-33218-0_1,

Keywords Data model · Disaster management · Evacuation simulation · Multi-agent systems · Social behaviour

1 Introduction

The management of natural or human disasters, such as the Hurricane Katrina and the tsunamis in East-Asia, requires urgent response and the coordination of human and material resources. Disaster management includes the actions that are taken in prevention of disasters, as well as those that are taken in response to an occurring disaster, and the recovery effort [1]. The efficiency of the disaster management process, especially the planning for response to disaster, can be improved notably by using evacuation simulations [2]. In particular, the GRIPS Project (GIS-Based Risk Analysis Information and Planning System)[1] has developed the multi-agent transport simulation toolkit MATSim[2] and used it to simulate regional evacuation [3]. In the current implementation of this evacuation simulation model, and similarly to other evacuation simulations models [4], less focus has been given to modelling the agents' behaviour and interaction, including interaction with authorities and agents with collective responsibility, their social characteristics and context and their reaction to changes in the environment. For example, authorities issuing a warning can affect differently the population with different beliefs, or crowd movement can have different impact on individuals [5]. A simulation can only be useful if it realistically represents the relevant social and environmental factors that impact the evacuation time into a single simulation model.

To support the development of such comprehensive simulation model, in this paper, we propose a generic data model for agent-based evacuation simulation. We aim at providing a data model that supports the representation of different levels of geometric or social detail, in order to perform simulations at different levels of detail. We also explore what key social factors should be taken into account and how they can be integrated into the simulation data model. The proposed generic data model describes three main parts, which are (1) the dynamic environment in which agents evolve; (2) the description of agents, their context, social characteristics and behaviour, as well as characteristics that are described in the disaster management literature as having an impact on decisions of agents regarding evacuation; (3) the parameters of the evacuation scenario.

The paper is organized as follows: Section 2 provides related work on multi-agents evacuation simulation. In Sect. 3, we give requirements for a generic data model on evacuation simulation, and we identify and discuss the relevant social parameters. In Sect. 4, the generic data model is presented and discussion is provided. Conclusions and future research avenues are given in Sect. 5.

---

[1] http://www.geog.uni-heidelberg.de/forschung/gis_grips.html

[2] http://www.matsim.org/

2 Multi-agent Evacuation Simulation

During recent years, multi-agent simulation has been improving and it can now support transportation simulation dealing with several million agents evolving in large regions [3]. As a result, multi-agent systems have increasingly been used in evacuation simulations (e.g., [3, 6, 7]). Evacuation simulation models based, for example, on cellular automata [8] or on flow dynamics [9, 10], are limited in their capacity to represent characteristics of individuals and the complexity of their behaviour. In contrast, multi-agent simulation allows incorporating more details and behaviour of agents to accurately represent the parameters that influence the time it takes for a population to evacuate a given region. For example, some approaches (e.g., [11, 12] model the travel demand (i.e., the number of people who will evacuate and when they will depart [2]) with a response curve that establishes the percentage of departures in each time interval, regardless of individuals' specific circumstances. In fact, several social and personal parameters influence one's decision to evacuate or not, for example, the perceived risk, socio-demographic characteristics, the distance to the disaster, whether an issuance order has been given or not, etc. [2]; or others more socio-structural and latent characteristics (normative structure, for instance) [13]. It is necessary to take into account these factors to understand the dynamics of the evacuation process and to obtain a simulation that accurately represents reality.

Numerous approaches of agent-based evacuation simulation have been proposed, among which only some can be mentioned here. Some approaches, such as the DYNASMART model [14], take into account the fact that people who choose to evacuate may make decisions to adapt their route during their evacuation. As another example, [15] define three different types of agents: agents who stick to the route they had initially chosen, agents who choose to modify their route every time they face congestion, and agents who change route only half the time they face congestion. Another kind of agent-based approach incorporates the impact of interaction between agents. For example, Murakami et al. [16] and Pelechano and Badler [17] include in their simulation model the behaviour of "leader" agents. Shendarkar et al. [18] employed the belief-desire-intention (BDI) framework to model the behaviour of agents during evacuation. The fact that agents can have complex behaviour also means that they can react to changes in their environment. This has motivated some researchers to incorporate a dynamic environment in their evacuation simulation. For example, Lämmel et al. [3] have used the notion of events to model the changes that affect the transportation network during the evacuation. More specifically, the attributes of roads, such as the flow capacity, can be modified at arbitrary points in time. As another example, Liu et al. [19] present an approach of evacuation caused by a flood where the depth of the water, which varies over time, affects the speed of the evacuating people. While existing evacuation simulation models focus on different aspects (the evacuees' decision-making process, interactions between evacuees, dynamic aspects of the environment, etc), to the best of our knowledge, there exist no comprehensive data model for evacuation simulation that would aim at representing a large spectrum of factors.

3 Data Requirements for Evacuation Simulation

The development of a generic data model for simulation is supposed to satisfy a wide variety of scenarios. The simulation is designed to support decision-makers in planning the evacuation of people from affected areas to safe destinations; the simulation also helps to assess whether the evacuation can be completed within the available safe evacuation time.

3.1 General Data Requirements

In order to support these decision-making and assessment processes, the data model underlying the simulation should satisfy the following general requirements:

R1: The data model should support the representation of different levels of detail, in order to adapt to coarse or microscopic simulations.
R2: The data model should support dynamic simulation, i.e., the representation of changes with respect to the level of risk, practicability and conditions of the road network, agent behaviour, etc.
R3: The data model should incorporate the major characteristics of agents that are known to influence their decision, and notably their decision to evacuate or not, when to evacuate, to which safe area and which route to take.
R4: The data model should support representation of different types of transport modes (pedestrian, by car, by bus, etc) and switching from one transport mode to another by agents during the evacuation.
R5: Despite computational considerations, the data model should not be over-simplified to the point where it is no longer reliable and does not represent the reality accurately enough.
R6: The data model should be interoperable with relevant ISO standards describing geospatial features, in order to support integration of input data from various databases relevant in disaster management (topography, hydrography, meteorological, demographics, etc) and sensor data that can provide information on current environmental conditions.

In this paper, we present a basis for the development of a generic data model for simulation that will comply with these requirements. Furthermore, our aim is to identify some of the more important social parameters that have an influence on the behaviour of evacuees.

3.2 Key Social Parameters to Understand the Behaviour of Evacuees

The field of major emergencies tends to be more and more interdisciplinary, and the importance of social elements is more and more recognized. The interest in social aspects in this field was already considered at its military origins [20]. In existing literature on emergencies, it is possible to distinguish two major types of social actors: *actors with collective responsibility* (referred to as ACR from now on) and *citizens*. ACR are governments and administrations that have responsibilities and competencies in the management of emergency situations. Despite the lack of specific literature on ACR in emergency contexts, it is generally recognized as one of the most important topics to deal with, because of the key role of ACR at decisional and operational levels [21].

With respect to citizens, studies on human behavior in emergencies share a *group-oriented* approach since the decade of 1980 [22]. Panic, as *individual-oriented* explanation of collective behavior, has been rejected as a probable behavior: people do not leave buildings or threaten areas immediately after knowing the event or risk, but they look for other relatives or partners before leaving [13]. Other features that also display analytical importance in times of emergency include normative bases of everyday life, as well as aspects such as group identity or the type and quality of intragroup linkages [23–25].

ACR are key actors in emergency situations, but few researchers have dedicated efforts to analyze their relations, responsibilities, capabilities, skills, competencies and needs [26]. That puts forward added difficulties for selecting social parameters for our evacuation model. Relevant social parameters can be found below, summarized and briefly characterized for an introductory model:

R7: **Authority for defining *threat* or *disaster*, or for giving the order of evacuation**. It can be observed that there are disastrous events that are not defined as a disaster. For instance, any recent heat wave during summer kills hundreds, and never a heat wave has been identified by the authorities as a "disaster". Also, a context of minor amount of human damage and greater magnitude of material damages can be defined as a disaster, or the order of evacuation of affected areas can be broadcasted by a government (typically, a typhoon, tornado, earthquake …). The definition of an event as *disaster* (real or potential) is a social construction, and it depends on the decision of a social actor with responsibility and political competencies in the matter; it can also be subject to external pressures (economic or political interests, for instance) [27]. In regard to our modeling exercise, the initial question of which risks or consequences of extreme events will activate the evacuation process must be considered.
R8: **Coordination between ARCs**. Responsibility and competences in emergency situations is often spread on several administrative levels. Different interpretations of risk, its effects, regulations, or interpretations about the speed with which the extreme event can occur are elements that point out the importance of coordination between authorities, as a necessary parameter in the model [28]. Notably, the presence or absence of coordination influences whether evacuation orders given in different regions are communicated simultaneously or sequentially.
R9: **Coordination of ARCs with operational and emerging organizations**. Depending on the magnitude of the event, international operational organizations (e.g., Red Cross), or at national level (e.g., Salvation Army, USA), or others locals,

could play a key role in the emergency and/or evacuation. In addition, literature agree the importance of "emergent organizations" [26] in these events. Emergent organizations are often formed by informal troops of volunteers that offer an interesting potential for evacuation. They appear spontaneously when the disaster occurred. The chain of communication and authority (ARCs—organizations) has to be clear and certain. These features will play an essential role in the evacuation of the population against the threat, so they have to be considered in our model.
R10: **Information management and mass media**. Mass media are the main reference for population to be informed, whether before, during or after the event [29]. The information management and diffusion made by authorities influences how citizens evacuate areas at risk.

With respect to citizens, our literature review suggests that the following elements should be considered in the evacuation model:

R11: **Type and size of the group**. The first reaction of people in emergencies is to look for the other members of their group, which can affect the evacuation process [24, 13]. For instance, persons in a massive concert search for the ones they came with when an emergency or a threat is perceived. The same, in a job-context, for job mates of the same department or section … Considering an extreme-event dimension for our model, two aspects would have to be taken into account: the type of ties between members of the group (primary or secondary) and the group size. Primary ties (family and friends, mainly) use to be stronger that secondary ties; they support an important part of the social identity of people. Secondary ties (ties between colleagues, for instance) can become important if the extreme event (or threat) is perceived whilst people are in the workplace. Numerous studies have shown that evacuation difficulties arise due to workers looking for each other before leaving [30]. Thus, the evacuation speed could be considered as inversely proportional to the size of group to be evacuated when the group is incomplete (that is, when some members are still missing). The worst case for evacuation would be incomplete primary groups.
R12: **Previous experiences on similar emergencies**. The uncertainty and threats inherent to emergency situations cause people to undergo high levels of stress. However, studies demonstrate that stress level is lower when people have experienced and survived to similar situations. Experience in other emergency situations encourages a general attitude of calm, rationality and organization in the process of evacuation [30, 31].
R13: **Socio-economic status**. It is difficult to find a sociological study that ignores the socio-economic status, because it is always assumed as one of the most explanatory characteristics of human behavior. In the field of disasters, studies have offered mixed results. To avoid this disagreement, we will here consider the "socio-economic status" parameter not in terms of degree (higher or lower status), but in terms of homogeneity or heterogeneity. Studies demonstrate that emergencies management in homogeneous communities is easier than in those where social inequalities are greater [29].

R14: **Other "individual" characteristics**. Very few human features are beyond the reach of social conditioning. However, for practical purposes, we consider here the age and physical ability as individual characteristics. We leave gender out of consideration, because the theoretical and empirical discussion does not offer clear conclusions. There seems to be a medical agreement in which the fullness of human capabilities is between 20 and 40 years of age. Therefore, it could be considered the optimal age to overcome an emergency situation and to behave properly in a hypothetical evacuation. The presence of dependent children in the group to evacuate has to be considered too, as that would make the process slower and more difficult. The existence of disabled people in the population would have a similar effect on the evacuation [32].

R15: **Crowd *supra-force***. Human density in evacuation processes can cause saturation at possible exits. It is also an important cause of increasing degree of stress and anxiety; and sometimes, if it is very high, this supra-force pushes people to non-desired directions. The crowd movements sometimes overwhelm persons and groups [30]. Incorporating the behavior of crowds in the simulation for evacuation is important, as it may suddenly change the evacuation pattern. Various works have studied the behavior of crowds, and existing research is reported in Challenger et al. [33]. Most models focus on the crowd as an entity. However, the behavior of the crowd depends on the behavior of individuals that compose it. Consequently, to capture the crowd's behavior, we need to relate a group's behavior to behavior at the individual level. This relationship must be integrated in both directions: the impact of individual's behavior on the crowd (bottom-up), and the impact of the crowd on the individuals (top-down). According to crowd models, a collective action (e.g., movement of a crowd) can only occur if the individuals in the group share the perception of a common social identity [34]. Therefore, possible social identities in a case of evacuation have to be defined, as well as the conditions for a group to start behaving as a crowd. Then, possible behaviors of the crowd have to be linked to aggregation of the behavior of the crowd's members. Finally, the simulation model should integrate the effect of the crowd on individuals (such as influencing the direction of evacuation).

4 Toward a Standard Generic Data Model with Social Parameters

The proposed model captures the necessary features to support an evacuation simulation representative of reality and incorporating social factors. It contains information on (1) the environment in which the agents evolve, and its dynamicity; (2) the agents, their generic structure, social characteristics, social contexts and actions, which model their behaviour; (3) the parameters of the evacuation scenario, which decision-makers can modify to assess the impact on the efficiency of the evacuation. Throughout the description of the model, we refer to the requirements of Sect. 3 to indicate which requirements are fulfilled by a given aspect of the model.

The environment model (Fig. 1) is designed around the *spatial entity* class, which inherits the properties of the *spatial feature* class of the ISO 19107 on geospatial features (note that classes from the other sub-models are reproduced to indicate relations between classes of different sub-models) (requirement R6 of Sect. 3). Spatial entities include the component of the transportation network, which is modelled with ways (streets, trails, roads, etc.) and nodes (intersections between ways). In order to support multiple levels of geometric details (R1), we have used the multi-representation notion [35], where a spatial entity can have multiple geometries, depending on the scale. For example, ways can be represented as GM: curve at a coarse level or with GM: surface at a finer level of detail (necessary for example, when considering pedestrians). Other spatial entities include obstacles (e.g., barriers, walls, waterways, etc.), and it should be noted that the obstacle class is associated with temporal validity period, which means that obstacles can be added or removed during the simulation (R2). The *Spatial entity* class also includes the *region* class, which can be associated with a risk assessment. The risk assessment is performed by actors with collective responsibility (ACR) agent, most likely a local or regional authority (R7). The level of risk is also varying spatiotemporally (meaning the risk assessment can be revised by the ACR agent) (R2). The *region* class is also associated with socio-demographic factors that are relevant due to their influence on the actions of agents (R3, R13).

All spatial entities are subject to events that can modify their existence (adding or removing a way, a node or an obstacle, modifying the capacity of a way, etc.) (R2). The events can be caused by agents (for example, ACR agent adding or removing a road blockade). The ways are also associated with modes of transport available on this way and changing transportation modes is possible at nodes (R4).

The agent model (Fig. 2) represents two main types of agents: the *population agent*, who are the people within affected areas, and the *ACR agent*, whose actions influence the population agents. ACR agents include authorities, but also other kinds of actors that have a responsibility role, such as emerging organizations. Each ACR agent is also assigned specific responsibilities that are related to their competencies, including the responsibility to issue warning, to perform risk assessment, to direct evacuating people or to issue evacuation orders (R7, R10).

Population agents are further distinguished into *evacuating agents* (those who have chosen to evacuate) and *non-evacuating agents* (those who have chosen to stay). Agents may dynamically change roles during the simulation. To support finer or coarser levels of detail, agents may be aggregated into *groups of agents* with similar behaviour (e.g., groups of agents going in the same direction) (R11, R15). This is necessary to efficiently support multi-modal simulation, i.e., a simulation in which evacuating agents can dynamically switch from one transport mode to another: for example, it is useful to consider individual agents when they are driving by car, but when 50 agents are on the bus, it is more efficient to consider them as a group of agents. Groups of agents can be formed based on various criteria, including relationships between the member agents. The class

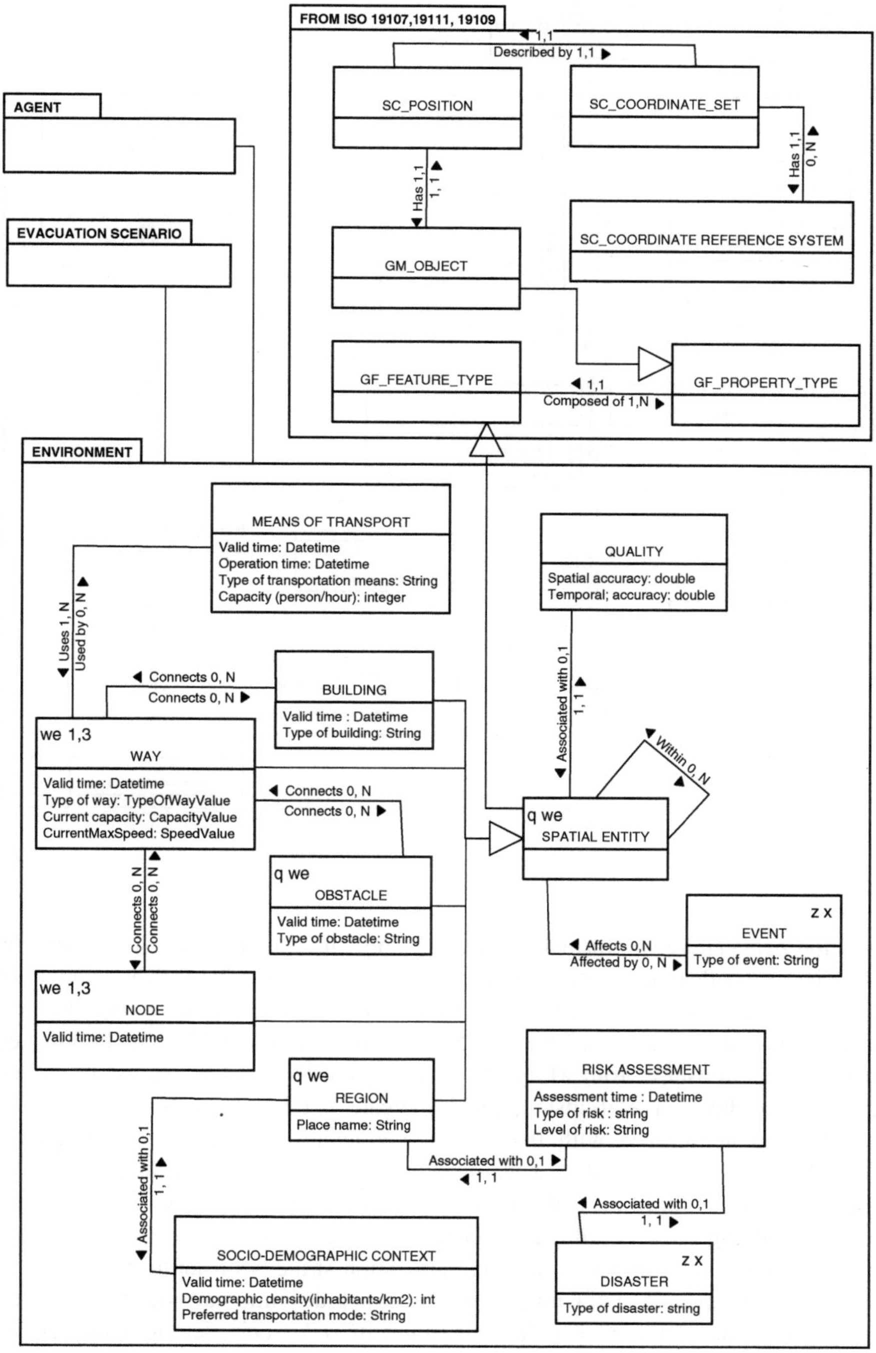

Fig. 1 Environment model

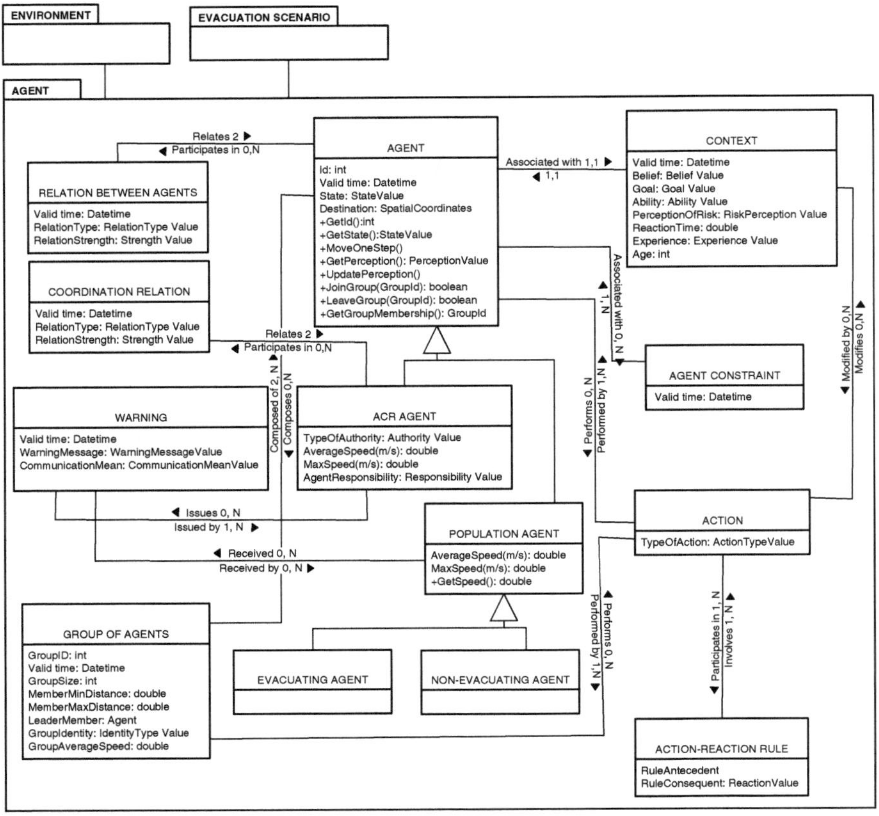

Fig. 2 Agent model

relation between agents formalizes this relation, and specifies the type of relation (family ties, colleagues, friends, etc) as well as its strength (R11). The strength of the relation can be used to determine whether the relation is strong enough to form a group. Groups are also assigned a group identity attribute, which can be inferred from the relationships from which the group was derived. ACR agents are related through coordination relationship, which indicate whether the actions of one ACR agent are synchronized with that of another agent or performed in sequence (R8, R9). For example, when a warning is issued by an official authority at the regional level, authorities at the local level can issue the same warning sequentially. Including these coordination relations into the model will allow evaluating how efficient, for example, is the communication strategy of authorities.

Each agent is associated with a *context* that includes time-varying attributes that influence its decision, such as its *perception of risk*, socio-demographic factors (R13) and knowledge of warnings and evacuation orders that were issued by ACR agents. Population agents are also associated with level of experience (R12), which can have an impact on their belief, and subsequently their reaction to events. The reaction of agents to such environmental factors, events on the transportation network or actions of other agents should be modelled with action-reaction rules that determine the actions they take (R3). For example, when an agent receives a warning to avoid a route, this agent can modify his evacuation route accordingly; but his reaction will also depend on his beliefs. Action-reaction rules also allow expressing the reaction of an individual agent following the action of a group of agents (e.g., a crowd) and conversely, the reaction of a group of agents following the action of an individual agent, such as a leader agent (R15). In the model, the association link between the classes *group of agents* and *agent* on the one hand, and the class *action* on the other hand, supports these types of reasoning. Agents can also be associated with constraints to express individual characteristics (e.g., an agent cannot drive a car, for example, which restricts the value of attribute *type of transportation mode* of the class *means of transport*) (R14).

The evacuation scenario model (Fig. 3) contains details of how the evacuation unfolds. The *evacuation strategy* class includes the attributes *available safe evacuation time* (that must be determined based on the extent of the disaster and other related factors) and *required safe evacuation time* (which must be determined by the simulation). The scenario includes the *areas to evacuate*, which can be dynamically determined during the simulation as the level of risk (in the environment model) evolves from a region to another. The *safe destination areas* are also a sub-class of the class *region* (from the environment model) and can also be modified dynamically according to the variations of the level of risk. *Safe destination areas* and *areas to evacuate* are associated with *evacuation routes*, which are composed of *ways* and *nodes* (from the environment model). Evacuation routes are associated to a cost, which is a time-varying attribute that reflects the level of practicability of the ways and nodes composing the route.

This practicability can depend on the capacity of the ways, their maximal speed and the type of way. We assume a general scenario regarding the evacuating agents' choice of a route: agents may choose an initial destination and route (not mandatory), and as this route may change during the evacuation, the class current route can be instantiated to reflect that actual route that an agent has followed. Agents may also not choose an initial route or destination, and having only a current route. In this case, the agent chooses a way (based on its attributes' values, such as max speed, capacity, etc.) at every node being encountered. In this way, the proposed data model is generic enough to encompass the different trip distribution modelling and traffic assignment modelling approaches mentioned by Pel et al. [2]. The choice of a route by an agent is a consequence of an action, which may result also from an event (from the environment model), from values of the agent's context, or can be influenced by the actions of other agents in the same group of agents (herding behaviour).

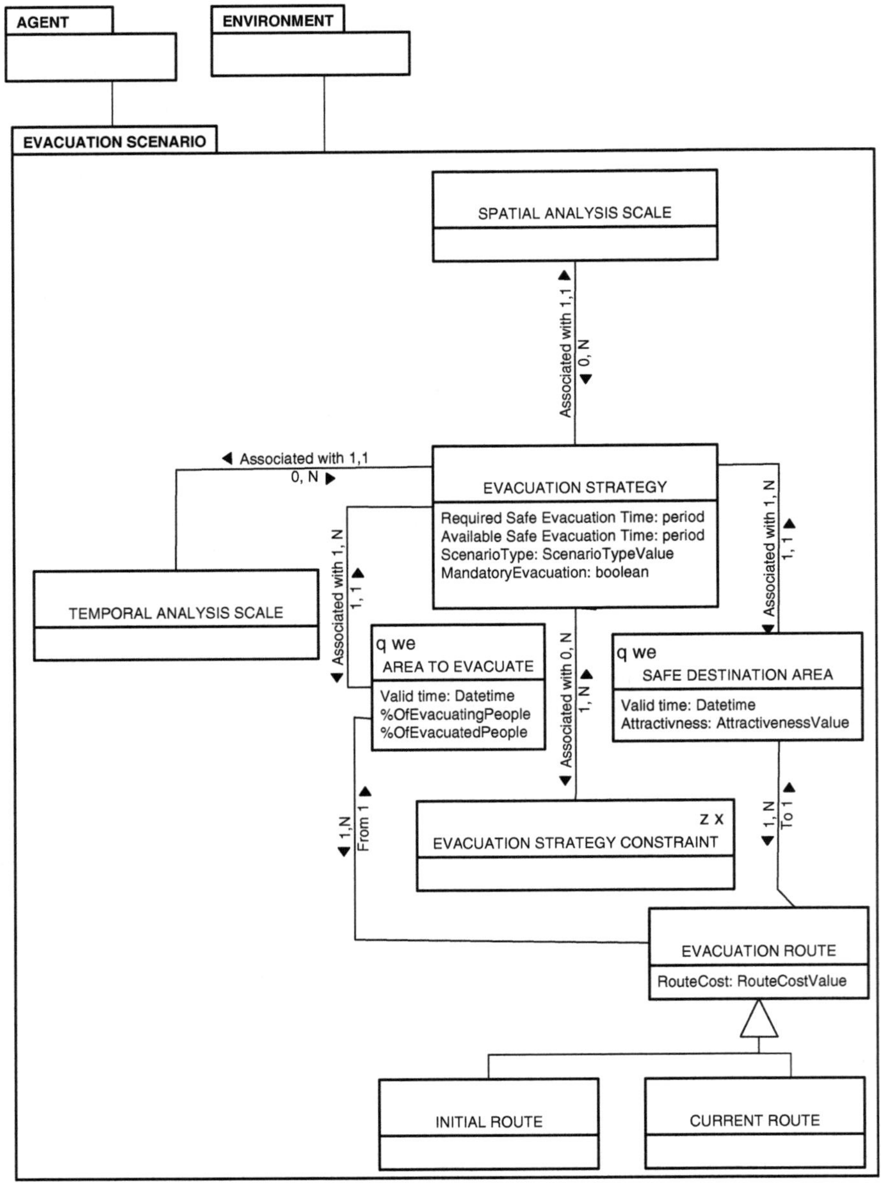

Fig. 3 Evacuation scenario mode

5 Conclusions and Future Work

We have presented a generic data model for multi-agent evacuation simulation. The data model aims at encompassing the characteristics that are known to influence the evacuation process and the required safe evacuation time, notably the social parameters identified in emergency literature. It describes the agents and their behaviour, the dynamic environment in which they evolve and make decisions regarding their evacuation plan, and the evacuation scenario. The inclusion of social parameters in the generic data model has for objective to support the enrichment of the evacuation simulation tool so that it represents more accurately the evacuation process has it happens in reality. Future work will include evaluation of the model with a case study scenario.

References

1. M. Bakillah, M.A. Mostafavi, J. Brodeur, Mapping between dynamic ontologies in support of geospatial data integration for disaster management. In *Proceedings of the Joint CIG/ISPRS Conference on Geomatics for Disaster and Risk Management*, Toronto, Ontario, 23–25 May 2007
2. A.J. Pel, C. Michiel, J. Bliemer, S.P. Hoogendoorn, A review on travel behaviour modelling in dynamic traffic simulation models for evacuations. Transportation **39**, 97–123 (2012)
3. G. Lämmel, D. Grether, K. Nagel, The representation and implementation of time-dependent inundation in large-scale microscopic evacuation simulations. Transp. Res. Part C **18**, 84–98 (2010)
4. H. Fu, Dissertation. Development of dynamic travel demand models for hurricane evacuation, Louisiana State University, 2004
5. M.-P. Kwan, D.M. Ransberger, LiDAR assisted emergency response: detection of transport network obstructions caused by major disasters. Comput. Environ. Urban Syst. **34**(3), 179–188 (2010)
6. X. Chen, F.B. Zhan, Agent-based modeling and simulation of urban evacuation: relative effectiveness of simultaneous and staged evacuation strategies. Washington, DC (2004)
7. X. Pan, C. Han, K. Dauber, K. Law, A multi-agent based framework for the simulation of human and social behavior during evacuations. Artif. Intell. Soc. **22**(2), 113–132 (2007)
8. H. Klüpfel, T. Meyer-König, A. Keßel, M. Schreckenberg, Simulating evacuation processes and comparison to empirical results, in *Traffic and Granular Flow'01*, ed. by M. Fukui, et al. (Springer, Berlin, 2003), pp. 449–454
9. M. Jafari, I. Bakhadyrov, A. Maher, Technological advances in evacuation planning and emergency management: current state of the art. Final research reports EVAC-RU4474, Center for Advanced Infrastructure and Transportation (CAIT), Rutgers University, March 2003
10. E. Kuligowski, *Review of 28 Egress Models. Technical report* (National Institute of Standards and Technology (NIST), Gaithersburg, 2004)
11. M.K. Lindell, EMBLEM2: an empirically-based large scale evacuation time estimate model. Transp. Res. A **42**, 14–154 (2008)
12. C. Xie, D.Y. Lin, S.T. Waller, A dynamic evacuation network optimization problem with lane reversal and crossing elimination strategies. Transp. Res. E **46**, 295–316 (2010)
13. N.R. Johnson, Panic and the breakdown of social order: popular myth, social theory. Empirical Evid. Sociol. Focus. **20**(3), 171 (1987)
14. P. Murray-Tuite, Perspectives for network management in response to unplanned disruptions. J. Urban Plan. Dev. **133**(1), 9–17 (2007)
15. B. Zhang, W. Kin, S.V. Ukkusuri, Agent-based Modeling for household level hurricane evacuation. In *Proceedings of the 2009 Winter Simulation Conference*, ed. by M.D. Rossetti, R.R. Hill, B. Johansson, A. Dunkin R.G. Ingalls (Institute of Electrical and Electronics Engineers, Inc Piscataway, 2009), pp. 2778–2784

16. Y. Murakami, K. Minami, T. Kawasoe, T. Ishida, Multi-agent simulation for crisis management. In *Proceedings of the IEEE on Knowledge Media Networking Workshop* (Washington DC, 2002), pp. 135–139
17. N. Pelechano, N. Badler, Modeling crowd and trained leader behavior during building evacuation. IEEE Comput. Graph. Appl. **26**(6), 80–86 (2006)
18. A. Shendarkar, K. Vasudevan, S. Lee, Y. Son, Crowd simulation for emergency response using BDI agent-based on virtual reality. In *Proceedings of the 2006 Winter Simulation Conference*, Monterey, 3–6 Dec 2006
19. Y. Liu, M. Hatayama, N. Okada, Development of an adaptive evacuation route algorithm under flood disaster. Ann. Disaster Prev. Res. Inst. **49**, 189–195, Kyoto University (2006)
20. E.L. Quarantelli, Disaster studies: an analysis of the social historical factors affecting the development of research in the area. Int. J. Mass Emerg. Disasters. **5**(3), 285–310 (1988)
21. C. Lalonde, Crisis management and organizational development: towards the conception of a learning model in crisis. Organ. Dev. J. **25**(1), 17 (2007)
22. B.E. Aguirre, E.L. Quarantelli, Methodological, ideological and conceptual-theoretical criticisms of the field of collective behavior: a critical evaluation and implications for future study. Sociol. Focus. **16**(3), 195 (1983)
23. R. Turner, L. Killian, Collective Behaviour, 3rd edn. (Prentice Hall, Englewood Cliffs, 1987)
24. J. Drury, C. Cocking, S. Reicher, Everyone for themselves? a comparative study of crowd solidarity among emergency survivors. Br. J. Soc. Psychol. **48**, 487 (2009)
25. M.J. Hornsey, Social identity theory and self-categorization theory: historical review. Soc. Pers. Psychol. Compass. **2**(1), 204–222 (2008)
26. T.E. Drabek, D.A. McEntire, Emergent phenomena and the sociology of disaster: lessons, trends and opportunities from the research literature. Disaster Prev. Manag. **12**(2), 97–112 (2003)
27. K.J. Tierney, From the margins to the mainstream? Disaster research at the Crossroad. Annu Rev Sociol. **33** 503–525 (2007)
28. S.K. Schneider, Governmental response to disasters: the conflict between bureaucratic procedures and emergency norms. Public Adm. Rev. **52**(2), 135 (1992)
29. E.L. Quarantelli, Basic themes derived from survey findings on human behavior in the Mexico City earthquake. Int. Sociol. **11**(4), 481–499 (1996)
30. B.E. Aguirre, S. El-Tawil, E. Best, K.B. Gill, V. Fedorov, Contributions of social science to agent-based models of building evacuation. Contemp. Soc. Sci. **6**(3), 415–432 (2011)
31. G. Proulx, Understanding human behaviour in stressful situations (2003) http://www.nrc-cnrc.gc.ca/obj/irc/doc/pubs/nrcc45394/nrcc45394.pdf
32. H.C.M. Vorst, Evacuation models and disaster psychology, in *1st International Conference on Evacuation Modeling and Management* (ICEM '09), Delft (2010), pp. 15–21
33. R. Challenger, C.W. Clegg, M.A. Robinson, *Understanding Crowd Behaviours*. UK Cabinet office. Crown (2009)
34. S. Reicher, Blackwell handbook of social psychology: group processes, Chap. In *The Psychology of Crowd Dynamics,* ed. by M.A. Hogg and R.S. Tindale (Blackwell Publishers, Oxford, 2001), pp. 182–208
35. E. Bernier, Y. Bédard, F. Hubert, UMapIT: an on-demand web mapping application based on a multiple representation database. In *Proceedings of the 8th ICA Workshop on Generalization and Multiple Representation*, A Coruna, Spain, 8–9 July 2005

An A*-Based Search Approach for Navigation Among Moving Obstacles

Zhiyong Wang and Sisi Zlatanova

Abstract Finding an optimal route in a dynamic transportation network affected by disasters is a critical problem for emergency response. Although many routing algorithms have been developed, and some of them show the ability to guide first responders around the static damaged infrastructure, there are few efforts devoted to the efficient routes avoiding moving obstacles. Emergencies caused by natural or man-made disasters can result in both static and moving obstacles in a transportation network, which poses a set of serious challenges for researchers in the navigation field. In this paper, we study the shortest-path problem for one moving object to one destination in a dynamic road network populated with many moving obstacles. Existing approaches, which are developed for stationary networks, are incapable of managing complex circumstances where the status of the road network changes over time. We propose a model to represent the dynamic network and an adapted A* algorithm for shortest path computations in the context of moving obstacles. Moreover, this paper presents a web-based application for route planning. It integrates an agent-based simulation tool for both analysis of the dynamic road network and simulation of first responders' movements, and web technologies for enabling the response community to easily and quickly share their emergency plans and to work collaboratively. We provide an experimental comparison of performance with the standard A* algorithm under different circumstances to illustrate the effectiveness of our approach.

Keywords A* algorithm · Navigation · Moving obstacles · Emergencies

Z. Wang (✉) · S. Zlatanova
Delft University of Technology, Jaffalaan 9, 2628 BX Delft, The Netherlands
e-mail: Z.Wang-1@tudelft.nl

S. Zlatanova et al. (eds.), *Intelligent Systems for Crisis Management*, Lecture Notes in Geoinformation and Cartography, DOI: 10.1007/978-3-642-33218-0_2,

1 Introduction

Disaster relief involves in a number of coordinated activities including searching and rescuing survivors, health and medical assistance, food and water distribution, and transferring injuries. Much of the successful and effective relief work relies on the safe and fast navigation. The complexity and dynamics of transportation network under disasters present considerable challenges to technology innovations related to Location-Based Service (LBS) and vehicle navigation. One challenging issue is that not only the position of the vehicle changes over time, but also the road network and its accessibility vary with the development of physical phenomena (flood, plume, fire, etc.) that cause disasters.

In a transportation network affected by disasters, road conditions could change drastically and many factors—such as plumes, landslides and floods—may cause one or more road segments to be unavailable during specific periods of time. For instance, in the context of a chemical plant explosion that results in many moving contaminant plumes, these moving plumes can be considered as obstacles with changing shapes and positions. Figure 1 presents an example of a moving obstacle and an edge *A_B* connected by two nodes *A* and *B*. As shown, the obstacle moves and intersects the edge *A_B* during the temporal interval $[t_2, t_3]$. During this period of time, the edge is temporarily blocked (i.e. out of service). In this case, the emergency response units should not be guided right through the edge during the time when it is affected by the toxic plume. For this purpose, rescue managers may need to know the movement of plumes and the spatio-temporal information of blocks in the road network in order to allow rescue vehicles to pass for quick response through affected areas. Therefore, how to obtain the safe relief route considering dynamic disaster-related information becomes important for emergency managers.

Route planning during disasters is a practical application that, in recent years, has attracted a lot of attention in the navigation field, but more work still needs to be undertaken. References [1, 2] investigate route planning services taking blocked areas or streets into account. Nevertheless since the status of the road network

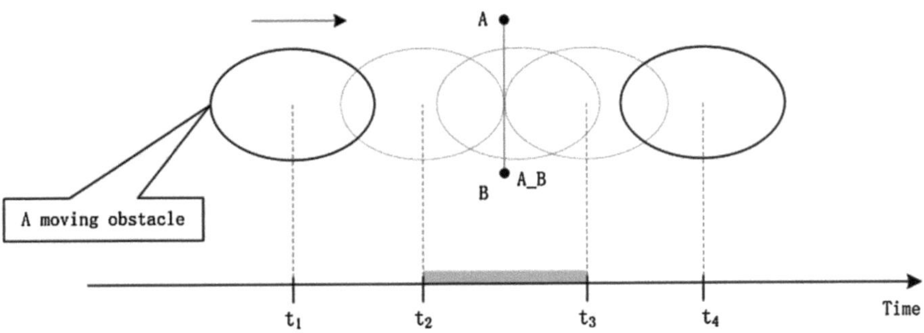

Fig. 1 Example of a block, and its corresponding temporal intervals when the edge *A_B* is blocked by a moving obstacle

varies with the disasters over time, it is necessary to take into consideration the moving obstacles in the path finding process. Similar research on navigation considering moving obstacles has been considerably investigated in the robotic field [3–5], but these related work mostly concern path planning in free space, and do not take into consideration constrains of the real road network. With the advance of disaster modeling and simulation technologies [6–9], some researchers try to incorporate the disaster simulation to improve the routing process. References [10, 11] both study the calculation of evacuation route under the flood disaster, considering vehicle types and the effect of water depth on walking speed respectively. However, they only focus on the routing in the case of flooding, taking its specific characteristics into consideration, which limits their application to other types of disasters, e.g. plumes.

In this research, we examine the problem of finding an optimal path for moving objects (first responders) to avoid moving obstacles. By considering multiplicity of moving objects and destinations, reference [12] tends to classify the navigation problem into 7 categories. With consideration of characteristics of obstacles, this classification can be extended to cases where one/many moving objects have to be navigated to one/many destinations avoiding many static/moving obstacles. In this paper, as the first step of the study of navigation among obstacles, we will focus our discussions on the shortest-path problem for one moving object to one destination in a road network populated with many moving obstacles. Since traditional techniques that are developed for static networks might not be applicable in dynamic scenarios, we propose our algorithm to solve the routing problem with moving obstacles. The proposed algorithm can compute the shortest path between a moving object and its destination in a dynamic network where road segments are blocked by moving obstacles. Because the A* algorithm [13] is generally more efficient than Dijkstras algorithm in terms of the running time and has been widely used in many applications [14–16], we develop the shortest path algorithm based on the classical A* algorithm. We extend the algorithm by incorporating the predicted information of moving obstacles. Besides, we also introduce the waiting option for the rescue vehicle to avoid moving obstacles, minimizing the total traveling time in the meantime. Although waiting has the clear disadvantage of using up precious time, it may be beneficial to allow the vehicle to remain at strategically favorable locations in some circumstances with moving obstacles, which can make their way faster than any other alternative route.

In the following sections, we also present a web-based application for navigation among moving obstacles, providing an effective way to make the shortest path calculation and application accessible to emergency managers and the public. The application can display the affected area on-line and provide the obstacle-avoiding route including waiting options. This application uses an agent simulation tool to compute the shortest path in the road network and to simulate the movement of the responder based on calculated route results. In connection with the simulator, this work also combines web-mapping technology where calculated routes and waiting information are visualized, enabling the response community to easily and quickly share their emergency plans and to work collaboratively.

2 The Model and the Shortest Path Algorithm

In a disaster response, it would be advantageous for the responder to have anticipation of the situation and include that anticipation in the path finding process to respond properly and effectively to challenges presented by the rapidly changing and dangerous environment. To provide a safer and maybe faster route to the destination, another improvement is expected to be made by introducing the possibility to wait in the route determination. For instance, in case of a chemical incident, a toxic plume moves across an edge, blocking the responder's way to the target point E temporarily, as shown in Fig. 2. When the fire truck arrives at grey point of the affected edge, it can either choose to find an alternative route or wait until the edge is available again depending on the time of arrival and changes of road conditions. In some circumstances, waiting at some specific points strategically might be the fastest and safest option. If the moving obstacle can move away from the determined route soon, the fire truck can wait for only a short period of time and continues its way to reach the destination point, saving more time than it would take to follow other alternative routes. To support dealing with changes of environment affected by disasters, we need a model to represent the dynamic information of the road network. The decision of determining whether to wait or not will also be made based on predictions, which can be done by extending the classical A* algorithm.

In following sections, we will first give some definitions and notations, and then discuss the basics of the model that is used to represent dynamics of the road network. According to this model, a modified A* algorithm is presented to incorporate the dynamic data of the road network affected by moving obstacles.

2.1 The Model

Let $G = (N, E)$ be a network consisting of a finite set of nodes N and edges between the nodes in N. For convenience, we denote the edge between two nodes

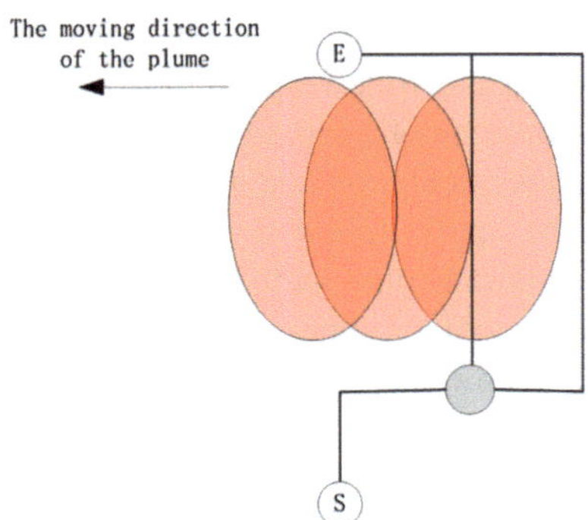

Fig. 2 An example of the waiting option for the responder

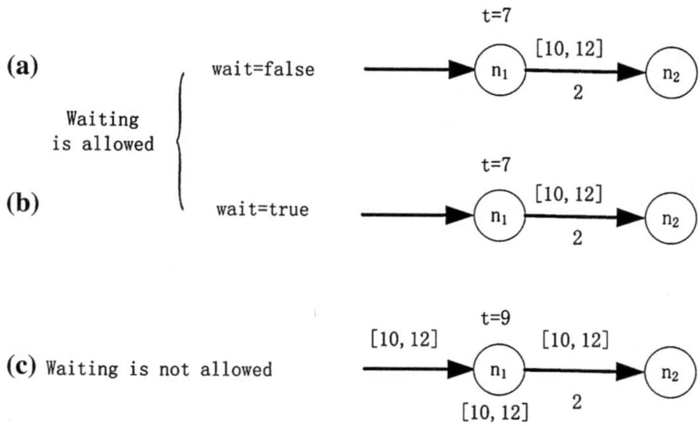

Fig. 3 **a** Not to wait if the object can safely pass through the next edge before the block starts. **b** Wait until the end of the block in the next edge. **c** Waiting is not allowed

u and v by uv. We also assume that each edge has a weight that is a non-negative real number. We denote it by $len(e)$ to represent the length of each edge e. To capture the possible changes of the availability of road segments, additional information is attached to the edges. In our approach, each edge in the road network is assigned a set of temporal intervals which can be represented as follows: $S_{uv} = (ID, b_1, ..., b_k, ..., b_m)$, $uv \in E$, $1 \leq k \leq m$, where ID uniquely identifies the edge uv, $b_k = (t_{ck}, t_{ok}), t_{ck} < t_{ok}$, indicates the time period in which the edge is inaccessible, t_{ck} represents the time when the closing starts, t_{ok} denotes the end time of the temporal block, and m is the total number of blocks in this edge. This allows for the multiple storage of road segment ID with different closing time series. In a similar way, we use $S_u = (ID, b_1, ..., b_k, ..., b_m)$, $u \in N$ to represent and store the state of the node. All these dynamic information of the availability of road segments in a disaster area are obtained through intersection computation between the road network and obstacles in the form of moving polygons. For simplicity, we do not differentiate points along the edge, i.e. once the edge intersects the obstacle, the whole edge is not available.

2.2 The Modified Shortest Path Algorithm

Following above discussions, we extend the standard A* algorithm by introducing waiting options to compute the shortest path from a given starting time in a dynamic network. The pseudocode of the whole algorithm is given in Algorithm 1. We assume the travel starts at time 0. The time instances when the edge is closed or open again are stored as an amount of time relative to the given starting time. The object moves at constant speed *moveRate*, starting from *startPoint* to *targetPoint*. We set a threshold to decide if the re-computation should be executed to allow for some beneficial waiting that is not considered in the previous computation. Our main adaptation to the A* algorithm is to introduce the waiting time as an additional cost attached to each edge, as shown in Algorithm 2. We propose three methods to compute the waiting time that needs to be added to the travel cost of the road segment. In the following, we will provide a brief description of how this algorithm works.

Algorithm 1: Modified A* algorithm

```
1: Initialize startPoint, targetPoint, moveRate, wait=false
2: Initialize closedSet // The set of nodes already evaluated
3: Initialize openSet // The set of tentative nodes to be evaluated
4: while openSet is not empty do
5:    x = the node in openSet having the lowest f_score value.
6:    if x = targetPoint then
7:       if totalTravelTime > threshold && iter < maxIteration then
8:          wait = true
9:          go to step 2
10:      end if
11:      return results: shortest route, total travel time, and total waiting time
12:   end if
13: remove x from openSet
14: add x to closedSet
15: for each y in neighbor_nodes(x) do
16:    if y in closedSet then
17:       continue
18:    end if
19: end for
20: arriveTime = x.g_score, nextEdgeID = get ID of Edge xy
21: calculateWaitTime(arriveTime, nextEdgeID) // main adaption
22: if y not in openSet then
23:    add y to openSet
24:    tentative_is_better = true
25: else if tentative_g_score < y.g_score then
26:    tentative_is_better = true
27: else
28:    tentative_is_better = false
29: end if
30: if tentative_is_better = true then
31:    y.came_from = x
32:    y.g_score = travelCost
33:    y.h_score = heuristic_estimate_of_distance(y, targetPoint)
34:    y.f_score = y.g_score + y.h_score
35:    x.waitingTime = waitingTime
36: end if
37: end while
38: end for
39: return failure
```

Algorithm 2: calculateWaitTime(arriveTime, nextEdgeID)

```
tentativeTravelCost = arriveTime + len(nextEdge)/moveRate
if the current node is affected then
   if arriveTime > t_om of the current node then
      if wait && waitLineList.ccontains(nextEdge) then
         WaitingTime = calcWTatEdgesInList(arriveTime, nextEdgeID)
      else
         WaitingTime =calcWTwithoutWaitOption(arriveTime, nextEdgeID)
      end if
   else
      WaitingTime = calcWTwithoutWaitOption(arriveTime, nextEdgeID)
   end if
else
   WaitingTime = calcWTwithWaitOption(arriveTime, nextEdgeID)
end if
travelCost = tentativeTravelCost + waitingTime
```

Algorithm 3: calcWTwithWaitOption(arriveTime, nextEdgeID)

```
if the next edge is affected then
   get blocking intervals of the next edge
   for k = 1 to m do
      if arriveTime <t_ck, then
         if tentativeTravelcost < t_ck then
            waitLineList.add(nextEdge); waitingTime = 0; break
         else
            waitingTime = t_ok − arriveTime; break
         end if
      else
         if arriveTime < t_ok then
            waitingTime = t_ok − arriveTime; break
         else
            waitingTime = 0; break
         end if
      end if
   end for
else
   waitingTime = 0
end if
return waitingTime
```

Algorithm 4: calcWTatEdgesInList(arriveTime, nextEdgeID)

```
get blocking intervals of the next edge
waitingTime = 0
for k = 1 to m do
    if arriveTime < t_ok then
        waitingTime = t_ok − arriveTime; break
    end if
end for
return waitingTime
```

Algorithm 5: calcWTwithoutWaitOption(arriveTime, nextEdgeID)

```
get blocking intervals of the next edge
for k = 1 to m do
    if arriveTime < t_ck then
        if tentativeTravelcost < t_ck then
            waitingTime = 0; break
        else
            waitingTime = inf; break
        end if
    else
        if arriveTime < t_ok then
            waitingTime = inf; break
        else
            waitingTime =0; break
        end if
    end if
end for
return waitingTime
```

In adaptions of the A* algorithm, we assign each edge a variable, called *waitingTime*, to represent the total waiting time associated with the road segment. The *waitingTime* is calculated based on the time of arrival and the timeframe that the road segment is closed. Since waiting is not always beneficial, two waiting policies are considered: waiting is allowed if the current node is not affected or the object arrives after the time the node is blocked by obstacles, and waiting is not allowed if the object arrives before the node is affected. Depending on the re-computation condition indicated by the boolean flag *wait*, two rules are applied in the first waiting policy: not to wait if the object can safely pass through the next edge before the block starts, and wait until the end of the block in the next edge. We also assume that the accessibility of the edge and the node follows a unidirectional relationship:

(i) If the edge is temporarily not accessible, one or both nodes belonging to this edge can still be accessible.

(ii) If the node is not accessible, all edges connected to this node are not accessible.

Following these assumptions, we present there examples in Fig. 3 to illustrate the calculation of the waiting time using three sub-algorithms 3, 4, 5 respectively. As shown in Fig. 3, there are two nodes n_1 and n_2, and it takes 2 time units for the object to move from node n_1 to node n_2. In example (a), waiting is allowed because the node n_1 is not affected by the obstacle. The object arrives at $t = 7$, can pass through the edge and arrive at node n_2 safely at $t = 9$ before the edge is affected by the moving obstacle. By applying the Algorithm 3, we obtain the waiting time 0, which means the object does not need to wait. The edge $n_1 n_2$ should be stored in the array *waitLineList* and will contribute to the next computation step if re-computation is invoked. In example (b), the scenario is the same as the first one but the object is asked to wait as indicated by $wait = true$, which means that the computed result by the first computation without considering beneficial waiting options exceeds the threshold, and the object should wait at this node. The waiting time calculated by algorithm 4 is $12 - 7 = 5$ and should be included to the re-computation process. In the third example, waiting is not allowed. The object arrives at $t = 9$, but it can not safely pass through the edge because the edge will be closed at 10 before the object arrives at node n_2 at time $9 + 2 = 11$. It can not wait at node n_1 either, since when the object is waiting at the interval $[9, 12]$, the closing will start at 10. Therefore the waiting time is set *inf* by the algorithm 5 to trigger the next computation step. As we obtain the waiting time of each edge, it should be added to the total travel cost of the edge to find the optimal route avoiding moving obstacles. After the first computation, the output total travel time is compared with the threshold to determine whether the calculated route is satisfied. If not, the re-computation will be executed taking some beneficial waiting at edges that are stored in Algorithm 3 into consideration.

3 Case Study

In order to assess the performance of our proposed approach as a feasible solution, we construct a web-based application and present an extensive experimental evaluation of the model and the modified A* algorithm. The application consists two main components: an agent-based simulation tool and a JavaScript web mapping library. The proposed model and the algorithm are realized in the multi-agent simulator, called Mason [17, 18], and are evaluated with a real road network. Moving obstacles are synthetically generated to simulate the disasters and visualized in the form of one or more moving polygons crossing the road network. First responder is modeled as a mobile agent who can follow the route calculated by the path finding algorithm. The road network dataset is obtained from the OpenStreetMap (www.openstreetmap.org) database for our testing. All the needed data are imported into a PostGIS database and are fed into the agent simulation tool via GeoTools (www.geotools.org). The simulation output data is displayed to users through Openlayers (www.openlayers.org), a tool for exhibiting spatial data on web pages.

As the first step of the case study, we apply our model and algorithm to the basic case that one moving object has to be routed to one static destination, avoiding many moving obstacles. The proposed algorithm has been tested with two different datasets: the urban area of Delft and the Rotterdam downtown area. A comparison with the standard A* algorithm without considering predictions is conducted to evaluate the practical application of our approach in path planning, as shown in Table 1.

3.1 Case 1: Delft

In the case of Delft the standard A* algorithm and our algorithm produce very different routes, which is depicted in Fig. 4. The upper line represents the shortest path through the city center, and the lower line is the route calculated by our algorithm, where points along the route indicates the waiting locations for the responder to avoid moving obstacles. Figure 5 shows the snapshot of the simulation of responder's movement. As shown in Table 1, even though the difference in anticipated travel time in this case is relatively small, the main difference here is that the route R3 calculated by our algorithm considering predictions introduces much less additional waiting time than the route R2 provided by the standard A* algorithm without considering predictions. Apart from slowing their movements, following the shortest path could also endanger responders themselves if they do not include predictions into their routing process. In some situations where moving obstacles are moving faster than the relief vehicle, it may be too late for responders to realize that it would be difficult for them to get out of the dangerous area if they continue with their shortest route.

3.2 Case 2: Rotterdam

We also test our approach in the case of Rotterdam with a different dataset of the road network. As can be seen in Fig. 6, the responder has to cross a river to reach the rescue point, and there are a number of tunnels and bridges between the start point and the target point.

Continuing with our analysis, Fig. 6 shows the routes calculated by the two algorithms and Fig. 7 shows the snapshot of simulation of movements of both the responder and obstacles. As can be observed from Table 1, similar to what is reported in the previous section, the generated routes differ significantly. This is caused by the fact that both moving obstacles block the shortest path, and the response unit, who follows the shortest route, has to wait for a long time until it

Table 1 Calculated results

Distance (km)		Anticipated travel time (min)	Waiting time (min)	Total travel (min)
Case 1	*(Delft)*			
R1	2.11	4.4	X	X
R2	2.11	4.4	13.9	18.3
R3	2.43	5.8	5.4	11.2
Case 2	*(Rotterdam)*			
R1	5.22	8.9	X	X
R2	5.22	8.9	18.4	27.3
R3	6.27	0	0	9.0

Notes

[a] R1: The shortest route calculated by the standard A* algorithm without considering predictions of obstacles

[b] R2: The shortest route calculated by the standard A* algorithm without considering predictions and followed by the object considering obstacles along the route (the distance of R2 equals the distance of R1)

[c] R3: Route provided by the modified A* algorithm considering predictions of obstacles

[d] X means no value

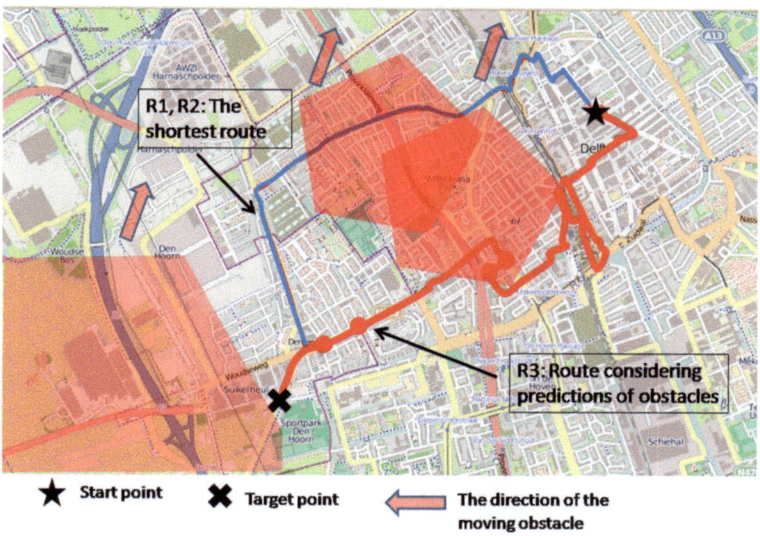

Fig. 4 Route considering predictions of obstacles versus the shortest route

can take the tunnel Maastunnel to reach the destination on the other side of the river. On the other hand, the proposed algorithm considering predictions computes a different and quicker route that crosses the bridge Erasmusbrug, introducing no waiting time and thus achieving a significant improvement of the performance of the response unit.

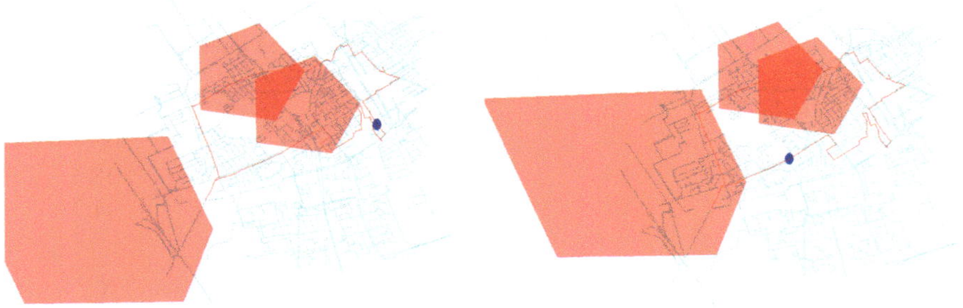

Fig. 5 Snapshot of simulation of movements of both obstacles (in *red*) and the responder (in *blue*)

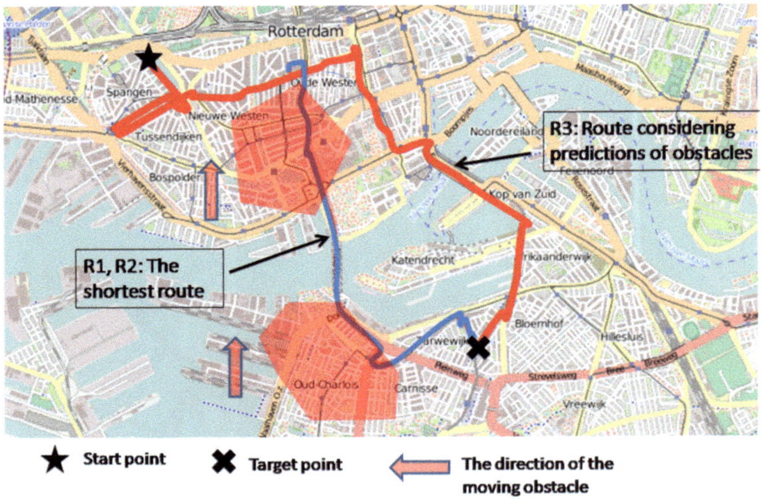

Fig. 6 Route considering predictions of obstacles versus the shortest route

Fig. 7 Snapshot of simulation of movements of both obstacles (in *red*) and the responder (in *blue*)

4 Conclusions and Future Works

Emergency navigation plays a vital role in disaster response and there is a great need for navigation support in the spatio-temporal road network populated by static/moving obstacles. However, despite the considerable amount of route guidance research that has been performed, investigations on navigation among obstacles are still sparse. The paper presents a model to represent a spatio-temporal network, and proposes an algorithm for obstacle-avoiding path computation considering predictions of obstacles' movements. Besides, a web-based navigation system, integrating the agent-based simulation tool and web-mapping technology, has been developed and tested in the real world network. We use both real-life and artificial data sets in our experiments. The real-life datasets are road systems in Delft and Rotterdam, and both are extracted from the OpenStreetMap database. The disaster data sets are artificially generated obstacles that are used to represent the physical phenomena during disasters. We also compare our algorithm with the standard A* algorithm. As demonstrated by experimental results, our approach provides a promising way for navigation among moving obstacles.

In our future work, we plan to adapt the proposed algorithm to deal with changes of the disaster-related information, in particular those that are not predicted for the environment. Due to the difficulty of collecting disaster-related data, there are always some gaps between the real situations and predictions provided by the disaster model, which creates a need for re-routing. A possible approach is to include the range of accuracy of the disaster model in obtaining the dynamic information of the road network to facilitate the re-evaluation of the calculated route. We also would like to introduce variable travel speed into the re-routing process, since the moving speed is an important factor considerably influenced by both traffic conditions and the infrastructure. Another next step would be to explore further some extreme cases (e.g., the obstacle covers the target point during the course of an incident, resulting in no available route until the incident is over) and the problem variants discussed briefly in [19] (e.g., one moving object has to be routed to many static destinations, avoiding many moving obstacles).

References

1. S. Schmitz, A. Zipf, P. Neis, New applications based on collaborative geodata—the case of routing, in *XXVIII INCA International Congress on Collaborative Mapping and Space Technology*, Gandhinagar, Gujarat, India (2008)
2. S. Nedkov, S. Zlatanova, Enabling obstacle avoidance for Google maps' navigation service, in *Proceedings of the 7th Geoinformation for Disaster Management*, Anltalya, Turkey (2011)
3. F. Kunwar, F. Wong, R.B. Mrad, B. Benhabib, Guidance-based on-line robot motion planning for the interception of mobile targets in dynamic environments. J. Int. Robot. Syst. **47**(4), 341–360 (2006)

4. H. Li, S.X. Yang, M.L. Seto, Neural-network-based path planning for a multirobot system with moving obstacles. IEEE Trans. Syst. Man Cybern. Part C Appl. Rev. **39**(4), 410–419 (2009)
5. E. Masehian, Y. Katebi, Robot motion planning in dynamic environments with moving obstacles and target. Int. J. Mech. Syst. Sci. Eng. **1**(1), 20–25 (2007)
6. F. Darema, Dynamic data driven applications systems: new capabilities for application simulations and measurements, in *Proceedings of the 5th international conference on Computational Science*, Atlanta, GA, USA (2005)
7. X. Hu, Dynamic data driven simulation. SCS M&S Mag. 1 16–22 (2011)
8. P. Pecha, R. Hofman, V. Šmídl, Bayesian tracking of the toxic plume spreading in the early stage of radiation accident, in *Proceeding of European Simulation and Modelling Conference*, Leicester, UK (2009)
9. G. Lu, Z. Wu, L. Wen, C. Lin, J. Zhang, Y. Yang, Real-time flood forecast and flood alert map over the Huaihe River Basin in China using a coupled hydro-meteorological modeling system. Sci. China Ser. E: Technol. Sci. **51**(7), 1049–1063 (2008)
10. D. Mioc, F. Anton, G. Liang, On-line street network analysis for flood evacuation planning, in *Remote Sensing and GIS Technologies for Monitoring and Prediction of Disasters*, ed. by S. Nayak, S. Zlatanova (Springer, Berlin, 2008), pp. 219–242
11. Y. Liu, M. Hatayama, N. Okada, Development of an adaptive evacuation route algorithm under flood disaster. Annuals of Disaster Prevention Research Institute, Kyoto University, vol. **49**, 189–195 (2006)
12. S. Zlatanova, S.S.K. Baharin, Optimal navigation of first responders using DBMS, in *Joint Conference of the 3rd International Conference on Information Systems for Crisis Response and Management/4th International Symposium on Geo-Information for Disaster Management* (2008)
13. P.E. Hart, N.J. Nilsson, B. Raphael, Correction to a formal basis for the heuristic determination of minimum cost paths. SIGART Newslett. **37**, 28–29 (1972)
14. B. Huang, Q. Wu, F.B. Zhan, A shortest path algorithm with novel heuristics for dynamic transportation networks. Int. J. Geogr. Inf. Sci. **21**(6), 625–644 (2007)
15. G. Nannicini, D. Delling, D. Schultes, L. Liberti, Bidirectional A* search on time-dependent road networks. Networks **59**(2), 240–251 (2012)
16. T. Ohshima, P. Eumthurapojn, L. Zhao, H. Nagamochi, An A* algorithm framework for the point-to-point time-dependent shortest path problem, in *Computational Geometry, Graphs and Applications*, ed. by J. Akiyama, et al. (Springer, Berlin, 2011), pp. 154–163
17. S. Luke, C. Cioffi-Revilla, L. Panait, K. Sullivan, Mason: a new multi-agent simulation toolkit, in *Proceedings of the 2004 SwarmFest, Workshop* (2004)
18. S. Luke, C. Cioffi-Revilla, L. Panait, K. Sullivan, G. Balan, Mason: a multiagent simulation environment. Simulation **81**(7), 517–527 (2005)
19. Z. Wang, S. Zlatanova, Taxonomy of navigation for first responders, in *Proceedings of the 9th International Symposium on Location-Based Services* Munich, Germany (2012)

A Two-level Path-finding Strategy for Indoor Navigation

Liu Liu and Sisi Zlatanova

Abstract In this paper, a two-level path-finding strategy is presented. It can derive a geometric indoor route according to different preferences. On the first level a room sequence is derived by means of non-metric criteria, such as the number of visited rooms or the number of obstacles in rooms. Based on the sequence, a geometric path with respect to obstacle shape and user size is computed for each traversed room. Then all of these paths compose a final path. The approach is illustrated with an example of a residential building. Compared to other related work, this strategy allows greater flexibility in providing the detailed path within changing indoor environments.

Keywords Indoor navigation · Path-finding · Two-level strategy

1 Introduction

Indoor navigation is a task consisting of three stages. They are path-finding, localization and guiding users. Path-finding is a key step to provide movement information based on knowledge about buildings. Thus path-finding should be conducted on certain model which represents a indoor environment. According to graph-based methods [1–3], certain types of graph can be derived from different

L. Liu (✉) · S. Zlatanova
GIS Technology Section, OTB Research Institute for the Built Environment, Delft University of Technology, Jaffalaan 9, 2628 BX Delft, The Netherlands
e-mail: L.liu-1@tudelft.nl

S. Zlatanova
e-mail: S.Zlatanova@tudelft.nl

S. Zlatanova et al. (eds.), *Intelligent Systems for Crisis Management*, Lecture Notes in Geoinformation and Cartography, DOI: 10.1007/978-3-642-33218-0_3,

building subdivision ways. Most of these graphs belong to Geometric Graph. It means that geometric information (metrics) is attached to nodes/edges of the graphs. Typically, the Combinatorial Data Model (CDM) proposed by Lee [2] is used to derive geometric network by making use of Medial Axis Transformation (MAT) [4]. Then it covers adjacency, connectivity and hierarchical and geometric relationships of a building. A framework named Multilayered Space-Event Model (MLSEM) [5] allows distinct links to be established between different space layers, such like between building geometry and sensor coverage layers. In this way, different representations of a building can be incorporated into the framework.

Another alternative is the Indicative Route Method (IRM) proposed by Karamouzas et al. [6]. It is built on the "corridor map" proposed by Geraerts and Overmars [7]. The "corridor map" structure offers a set of collision-free spaces. IRM can generate routes as smooth skeletons in a "corridor map". Moreover, there is another type of solution that can be named triangulation-based method [8, 9]. Based on a certain triangulation strategy, (indoor) environments are represented by different triangle areas. Thus some sort of a medial axis network can be derived and it would be used for path-finding. Though IRM is a nice method for considering obstacle avoidance, currently the network of indicative route can not be automatically created. Similarly, various triangulation-based methods are also attractive, yet much more artificial nodes may be added in for path-finding compared to graph-based methods.

In our research, we concentrate on the graph-based method for path-finding. However, we intend to avoid constructing the entire geometric graph of a building, because there may be considerable increment of a graph's node/edge number when the graph represents a complex building with plenty of rooms and openings. For instance, a respectable amount of nodes and edges may be added in when the medial axis of a space is computed. Furthermore, the conventional geometric graph is a bit feeble to take indoor obstacles into account. Usually the wide-used geometric network [2] representing the interior of a building lacks a proper representation of obstacles. Additionally, it's not adaptable to dynamically changing scenarios. Even if an indoor environment changes a little, yet a new up-to-date graph needs to be generated. The process would be appreciably time-consuming.

Therefore, it is necessary to develop a strategy to overcome the deficiencies of these widely-used graph-based methods (e.g. path-finding on CDM). A pivotal foundation of our two-level strategy is an appropriate data model representing buildings [10]. The model we used is a logical graph. It only represents the building connectivity. The logical graph (i.e. connectivity graph) could be automatically derived from the building semantics and thus metrics would not be introduced. In our two-level strategy, it will be used in the first level (i.e. the *Rough* level). Then a space/room sequence to be passed would be provided. On the second level (i.e. the *Detailed* level), "door-to-door" paths will be computed for single spaces of the previous sequence.

This paper is organized as follows. In the next section, methods/technical aspects will be introduced. Then the two-level strategy will be presented and exemplified. Finally, we will elaborate on conclusions and future developments.

2 Related Issues

The purpose of indoor path-finding is to obtain different indoor paths according to distinct requirements of various users. A specific path-finding task is labeled by its purpose, such as the shortest path or the fastest path. According to typical graph-based methods (where nodes represent spaces and edges represent openings), a classification of diverse indoor paths is given:

- *Least-effort*: it minimizes the total required travel time for reaching a destination [11].
- *Shortest (shortest-distance)*: it minimizes the required distance between a start and a destination [2, 3].
- *Central point strategy*: finding a path by trying to transit well-known locations (e.g. landmarks) of buildings [12].
- *Direction strategy*: heading to the horizontal position of a destination as directly as possible and regardless of the level-changes [12].
- *Floor strategy*: firstly finding a path to the storey of a destination, then horizontally finding the destination on the storey [12].

In our two-level strategy we apply different criteria for path-finding on the first and the second levels. We will elaborate them in the following subsections.

2.1 Criteria of Path-finding on the First Level

We distinguish between two path types as follows:

- *Least-space-visited*: finding a path by visiting the least number of indoor spaces between a start location and a destination.
- *Least-obstruction*: finding a path guaranteeing the least degree of obstruction between a start location and a destination.

We believe much more other path types could be defined according to specific preferences of various users and the strength of preference can be estimated.

Table 1 lists the merits and drawbacks of indoor pathfinding approaches. As we mentioned before, a *Shortest-distance* path derived from conventional geometric graph usually does not concern obstacles due to the insufficient representation. In this sense the path may not be real closest path between a start and a destination. To get a *Least-effort* path, the type and walking speed of specific users would be reckoned. Yet these estimated data might not be accurate. The *Central point*

Table 1 Comparison of different path types

Path type	Merit	Drawback
Shortest-distance	It is a path providing the minimum distance between a start and a destination	Full building metrics is required to get an accurate shortest path. Sometimes, the shortest path isn't the one that is passable or easy to be followed
Least-effort	It is a path considering the minimum travel time as the 'optimal' criterion	It needs users' type and speed as parameters for computation. Yet usually these parameters are just rough estimations
Central point strategy	It's an easy and practical way for users to find a path by themselves	It may cause considerable detours
Direction strategy	Finding a path by sticking to the direction between the start and the destination	It may result in winding path which can get unfamiliar users lost
Floor strategy	It is a easy-followed pattern to arrive the floor of destination	It may not be the minimum distance path
Least-space-visited	It is a path presenting the least number of visited rooms between assigned start and destination	To follow it may result in more travelling time and distance
Least-obstruction	It is a path considering the least degree of blockage between a start and a destination. Moreover, it has no request for building metrics	Gaining an accurate path of such type needs accurate dynamic information. The information is difficult to be exactly collected in practice

strategy, *Direction strategy* and *Floor strategy* are proposed by Hölscher et al. [12]. They mainly act as strategies to help users find paths by themselves.

Liu and Zlatanova [10] classify three types of indoor obstacle. They are moveable, fixed and dynamic obstacles. The *Least-obstruction* path concerns all these obstacles and it's the potentially least-interrupted route. The *Least-space-visited* path may avoid detours by controlling the number of passed spaces. In this paper, we initially decide to compute the *Least-obstruction* and/or the *Least-space-visited* path on the first level.

2.2 Visibility Graph Construction on the Second Level

In our research we select Visibility Graph (VG) method for path-finding with respect to indoor obstacles [13]. The nodes of a VG correspond to geometric vertices of all polygonal obstacles on 2D plane. A pair of mutually visible nodes is corresponding to a visible edge. Then all of nodes and visible edges compose a

VG. When a Euclidean shortest path is to be computed on 2D plane, the basic approach is to build a VG and apply certain shortest-path search algorithm (e.g. Dijkstra algorithm, [14]) on the graph. Specifically, the weight of each visible edge is the distance of related two nodes.

Alt and Welzl [15] and Overmars and Welzl [16] give an algorithm to support visibility graph construction for polygons in $O(n^2)$ time. This algorithm is better than traditional ones with $O(n^2 \log n)$ time complexity [13]. So it's introduced in this paper. Figure 1 illustrates a VG in a 2D single space/room. The nodes of the VG involve both geometric vertices of obstacles and concave vertices of the space/room polygon.

2.3 Constraint of User Size on the Second Level

A user's size and shape may be critical when a path is computed in a room with many obstacles. In robot motion planning if a robot is represented by a polygon on a 2D enviroment, the real space for its movement is named Work Space (WS) [13]. Reference point is a notion to depict a robot's location. The robot's state can be reflected by the reference point and translation and rotation parameters. A space related to the two parameters is named Configuration Space (CS). Apparently a robot in a location of WS (i.e. real world) may intersect certain obstalce area. Thus the non-obstacle part is named Free Space. In WS the obstacle-free movement of a robot can be represented by certain curve in Free CS. In this case, the Minkowski Sums are used to derive obstacle-free paths in robot motion planning [13]. Moreover, Yuan and Schneider [17] presents another approach. They claim it can test the physical traversability of paths for arbitrarily-shaped users in a 3D environment.

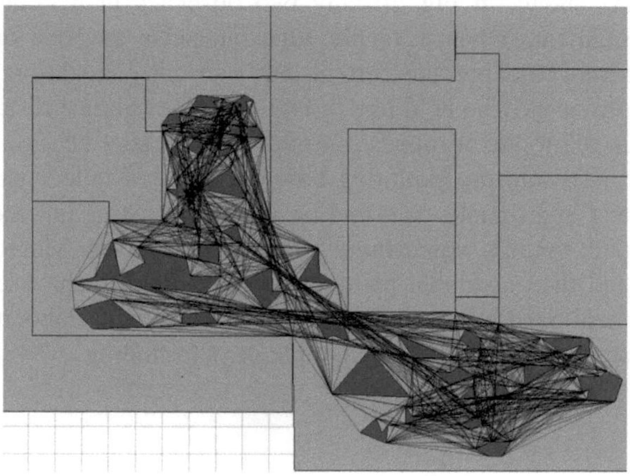

Fig. 1 A visibility graph in a single room

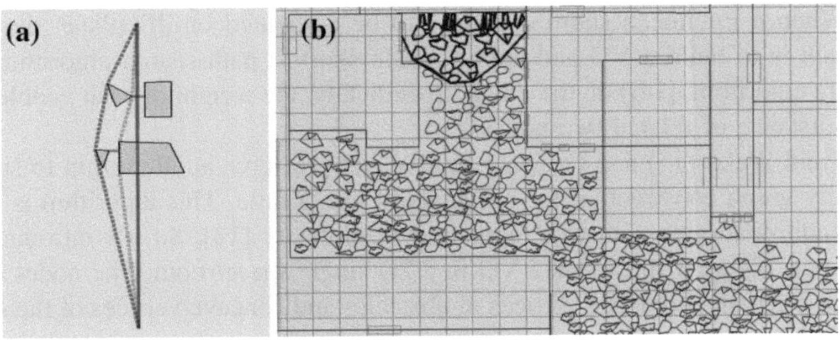

Fig. 2 **a** Paths with respect to user sizes; **b** a path complying with a constraint of user size in a single room

As people are more flexible than robots, it's sufficient to only concern the size constraint of users to report "bottlenecks". These "bottlenecks" involve the minimum gap between any two obstacles. On a 2D plane, the size constraint can be specified as the minimum width of a user. There are different size constraints for diverse users. For instance, the solid line in Fig. 2a is the path without regarding of obstacles, and the middle dash line is the path to avoid obstacles for a child. In Fig. 2a the leftmost dash line indicates the path for an adult. In order to compare user's widths to the "bottlenecks", we detect the minimum distance between any two obstacles. Rotation Calipers algorithm [18] is selected to calculate the minimum distances. Yet it only can be applied to convex polygons. There is an assumption that people would like to move around an obstacle. It means the shortest way for moving around an obstacle is its convex hull. Thus we concern convex hull of obstacles and Rotation Calipers algorithm can be applied.

In Fig. 2b, obstacles in a room and their convex hulls are presented. Minimum distances between all obstacle pairs are computed. If a minimum distance is less than a given tolerance then the related two obstacles are in a same group. As shown in Fig. 2b, any two obstacles in a same group are connected by a segment. When a visible edge intersects any one of these segments, it will be removed. Thus the current user can not pass through the interior of a group of obstacles. The bold line in Fig. 2b is the shortest-distance path between two doors in the room. It rounds a group of obstacles (the highlighted part).

To sum up, exploring different types of indoor path could help us design and test appropriate path-finding results; constructing the visibility graph can assist path-finding which involves indoor obstacles. Moreover, user size is concerned and then available paths can be provided to the user. Based on related aspects mentioned above, in the next section we will elaborate the two-level path-finding strategy and show an example of path-finding.

3 Two-Level Path-finding Strategy

As mentioned previously, the two-level path-finding strategy is a graph-based method. It uses two kinds of graph for path computation. On the *rough* level the logical graph of a building is required. Nodes of the logical graph represent spaces/ rooms, while its edges denote openings in the building. On the *detailed* level, a "door-to-door" path in a single space is computed. Thus the nodes represent all openings and transition locations among obstacles in the space, and the edges imply the Euclidean distances.

Compared to the work of Liu and Zlatanova [10], we improve the two-level strategy mainly in following aspects:

1. Metrics are not introduced to the *rough* level path-finding (routing);
2. Obstacles are taken into account in the *detailed* level routing. Related path-finding results are still "door-to-door" paths.
3. The size and shape of users are considered.

As connectivity is the only required information on the *rough* level, it would facilitate an automatic derivation of the logical graph from building semantics. In this case, distances between nodes are not available. Routing result of *rough* level is a space sequence. It implies the spaces to be passed through.

Point 2 indicates the resulting path is the shortest way between two doors of a space. The door-to-door paths are geometric routes in single spaces. According to the space sequence, door-to-door paths in those spaces compose the final path. Therefore, geometric paths are computed on the fly. Point 3 guarantees users can follow the derived paths. In the following subsections, we will illustrate how to apply the two-level strategy for path-finding.

3.1 Rough Level

At first, the connectivity graph of a building should be given. Floor plans of a building with four stories are shown in Fig. 3. Some spaces of the building contain obstacles. There are two elevators and one stairwell as vertical movement spaces.

With the logical graph, paths can be computed according to different attributes of edges. It is also common to compute the value of edges by some weighted function [19]. In order to facilitate the explanation of our method, in this paper we just consider the attribute named *obstruction degree* of nodes. More specifically, the logical graph can be a directed graph to denote available passing directions. If a space (i.e. a node) leads to another space, then its *obstruction degree* value is the value of related directed edge. Finally, the *Least-obstruction* path could be acquired as a space sequence from the first level routing.

Figure 4 expresses the connectivity graph of the building. As shown in Fig. 4, the central part includes the vertical movement spaces. And the centrifugal clusters

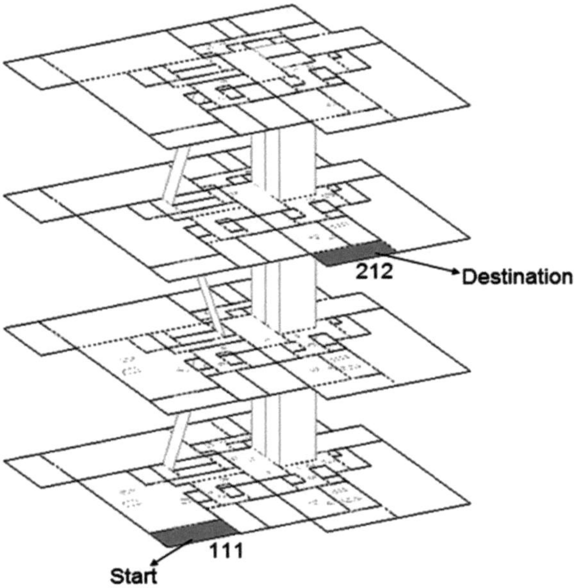

Fig. 3 Floor plans of a test building

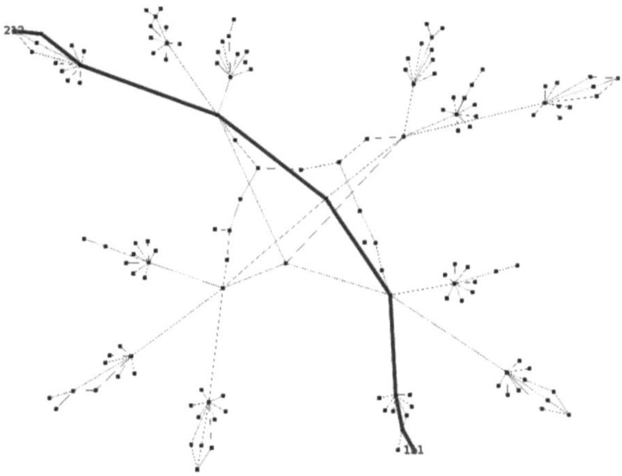

Fig. 4 The connectivity graph of the test building

denote horizontal spaces in different stories. The bold line denotes the *Least-obstruction* path from the node 111 to the node 212. It reveals that an elevator is selected for a vertical transition. It is the conformable and maybe prompt way for users when there is no emergency. Besides, we could eliminate some apparently useless nodes. For instance, if a space is only connected to one other space and it's not the start or destination, then it can be removed. In this way the connectivity graph in Fig. 4 can be simplified, which is shown in Fig. 5.

It is apparent that the number of nodes and edges of a logical graph is almost constant. Yet more nodes and edges would be removed (or added) on the graph

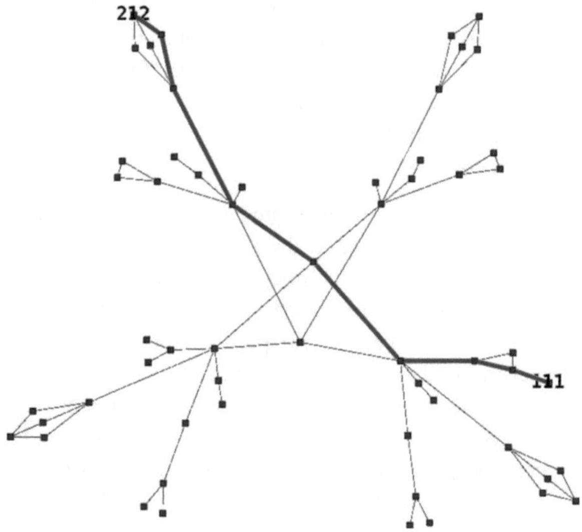

Fig. 5 The connectivity graph after selection

according to dynamic changes in a building. The updating process can be finished on one logical graph, so there is no need to derive a new graph of the building every time. The path derived from the first level is just an indication of spaces to be passed. Yet the movement in the interior of spaces is not clear. In the following subsection, we will introduce the detailed path.

3.2 Detailed Level

After the space sequence is determined at the first level, a detailed path would be searched orderly in each space. Generally, if a space has multiple openings (which mostly are doors) to the next space, then we should compute the door-to-door route for each opening to select the shortest one. For any two openings in a single space, the concrete computation steps are described as follows.

1. Use Rotating Calipers algorithm to compute the minimum distance between any two of obstacles.
2. Construct a visibility graph in the space.
3. Use a given tolerance to filter visible edges. We should have detected the obstacles whose minimum distance is less than the tolerance before that.
4. If the start and the end nodes are invisible, we check whether there are visible edges connecting the start/end nodes. If it's not, there is no path from the start node to the end node.
5. If the path exists, compute the *shortest-distance* path on the visibility graph. Otherwise, return to the previous space and re-compute the space sequence on the first level.

If there is no passable path in the current space, iterative computation of the space sequence starts from the previous space. For each passable space, its detailed path is the shortest path in it. The iteration computation stops when the detailed path in destination space is obtained.

According to different scenarios, several situations in a single space could be categorized:

- No obstacles. If there is only one door to the next space, we just compute the door-to-door path with the shortest distance.
- With obstacles. A visibility graph will be constructed at first. Also, impassable visible edges are removed. Then the *shortest-distance* path to the next space is computed. If there is no path to the next space, then the space sequence is re-computed from the last space.
- With emergency. There may be some unexpected dynamic or new static (like collapsed cupboards) obstacles in current space. The processing is similar to the scenarios with obstacles: if the dynamic obstacles block all possible openings, users should return the previous space and the space sequence is re-determined.

All the detailed paths in these spaces on the space sequence will be put together. This is the final path. Except for the start and the destination spaces, any other space on the sequence has a determined start location as the end location in the path of the previous space. As shown in Fig. 6, the bold line is the final path between the start and the destination spaces. From the top view of its horizontal parts, obviously the path is a door-to-door and obstacle-avoided path.

In brief, the two-level path-finding strategy is feasible and flexible. Compared to the widely-used geometric graph method, it has preferable ability to take indoor obstacles into account.

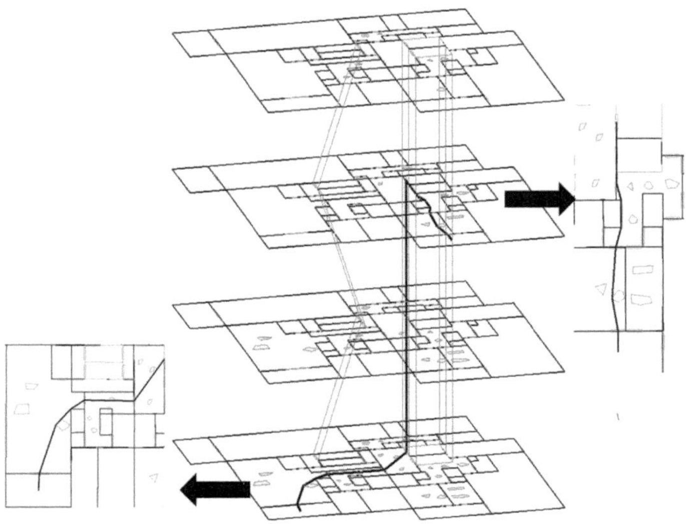

Fig. 6 The detailed path from the start to the destination in the building

4 Conclusions

This paper presents our two-level indoor path-finding strategy. In general, the two-level routing strategy has several primary merits:

1. Path-finding on the first level can be conducted on connectivity graph without metrics and with the limited number of nodes and edges. The number is determined by the structural subdivision of buildings;
2. The detailed geometric network per room is computed only to define the path in the room. Thus extra nodes/edges will not be added to the connectivity graph derived at the first level. Also, the method can support path-finding in dynamic environments as detailed paths are derived on the fly.
3. Obstacles and the shape and size of users can be considered in this method, which is seldom taken into account by other related researches.

When the two-level path-finding strategy is used with a data model of buildings, extra subdivision on buildings would not be introduced (as with the triangulation or MAT methods). That means it's unnecessary to adjust the size of building spaces. Even a spacious room would not be subdivided if it's a basic space in a building. Therefore, the strategy has potentially a broad range of applications.

In this paper, we only show an initial test about single-attribute path on the rough level. It concerns merely one preference. In the future, more criteria of path-finding would be used together to get a space sequence. In this case, weighted functions can be devised with preference attributes, and then the weight value of each edge of a logical graph will be computed with respect to every attribute. The anticipated space sequence is a weighted result as well. On the detailed level, currently only visibility is taken into account for computing a geometric path. Yet it's a purely geometric criterion. In order to concern diverse capabilities of the navigation users (e.g. vision-impaired or wheelchair users), some other factors may be considered to imply traversability in the future.

Currently all indoor spaces are inputted as 2D polygons. In the next step we will investigate how to apply the two-level strategy to 3D solid indoor spaces. Furthermore, its application in emergency scenarios will be explored. On the rough level, connectivity is impacted by emergencies, and then we need an updating mechanism to maintain nodes and edges of the connectivity graph. On the detailed level, the number of dynamic and static obstacles will sharply increase in a short time. Then temporal factors could be introduced to computing detailed paths.

References

1. G. Franz, H. Mallot, J. Wiener, Graph-based models of space in architecture and cognitive science—a comparative analysis, in *Proceedings of the 17th International Conference on Systems Research, Informatics and Cybernetics*, Windsor, Canada, 1–7 Aug 2005, pp. 30–38

2. J. Lee, A spatial access-oriented implementation of a 3-D GIS topological data model for urban entities. Geoinformatica **8**(3), 237–264 (2004)
3. M. Meijers, S. Zlatanova, N. Preifer, 3D geoinformation indoors: structuring for evacuation, in *Proceedings of Next generation 3D City Models* , Bonn, 21–22 June 2005
4. H. Blum, A transformation for extracting new descriptors of shape, in *Proceedings of the Symposium on Models for the Perception of Speech and Visual Form*, ed. by W.W. Dunn (MIT Press, Cambridge, 1967), pp. 362–380
5. C. Nagel, T. Becker, R. Kaden, K. Li, J. Lee, T.H. Kolbe, *Requirements and Space-Event Modeling for Indoor Navigation* (Open Geospatial Consortium Discussion Paper, Open Geospatial Consortium, 2010)
6. I. Karamouzas, R. Geraerts, M. Overmars, Indicative routes for path planning and crowd simulation, in *Proceedings of International Conference on the Foundation of Digital Games*, Orland (2009), pp. 113–120
7. R. Geraerts, M. Overmars, The corridor map method: real-time high-quality path planning, in *Proceedings of the IEEE International Conference on Robotics and Automation* (2007), pp. 1023–1028
8. M. Kallmann, Path planning in triangulations, in *Proceedings of the IJCAI Workshop on Reasoning, Representation, and Learning in Computer Games* (2005), pp. 49–54
9. W.V. Toll, A.F. Cook IV, R. Geraerts , Navigation meshes for realistic multi-layered environments, In *Proceedings of IEEE/RSJ International Conference on Intelligent Robots and Systems* (San Francisco, 2011), pp. 3526–3532
10. L. Liu, S. Zlatanova, A door-to-door pathfinding approach for indoor navigation, in *Proceedings of GeoInformation for Disaster Management Conference 2011*, Antalya, Turkey, 3–8 May 2011
11. J.-C. Thill, T.H.D. Dao, Y. Zhou, Traveling in the three-dimensional city: applications in route planning, accessibility assessment, location analysis and beyond. J. Transp. Geogr. **19**, 405–421 (2011)
12. C. Hölscher, T. Meilinger, G. Vrachliotis, M. Brösamle, M. Knauff, Up the down staircase: wayfinding strategies in multi-level buildings. J. Environ. Psychol. **26**, 284–299 (2006)
13. M.D. Berg, M.V. Kreveld, M. Overmars, *Computational Geometry: Algorithms and Applications* (Springer-Verlag, New York, 2000)
14. E.W. Dijkstra, A note on two problems in connexion with graphs. Numer. Math. **1**, 269–271 (1959)
15. H. Alt, E. Welzl, Visibility graphs and obstacle-avoiding shortest paths. Z. Oper. Res. **32**, 145–164 (1988)
16. M.H. Overmars, E. Welzl, New methods for computing visibility graphs, in *Proceedings of the 4th ACM Symposium on Computational Geometry* (1988), pp. 164–171
17. W. Yuan, M. Schneider, 3D Indoor route planning for arbitrary-shape objects, in *Database Systems for Advanced Applications,* ed. by J. Xu, G. Yu, S. Zhou, R. Unland. Lecture notes in computer science, vol. 6637 (Springer, Berlin/Heidelberg, 2011), pp. 120–131
18. G. Toussaint, Solving geometric problems with the rotating calipers, in *Proceedings of IEEE MELECON'83* (Athens, 1983), p. A10.02/1–4
19. F. Lyardet, D.W. Szeto, E. Aitenbichler, Context-aware indoor navigation, in *Proceedings of the European Conference on Ambient Intelligence* (2008), pp. 290–307

An Approach to Qualitative Emergency Management

Rami Al-Salman, Frank Dylla and Lutz Frommberger

Abstract Emergency Management Systems (EMSs) are playing an important role to save people's life's and to reduce the effects of disasters such as earthquakes or floods. In this paper, we propose an approach to qualitative emergency management. This will empower emergency managers to query spatial databases using qualitative terms used in spoken language, such as 'near' or 'north of'. By providing a qualitative DBMS layer that covers the three qualitative aspects topology, distance, and direction, our system is able to handle qualitative spatial queries. Qualitative spatial queries are translated into formal Structured Query Language (SQL) database queries which are used to query and retrieve results from spatial databases.

Keywords Emergency management system · Volunteered geographic information · Qualitative spatial representations · Matching · Spatial indexing

R. Al-Salman (✉)
Initial Training Network Geocrowd, University of Bremen,
P. O. Box 330 440, 28359, Bremen, Germany
e-mail: rami@informatik.uni-bremen.de

F. Dylla
SFB/TR 8 Spatial Cognition, University of Bremen,
P. O. Box 330 440, 28359, Bremen, Germany
e-mail: dylla@sfbtr8.uni-bremen.de

L. Frommberger
Capacity Lab, University of Bremen, Enrique-Schmidt-Str. 5
28359 Bremen, Germany
e-mail: lutz@capacitylab.org

S. Zlatanova et al. (eds.), *Intelligent Systems for Crisis Management*, Lecture Notes in Geoinformation and Cartography, DOI: 10.1007/978-3-642-33218-0_4,

1 Introduction

Every year thousands of people are killed and hundreds of thousands more have to escape their homes due to natural disasters like earthquakes or floods. From 1970 up to 2007, 860,000 people were killed by earthquakes and 535,000 by storms and floods [8]. However, immediately after natural hazards such as the earthquakes in Vientiane/Laos (2011), Tabriz/Iran (2012), and Sulawesi/Indonesia (2012)[1] fast response and recovery capabilities of Emergency Management Systems (EMSs) play a crucial role to save people's lifes. People who are stranded by landslides and partially or totally damaged buildings in the disastrous areas have only a good chance to survive if they are rescued within 72 h—the so-called 'Golden 72 h' [9].

Especially in emergency situations where the scenarios can change rapidly, exact geometric information is often not (yet) available (for example, detailed information about flooded areas). Especially in less developed countries, up-to-date information about emergencies can be impossible to acquire in short time, in particular from remote areas. Thus, we can expect that much of the information is coming from oral reports. Humans rather communicate by means of qualitative terms such as 'near', 'big', or 'away from' than in terms of quantitative values [2]. Hence, a major challenge to EMSs is to allow for coping with qualitative data in geo-spatial databases. In particular, this enables an EMS to offer more natural and intuitive interfaces. Queries and instructions like "Spread goods in areas that are near to Main Street, but far away from damaged buildings and water" should be handled appropriately.

In this paper, we present an approach to integrate qualitative data into an EMS. We claim that emergency managers will benefit from our approach to query geo-spatial databases of EMSs by means of qualitative spatial terms. To handle qualitative spatial data, a qualitative layer that covers the three qualitative aspects, features respectively, topology, distance, and direction needs to be integrated into the DBMS. We show the applicability of our approach by implementing a prototypical system query system that allows for querying qualitative information from a spatial database. Simplified, qualitative queries are translated into formal geospatial SQL queries which are then matched against precomputed qualitative information within the geospatial database. Finally, a list of matches is generated and presented to the emergency manager.

2 A Framework for Qualitative Emergency Management

In this section, we describe the overall architecture of a *Qualitative Emergency Management System* (QEMS), which is illustrated in Fig. 1. Initially, we assume that databases of QEMS contain *n* objects in a vector format which are indexed by

---

[1] See www.emdat.be for details.

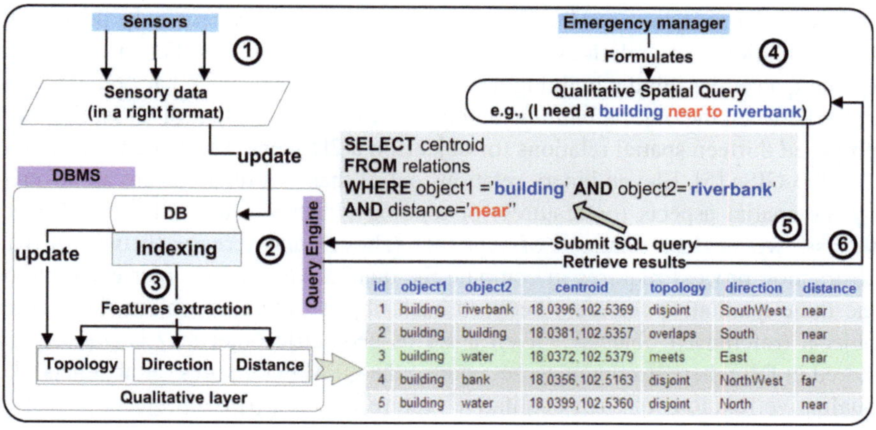

Fig. 1 Architecture of the qualitative emergency management system (QEMS)

an R-Tree [6]. The QEMS indexes and updates new objects (coming from step 1) in the database in step 2. Qualitative spatial information is achieved by abstraction and brought to the qualitative layer of the DBMS in step 3. The querying operation is performed in the remaining steps.

2.1 Indexing

Spatial databases contain a vast amount of objects. In general, answering qualitative spatial queries requires exponential time [14], because all objects and their relations towards each other must be considered to answer a query. Several data structures were proposed to cope with a huge amount of spatial objects [11]. R-tree supports spatial access methods by indexing multi-dimensional data (e.g., polygons or geo coordinates). The key idea of R-trees is to compute the Minimum Bounding Box (MBB) of objects which are then grouped into clusters in the next higher level of the tree where the MBB of all objects contained is calculated. If a spatial query does not intersect the MBB of a cluster then it cannot intersect any of the contained objects in this cluster. However, R-trees provide efficient update, add, and delete operations which allow to update and integrate incoming data (from step 1) into the database.

2.2 Feature Extraction

Geo-spatial databases store the data in quantitative vector format. To enable qualitative usage, all possible qualitative spatial relations in the 2D domain need to be identified [1]. These relations need to be abstracted from the database objects

and administrated in a qualitative layer. Currently, our system only deals with binary relations, i.e., relations that hold among pairs of objects. For instance, in a spatial query like 'Find a building near to riverbank', *near* is a binary relation that holds between the objects *building* and *riverbank*. In the context of GISs, Freeman proposed thirteen spatial relations for objects in a 2D scene which are beneficial to develop GISs [5]. Eleven binary relations out of these 13 relations are all covered by the spatial aspects (or features) of topology, distance, and direction. Nevertheless, we neglect relations of type 3D (above) and ternary (between). The abstraction process requires to compute the qualitative relations (per each one of the three qualitative features) between each object (reference objects) and other objects (referenced objects). Given a set of geometric objects $O = \{o_1, \ldots, o_n\}$ stored in DB where o_i contains a set points , the QEMS computes (or abstracts) the qualitative relations of each qualitative feature for each pair of objects (o_i, o_j) in the database, where $i \neq j$.
Topology relations: To abstract topological relations, the 9-intersection model [3] that abstracts 8 topological relations {equal, disjoint, meet, overlap, contains, covers, inside, and coveredBy} is used.
Direction relations: We apply the Cardinal Direction Model [13] for extended objects to abstract direction relations. Based on the minimum bounding box (MBB) of a reference object the space is partitioned into nine regions: (*S*outh, *S*outh*W*est, *W*est, *N*orth*W*est, *N*orth, *N*orth*E*ast, *E*ast, *S*outh*E*ast, and same location *B*). A referenced object (referent) may be completely contained in one of the nine regions. This is called 'single (tile) relation'. However, as the model deals with extended objects a referent may cover more than one region (partially or totally) which leads to 512 'multi-tile' or conjunctive direction relations wrt. a reference object.
Distance relations: Based on the qualitative distance model [7], we propose a distance model that assigns one of the four distance relations *equal*, *near*, *medium*, and *far* to pairs of objects in $\mathbb{R}^2$. Again, this representation is based on the MBB. Let P denote the centroid of the MBB. In addition, $d_{max} = d_3$ denotes the maximum distance between P and one of the corners of the MMB. In order to define the four relations we need two additional distance values d_1 and d_2 with $d_0 < d_1 < d_2 < d_3$. d denotes the distance between the object of interest and P. The relation is considered *equal* iff $d = 0$, *near* iff $d_0 < d \leq d_1$, *medium* iff $d_1 < d \leq d_2$, and *far* iff $d_2 < d \leq d_3$. In Fig. 2 we depict a distance model regarding a fixed region of interest with an equidistant partition scheme regarding all d_i.

The bottom-right area of Fig. 1 shows a snapshot from the qualitative layer as a qualitative spatial relation table. Field 2 (Object1) and field 3 (Object2) represent values for pair of objects (e.g., building as a reference object relates to other referenced objects). The fourth field (centroid) represents the location of the geometric center of objects in the 2D plane by using latitude and longitude coordinates. The remaining fields store three types of qualitative spatial relations that hold between reference objects and other referenced objects. Practically,

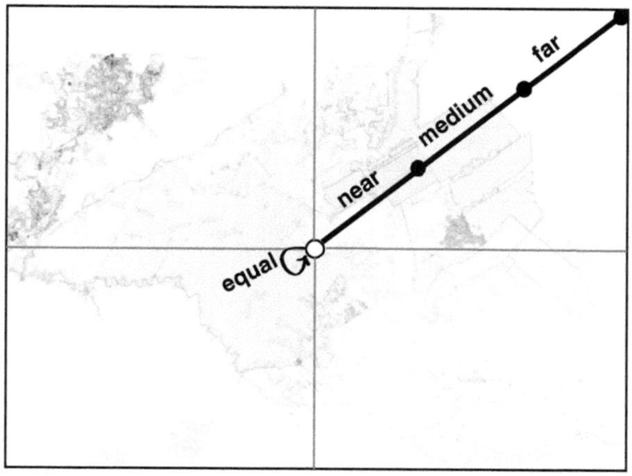

Fig. 2 The distance model with four relations

instead of computing the qualitative spatial relations at run time, they can be precomputed in advance and stored in a database which allows near real-time answers of spatial queries. However, the abstraction process still leads to an explosion of data which needs to be stored in a database. To deal with this problem, reduction techniques are proposed in [4]. By using reduction techniques, some relations can be inferred from others and suggest a reduction that exploits the superimposition of a uniform grid over the spatial dataset to identify a set of inferable relations. Such relations are not stored and are reconstructed at query execution time. When a database is updated (step 2), the DBMS triggers the qualitative layer to recompute the qualitative relations between updated objects and the remaining objects in the database. Hence, the update operation is a time consuming process. We leave this challenging point for future work.

2.3 Querying

To allow for handling qualitative spatial queries formulated by an emergency manager (step 4), these queries are translated into an SQL query (step 5). The structuring operation is done automatically by detecting each pair of an EM query. Then a reference object, a referenced object, and a symbol of the qualitative feature of each pair are assigned to appropriate fields in SQL which are related to fields in database tables (an example of an EM query in SQL is shown in step 5). Via the interface of the DBMS the query is processed (step 5 and 6). Finally, the results are presented to the EM in a graphical user interface. Due to reasons of space we point the interested reader to [14].

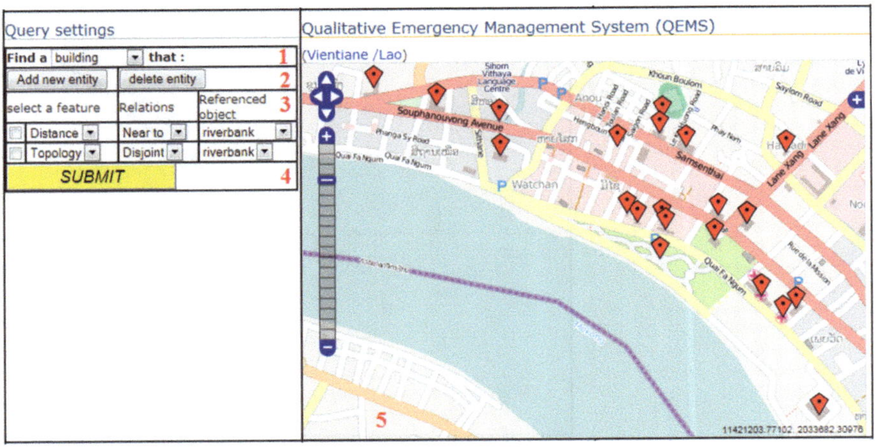

Fig. 3 A snapshot of the graphical user interface of the QEMS, showing the result for the query "buildings near to the riverbank, but not directly at the riverbank"

3 Implementation

Based on the architecture in Fig. 1, we implemented a prototype of an QEMS that enables emergency managers to query interesting places based on qualitative features. We focus on the task to find geo-coordinates of locations that fulfill some specific qualitative spatial features.

A snapshot of the graphical interface of QEMS is shown in Fig. 3. As a testbed we use a geo-referenced dataset of Vientiane (Lao PDR) which was extracted from OpenStreetMap.[2] It contains more than 500 objects. The extracted dataset is stored in a Postgres/PostGIS data-base, rendered by Osmarender and viewed by OpenLayers. However, not all extracted objects are annotated by volunteers, hence a preprocessing step is applied to eliminate non-annotated objects. We marked different spots in the figure from one to five. In the input field denoted by '1', a reference object (e.g., a building) can be selected from a drop-down list. Directly below objects which are supposed to be referenced to the reference object can be added or deleted (field '2'). In connection to this object a qualitative feature (e.g., distance) and a qualitative relation (e.g., "near to") needs to be selected (field '3'). The user of the system does not need to specify any quantitative value for any spatial feature. Figure 3 illustrates that the output of such queries could be helpful for EMs. As an example, it shows the answer to a query for buildings that are near to the riverbank, but disjoined from it (not directly at the riverbank) Emergency managers can formulate such queries to find coordinates of buildings which may, for example help rescue teams to spread goods in right places quickly. From this, an according query is generated (field '4'). Based on this query the QEMS

[2] www.openstreetmap.org

retrieves a set of matches which are displayed (as red markers) on a map (field '5'). Red markers denote the result: a set of matches that satisfy the spatial constraints that hold among pairs of objects in the EM query.

4 Related work

Recently, several systems and methods have been developed to response to natural disasters. Google Crisis Response[3] is a platform for empowering people to contribute to the status of disastrous areas (e.g., people can send test messages via mobile phones). Based on this data a spatio-temporal database of events is created, which can be queried by emergency managers.

A system that deals with natural hazards during the phase of disaster responding is presented in [10]. The authors propose a new spatio-temporal indexing method to speed up the query process. However, qualitative aspects are not tackled in this work.

A framework that detects damaged buildings in high resolution satellite imagery is proposed in [12]. The authors use clustering-based techniques to detect damaged buildings by comparing the roof texture of buildings in satellite imagery before and after the actual disaster. The focus of this work was not on querying databases. Other systems and researches were proposed to deal with response and recovery phases of emergency management. However, they mainly do not concentrate on extending DBMSs to handle qualitative spatial queries.

5 Conclusions

Handling qualitative data as coming from natural language is a crucial aspect of Emergency Management Systems. We proposed the integration of a qualitative layer into an EMS that covers topology, distance, and direction. In a prototypical implementation we presented a query system that translates qualitative statements to formal SQL queries and matches them to precomputed relations between objects in the spatial database. Thus, it is possible to answer qualitative queries by presenting a set of matches to the emergency manager for further inspection. This shows the general feasibility of our approach for extending EMSs with the ability to handle qualitative information.

Acknowledgments Funding by the German Research Association (DFG), SFB/TR 8 Spatial Cognition, and by the European Commission Initial Training Network geocrowd (FP7 - People Marie Curie Actions) is gratefully acknowledged.

[3] http://www.google.org/crisisresponse

References

1. A. Cohn, J. Renz, *Qualitative Spatial Representation and Reasoning, Handbook of Knowledge Representation* ed. by F. Harmelen, V. Lifschitz and B. Porter (Elsevier, 2007), pp. 551–596
2. M.J. Egenhofer, Query processing in spatial-query-by-sketch. J. Vis. Lang. Comput. **8**(4), 403–424 (1997)
3. M.J. Egenhofer, R.D. Franzosa, On the equivalence of topological relations. Int. J. Geogr. Inf. Syst. **9**(2), 133–152 (1995)
4. P. Fogliaroni, G. De Felice, F. Schmid, J.O. Wallgrün, Managing qualitative spatial information to support Query-By-Sketch, in *Understanding and Processing Sketch Maps (COSIT'11)*, vol 42 (IOS Press, Amsterdam, 2011), pp. 21–32
5. J. Freeman, The modelling of spatial relations. Comput. Graph. Image Proc. **4**, 156–171 (1975)
6. A. Guttman, R-Trees: A Dynamic Index Structure for Spatial Searching. SIGMOD Conference,47–57 (1984)
7. D. Hernández, E. Clementini, P. Di Felice, Qualitative distances, in *COSIT'95* (Springer, Berlin, 1995), pp. 45–57
8. K.D. Hurley, Planning and mitigation for emergency situations and natural disasters in hennepin county, minnesota utilizing GIS, in *Resource Analysis* (Saint Mary"s University of Minnesota Central Services Press, Winona, 2009)
9. H.C. Jang, Y.N. Lien, T.C. Tsai, Rescue information system for earthquake disasters based on manet emergency communication platform, in *Proceedings of the 2009 International Conference on Wireless Communications and Mobile Computing: Connecting the World Wirelessly*, IWCMC'09 (ACM, New York, 2009), pp. 623–627
10. R. Laurini, S. Servigne, G. Noel, Soft Real-Time GIS for Disaster Monitoring, ed. by P. Oosterom, S. Zlatanova, E.M. Fendel, in *GI4DM*, (Springer Berlin , 2005), pp. 465–479
11. Y. Manolopoulos, Y. Theodoridis, V.J. Tsotras, Spatial indexing techniques, in *Encyclopedia of Database Systems* ed. by L. Liu, M.T. Zsu, (Springer, New york, 2009), pp. 2702–2707
12. F. Samadzadegan, H. Rastiveisi, A clustering-based technique for automatic damage map generation using high resolution satellite imagery, in *Gi4DM*, Antalya, Turkey, 2011
13. S. Skiadopoulos, M. Koubarakis, Composing cardinal direction relations. Artif. Intell. **152**(2), 143–171 (2004)
14. J.O. Wallgrün, D. Wolter, K.-F. Richter, Qualitative matching of spatial information. in *Proceedings of the 18th SIGSPATIAL International Conference on Advances in Geographic Information Systems, GIS'10* (ACM, New York, 2010), pp. 300–309

Smoke Plume Modeling in Crisis Management

Hein Zelle, Edwin Wisse, Ágnes Mika and Tom van Tilburg

Abstract The current state of numerical modeling allows detailed, accurate weather forecasting, as well as forecasting of the transport and dispersion of smoke plumes and gas clouds. However, emergency workers do not have access to such state-of-the-art information during large incidents. BMT ARGOSS and Geodan have developed an integrated system to provide emergency workers with weather and smoke plume forecasts, in a modern crisis management system based on a geographical information system.

1 Introduction

In February 2011, a large fire in the Netherlands illustrated the need for better information during incidents. A fire broke out in a chemical storage facility near Moerdijk. The smoke plume extended well into the neighboring regions, requiring the participation of fire fighters from different fire brigades (crisis response in the Netherlands is organized into regions). As the smoke plume cooled downwind,

H. Zelle (✉) · Á. Mika
BMT ARGOSS, Voorsterweg 28, 8316 PT Marknesse, The Netherlands
e-mail: hein.zelle@bmtargoss.com

Á. Mika
e-mail: agnes.mika@bmtargoss.com

E. Wisse · T. van Tilburg
Geodan, President Kennedylaan 1, 1079 MB Amsterdam, The Netherlands
e-mail: edwin@kandedifang.com

T. van Tilburg
e-mail: tom.van.tilburg@geodan.nl

S. Zlatanova et al. (eds.), *Intelligent Systems for Crisis Management*, Lecture Notes in Geoinformation and Cartography, DOI: 10.1007/978-3-642-33218-0_5,

it descended and touched the ground at a considerable distance from the fire. The smoke potentially contained several chemical species from the storage facility, as well as the products of the chemical reactions in the fire. As the species and concentration levels were not well-known, estimating the health risks for people in the affected regions was difficult or impossible.

The need for better information about the movement and dispersal of a smoke plume was emphasized by this particular incident. Authorities in the municipalities where the plume reached ground level had no information about the pollutants in the cloud and the concentration levels. The crisis management team at the incident location of the fire could not adequately predict where the cloud was moving. Also, there was a lack of ability to share the information effectively between affected regions and the involved crisis management teams.

1.1 Current State

During an incident, an expert on dangerous chemical species is made responsible for providing information to other emergency workers and decision makers. In the Netherlands, this expert is called the "advisor on hazardous substances" (AGS). The expert collects information about the substances present at the incident location and, using standard procedures, reports about the consequences and gives advice on the actions that should be taken (e.g. evacuation). During a typical incident with a large fire, the following events and actions would occur:

1. Fire detected and reported to central operator (emergency number)
2. Fire crew alerted, present at location
3. Chance of emission of dangerous chemicals asserted
4. Specialist in chemical species called to incident, arrives within 30 min.

At this point, the specialist will assess the weather conditions, the state of the fire and the possibility of dangerous chemicals being released. He uses a standard worksheet to estimate wind, emission strength and species, and obtains a template representing the possible spread of the plume. This template can then be placed on a paper map or applied on a digital map.

5. Wind direction and strength are obtained from the nearest weather station or estimated.
6. The specialist estimates fire parameters using a standard worksheet: surface area, temperature, emission strength.
7. A "gas template" is selected using the standard worksheet. This is an outline of a smoke plume, printed on transparent plastic, which can be overlayed on a map.

The specialist can now draw affected areas into the map. He will decide where to send observation teams, which areas and access roads to block, and which areas should potentially be evacuated.

One of the main disadvantages of this approach is the lack of real-time information and forecasts about the weather and the movement of the smoke plume.

1.2 Proposed Solution

The aim of this project is to provide emergency workers access to real time, sophisticated information about the weather and the movement of smoke and chemical species. To achieve this, a meteorological model is coupled to a chemistry-transport model. The models are integrated with a state-of-the-art crisis management system: Eagle [2, 4]. A user interface is developed which conforms to existing procedures in crisis management. Network communication is used to give all the parties involved in an incident access to the same information [3]. This also allows the fire brigade's chemical specialist to share relevant information about a smoke plume with the other parties involved, in an understandable form.

1.3 Challenges

Scientific institutes that create and run numerical models work in a very different environment compared to emergency workers that deal with the direct effects of a fire or gas cloud. Numerical models are typically run in a computer lab with the availability of large meteorological data sets and strong computing power. A typical weather forecast takes several hours to compute, and can only be performed 2–4 times per day. In contrast, during a crisis situation near-real-time access to information is required. In practice this means that an information request must be answered within minutes and all model data must be readily available. In addition, the quality of a model result will be judged differently: a scientist will emphasize on physical accuracy, whereas a fire brigade may focus on timely delivery, useful presentation and simplicity of the results.

The challenge of this project is to bridge the gap between models in the lab and real time situations in the field, without compromising the quality of the model results or the reliability, speed and usability of the system.

2 System Description

The basic system design consists of three clearly defined components: a weather model, a smoke plume model and a crisis management system. These components are linked together, and presented to the user with a dedicated user interface.

In this section we first describe the intended use of the integrated system. Each individual component is then described separately. The integration of the components is described in Sect. 2.9.

2.1 Scenario with Integrated System

The capabilities of the integrated system are demonstrated with a scenario of a large fire in a chemical factory. Within the crisis management system, an incident is opened and other participants are invited (e.g. the commander in charge). Opening an incident initiates a map for that incident which will be shared among the involved parties. Since the location is a chemical plant the commander decides to call the specialist: the adviser hazardous substances (AGS) to the incident location.

The specialist arrives at the scene and determines it is likely that dangerous chemicals are being released into the smoke plume. The specialist opens the crisis management system and enters the fire and emission details in the input screen of the smoke plume module. To do this, he follows standard procedures to estimate the type of fire, the size of the source and the temperature of the fire. Based on what materials are known (or guessed) to be burning, the chemical species that are emitted are chosen. The specialist will then arrive at a rough estimate of the strength of the emission source: for example, 1 kg CO is being emitted per second.

1. Specialist enters fire parameters: location, size, temperature, emission strength. Emission strength and type are estimated using the standard worksheet and methods.
2. Model simulation is automatically started.
3. Simulation process is monitored and shown to the specialist ($\sim$3 min).
4. Simulation results are shown on the map as concentration contours and wind vectors.

The specialist now has access to all information regarding the smoke plume forecast. He can verify that the observed wind direction and strength match with the weather forecast. The smoke plume can be visualized in different ways, looking from above or in the form of a vertical cross-section. The plume is drawn using 3 contours representing "alert", "evacuate" and "lethal" levels.

In the crisis management desktop, these contours can be combined with information about sensitive buildings, number of people affected and whether the plume affects evacuation routes.

5. Specialist draws affected areas and evacuation zones into the map, using the plume forecast as a guide. Typically this is done with streets or blocks as logical boundaries, keeping access possibilities for vehicles in mind.
6. The affected areas are shared with all emergency workers that should have access. The smoke plume contours are not shared with the other (non-specialist)

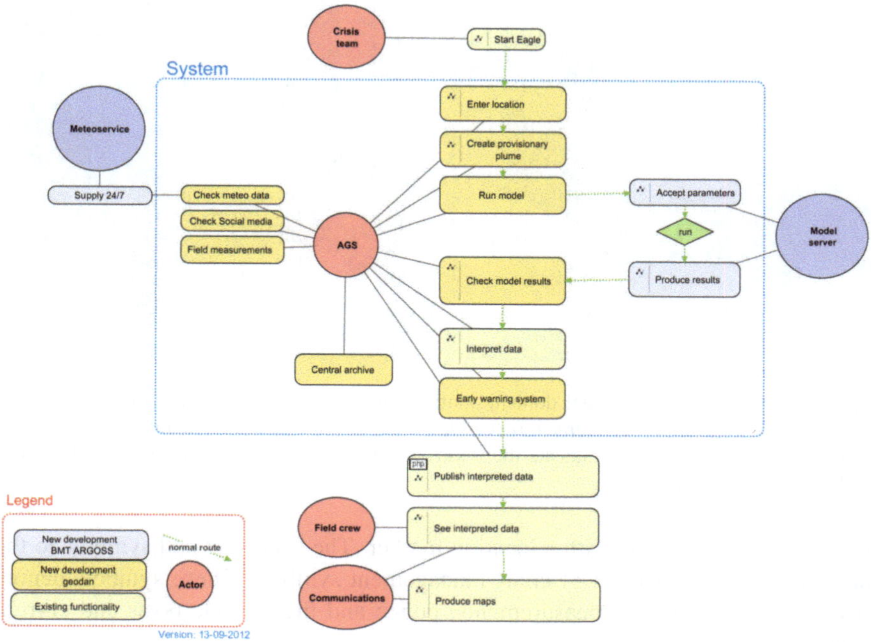

Fig. 1 Schematic for the smoke plume use case

parties involved in the incident like municipalities, the police and medical services.

A conscious decision was made not to share all information between all parties involved. The output of a smoke plume model requires expert knowledge about the processes involved, interpretation of the output and the model limitations. Therefore, the chemical specialist will only share his conclusions with the other parties involved, not the model results themselves. These conclusions (which areas are safe, which should be evacuated, which roads should be blocked) are shared as annotations and marked areas on the map. Communication with measurement teams in the field may also take place through the shared map.

Figure 1 gives a schematic representation of the system components and the use cases during an incident.

2.2 Crisis Management System

The Eagle crisis management system is used to share information during an incident. Eagle uses a map as the shared communication layer. An incident is opened, participants are invited and from that moment on all invited parties see a shared map of the incident location. This map will show all communication

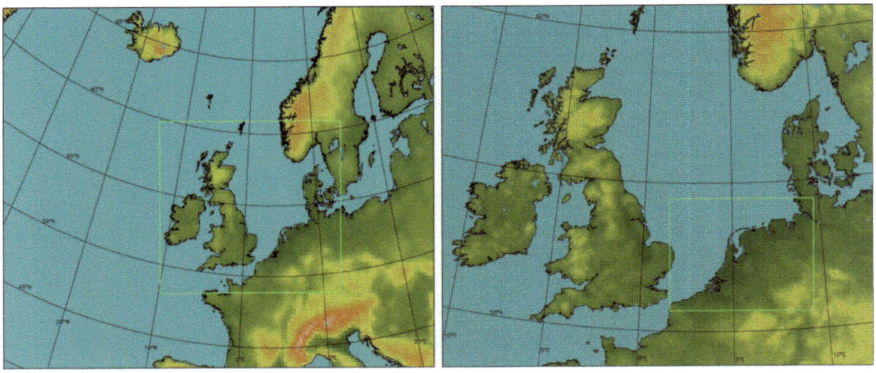

Fig. 2 The WRF weather model domain configuration. *Left panel* the outer domain grid has 27 km spatial resolution. The *green rectangle* marks the location of the intermediate, 9 km domain. *Right panel* the *dots* indicate the 9 km grid, the *green rectangle* over the Netherlands marks the finest, 3 km grid

messages as annotations with a location marker. The users can add symbols to the map from a predefined set of crisis management symbols. These symbols denote road blocks, location of measurements, police and fire brigade posts and current locations of vehicles.

Eagle is implemented in a GIS system. Therefore a geo-spatial analysis of available information can be made and shared immediately. Different extensions have been built in Eagle, for example to estimate the number of people within a given area. Other extensions are available for the use of social media as information sources and for the display of video feeds from (unmanned) aerial vehicles. The smoke plume simulation interface has been implemented as an extension to the Eagle system.

2.3 Weather Model

BMT ARGOSS operates the WRF (Weather Research and Forecasting) atmosphere model for daily weather forecasting. WRF is an open-source model, developed in cooperation between several institutes and universities in the USA [5, 6]. The model is state-of-the-art, used both in research and operational weather forecasting.

For Europe and the Netherlands, the model is configured with 3 nested domains, going from a large, coarse (27 km resolution) domain covering north-west Europe to a small, fine (3 km resolution) grid over the Netherlands. The nested grids allow us to refine the model quality, while keeping the required computational power limited. Figure 2 shows the WRF domain configuration.

Model forecasts are produced 4 times per day, with the following parameters:

- forecast times: 00, 06, 12 and 18 UTC
- forecast window: from +0 to +120 h (5 days into the future)
- boundary forcing: NCEP GFS global forecast model
- nesting: fine-scale results are fed back to the coarse-scale model
- topography: SRTM (Shuttle Radar Topography Mission) data with 90 m spatial resolution
- land use: MODIS satellite imagery classified in 20 categories
- vertical structure: 31 levels from the surface up to approximately 20 km. Extra levels were added for more detail in the lower 200 m.
- convective precipitation: clouds and rainfall are explicitly resolved in the 3 km domain.

The result of the model forecast is a 4-dimensional database of all atmospheric parameters. The data is post-processed to compute any derived variables, to standardize the NetCDF output files and to visualize the most important parameters on a web site. Figure 3 shows a sample of the weather charts generated from the operational weather forecast.

2.4 Smoke Plume Model

We now have an up-to-date weather forecast, at most 6 h old, which provides weather information up to 5 days in the future. The next step is to couple the weather model to a chemistry-transport model. Given the state of the weather and an emission source (e.g. a fire), such a model will calculate the transport of chemical species, as well as their transformations due to chemical reactions. The following characteristics are required:

- Capable of very high detail levels: street level or better. The target resolution is 50 m.
- Computationally very fast. Target response time: <5 min.
- Capable of simulating extreme situations such as high concentrations, strong flows, high temperatures.
- Capable of simulating dynamic situations with changing weather conditions and variable emission sources.
- Capable of simulating vertical plume motion due to heat transport, and providing 4-dimensional output.
- Preferably open source with a sound scientific basis.

Out of several candidates, the CalPuff model was selected [1]. This model complies with all the requirements above. CalPuff is a so-called "puff" model, which means it simulates a gas cloud as a collection of puffs, small packets of chemical species. These packets are transported by the wind, dispersed or affected by chemical reactions. The advantage of this system, compared to a Eulerian model

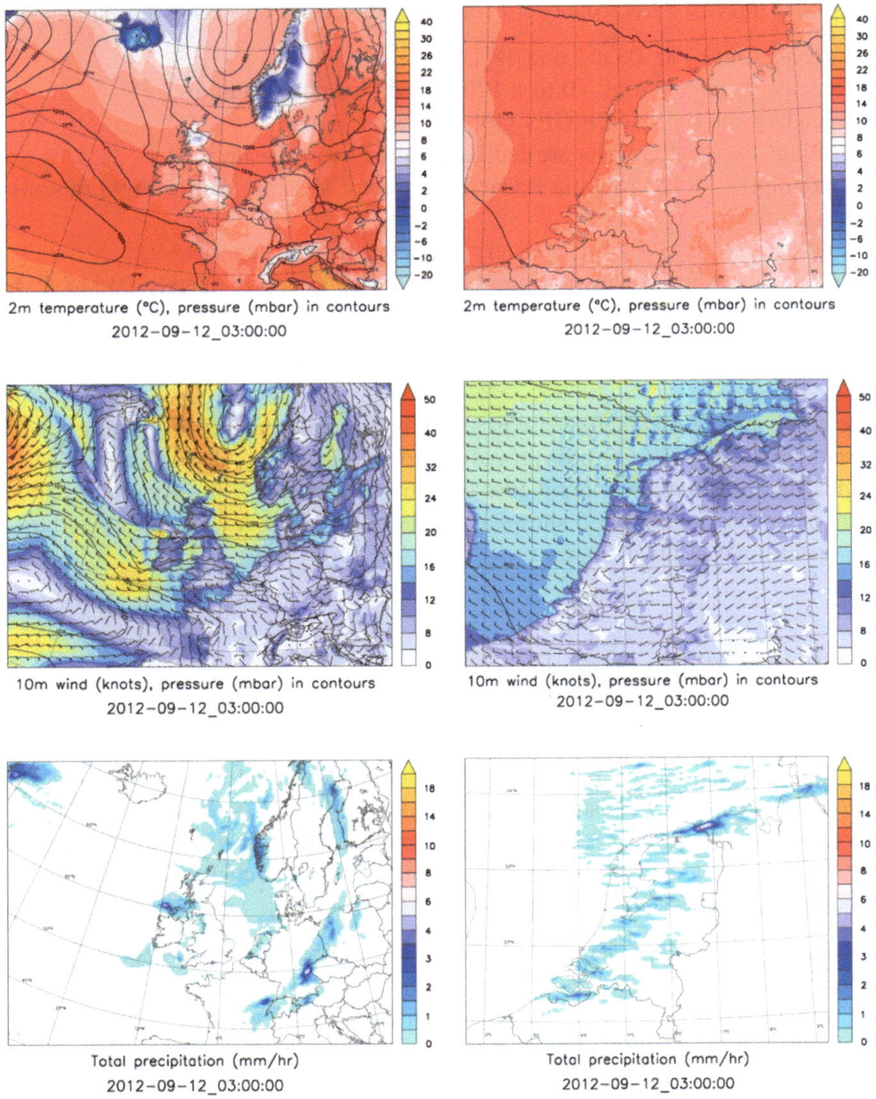

Fig. 3 Sample weather forecasts for north-west Europe (*left*) and for the Netherlands (*right*). *Top* 2 m temperature (°C). *Center* 10 m wind speed (m/s). *Bottom* total precipitation (mm/h)

with a regular "box" grid is that it allows very localized emission sources and it is more suitable for highly dynamic situations, such as the plume coming out of a chimney. The puff model is also much faster computationally, a big bonus for this application.

CalPuff is well tested and is used operationally as well as for regulatory purposes. For specific applications such as the modeling of explosions or the dispersion of heavy gases, the use of a different, specialized model may be more appropriate. The system design is modular and allows replacement of individual components so this an be easily achieved. An extensive comparison of Calpuff with other fire models is outside the scope of this paper but the main reason for not using them is that they are computationally too slow if operated at the required space-time resolution. This is especially true for the Eulerian models.

2.5 Model Coupling

The coupling between the atmosphere model and the smoke plume model is relatively straightforward. The output of the WRF model consists of full 4-dimensional data fields, in NetCDF format. The only non-trivial aspect is the projection of the model grid, which is a Lambert Conical Conformal projection. The CalPuff model requires 4-dimensional wind, temperature and moisture fields. From these data it can compute all required information, notably the stability of the atmosphere, the horizontal and vertical transport and the diffusion. The CalPuff model grid uses a different map projection (Rijksdriehoek, a variant of UTM zone 31 specifically designed for the Netherlands). A custom coupling program reads each of the required variables in from the WRF data files, transforms the source grid to the target grid, and writes output files in a format specified by CalPuff.

At this point the CalPuff model can be operated normally by providing it with information about the emission source, a reference to the pre-processed weather data, and other required configuration information about the simulation such as start and end times.

2.6 Validation of the Coupled Modeling System

The WRF and CalPuff models have both been validated extensively. Both models are used by the research community, where validation studies are regularly performed and published.

The operational WRF configuration at BMT ARGOSS has been validated in previous projects using all available observation stations from the national observation network in the Netherlands [7]. This was performed for the period 2000–2010. Extensive statistics were computed and presented. Overall, we can conclude that the model performs equally well as the HIRLAM model when compared on a large scale. At smaller scales the WRF model shows clear improvements, notably near coastal boundaries, in cities and in more difficult terrain such as hills.

The coupled smoke plume model was validated using data from a fire on the fish trawler Willem van der Zwan (SCH 302) which took place on January 31, 2007 near Velsen in the west of the Netherlands (Fig. 4). The incident report produced by the chemical expert during the initial stage of the fire was used to

Fig. 4 Fire fighters at work on the fire at the trawler Willem van der Zwan, 2007 January 31

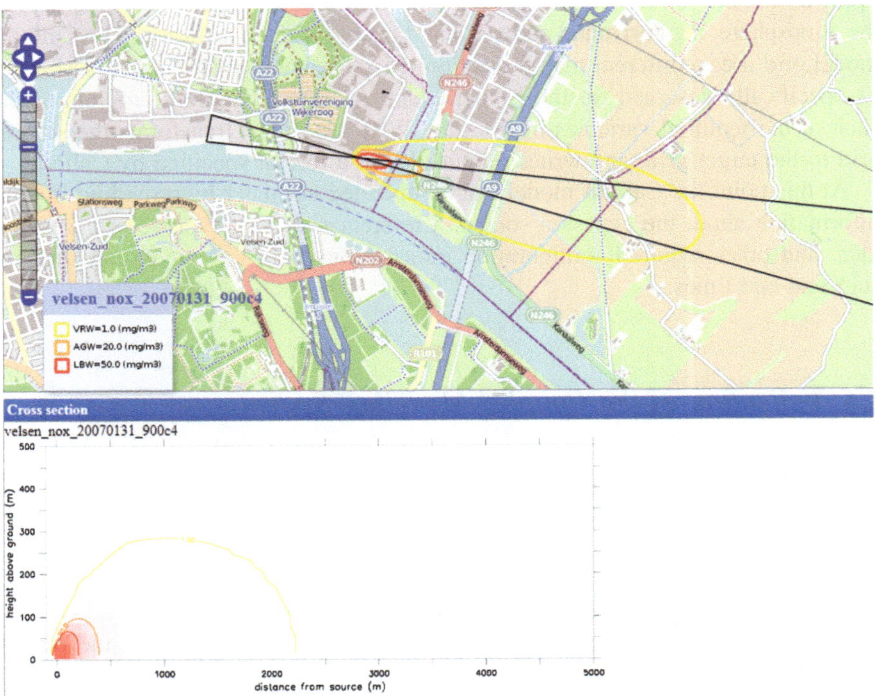

Fig. 5 Estimated NO_x concentration at the start of the fire incident at the trawler Willem van der Zwan, 2007 January 31. *Top panel* concentration contours at ground level. *Bottom panel* vertical cross-section along the plume *center line*

assess the emission rate of NO_x. As the fire fighters arrived, measurements indicated a concentration of 20 mg/m^3 NO_x in the vicinity of the vessel. Figure 5 shows this value (orange contour) as predicted for the time of measurement. The top panel shows the concentration contours at ground level and the bottom panel the concentration contours in the vertical along the plume center line. More measurement/prediction comparisons at different times show the same similarity indicating that the model provides plausible results. In cooperation with the regional fire brigades more validation data are collected of other incidents.

2.7 Dealing with Uncertainties

It is possible, even likely, that the weather and smoke plume forecasts deviate from reality. This can be due to the general uncertainty of the weather or local effects such as trees, water bodies and buildings. Even more likely are errors in the model input parameters. The emission strength may easily be wrong by a factor of 10. The size and temperature of a fire are all estimated and may contain significant errors. This means a specialist will always have to interpret the information, adding uncertainty margins when drawing affected work areas before sharing them.

In case of weather forecast errors, the chemical specialist may contact a meteorologist to ask for advice. The meteorologist can verify the forecast error and explain the difference (e.g. a low pressure system arrived earlier than expected). He can then provide advice on the expected developments: e.g. the real plume will likely deviate to the south compared to the forecast.

Errors in the estimated fire temperature and source strength are best dealt with by the chemical expert. The current system is designed in such a way that it is easy to generate multiple forecasts: the specialist can re-do a forecast with a modified emission strength or a different fire temperature. The results can then be inter-compared to make a decision based on multiple possible outcomes.

2.8 Operation and Reliability

When applying a modeling system toward emergency services, it is vital that the system is always available. BMT ARGOSS faces similar requirements for its regular forecasting services. Service reliability is guaranteed by the following means:

- a cluster computer with resilient components
- partial backup computing capacity at an off-site location
- a job scheduling system (SMS) which plans, submits and checks each task at regular times. SMS was developed at the European Centre for Medium Ranged Weather Forecasting (ECMWF) for this purpose. It automatically deals with scheduling jobs at the right time, inter-job dependencies (job *b* can only run after job *a* is complete), error handling, sending email notifications, automatic re-running of failed jobs.
- Round the clock system monitoring by meteorologists and technical support.

These systems have proven very reliable, with very low failure rates and fast recovery in case of failure. However, for the purpose of an operational deployment for emergency services the level of reliability should likely be increased even further: a fully redundant system at a different site may be required to deal with power outage, complete internet failure or a fire.

2.9 System Integration

The Eagle crisis management system is used as an interface to the smoke plume modeling service. The aim is to allow the user seamless access, to enable geographical analysis combining smoke plume information with other information sources and to allow the user to share information resulting from his analysis with other involved parties. To achieve this integration, the Eagle system must be able to do the following:

- enter fire parameters (user interface component)
- start a remote task (run a remote simulation)
- request model information (smoke plume model results)
- display this information in its own map (contours in GIS map).

Entering the fire parameters is achieved through a user interface component which was implemented as an extension to Eagle. Figure 6 shows the input form in Eagle.

Displaying the model results is made possible through the use of open geographical standards: web map services (WMS) and web feature services (WFS). Performing analyses requires access to the numerical model results: this is achieved via web coverage services (WCS). These protocols are supported by most GIS clients. The model results (contours) are delivered via the WMS or WFS protocols, and can be instantly displayed in the Eagle map. Figure 7 shows a sample smoke plume in Eagle.

Finally, Eagle must be able to start a model simulation or request status information. Such communication between Eagle and the model server is implemented using the open Web Processing Standard (WPS) protocol. The advantage of the WPS protocol is that it allows a clear division between user interface and model implementation. This allows the implementation of alternative user interfaces or model components speaking the same protocol. A web interface with similar functionality to the Eagle map has been developed to demonstrate this, and to perform tasks such as archiving of previous model simulations. Figure 8 shows the system components and communication protocols.

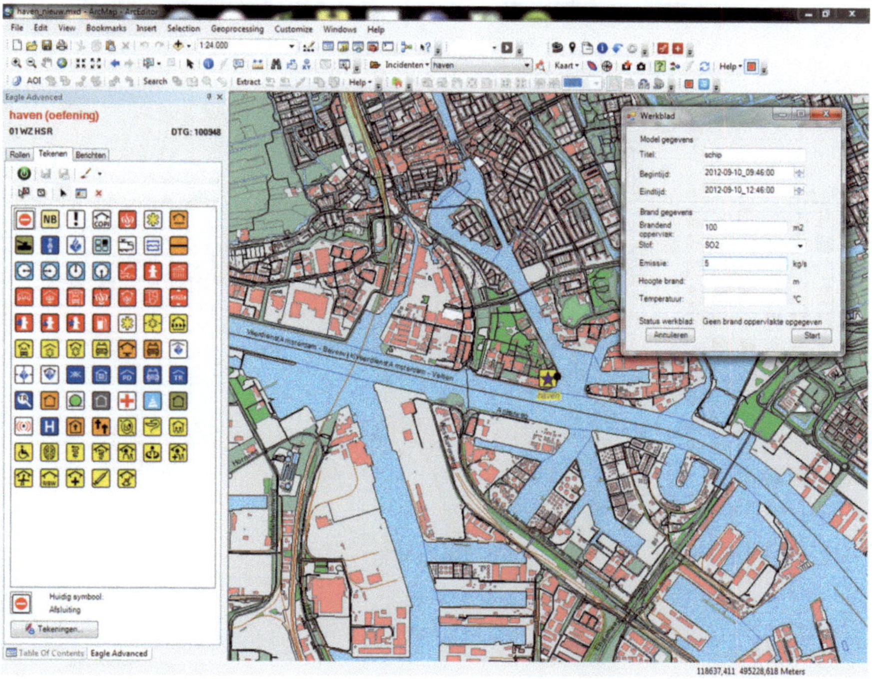

Fig. 6 The input form in the Eagle crisis management system

3 Future and Ongoing Work

The service is currently being evaluated by several fire departments. The next step will be to evaluate the system during live exercises where incidents with dangerous chemical species are simulated. These are planned for the end of 2012.

The smoke plume is currently represented as a set of contours on a 2D map and as a vertical cross-section. There are cases in which the height of the plume is important, for example in a city with tall buildings. Full 3D visualization of the smoke plume is being explored in the project. A limiting factor lies in the representation of the common map (top down view) in the crisis management system: 3D information can not be shared and 3D visualization will remain a separate option, for now. Other options such as visualization in Google Earth are being developed.

There are several highly specific types of incidents which are not yet covered by the system. These include explosions, the release of gases heavier than air and high-pressure gas leaks. Work is ongoing on the implementation and validation of some of these incident types. Validation may be required in cooperation with scientific institutes. Some incident types may require specialized models for proper representation of the physical effects. These types of incidents will be added in the future.

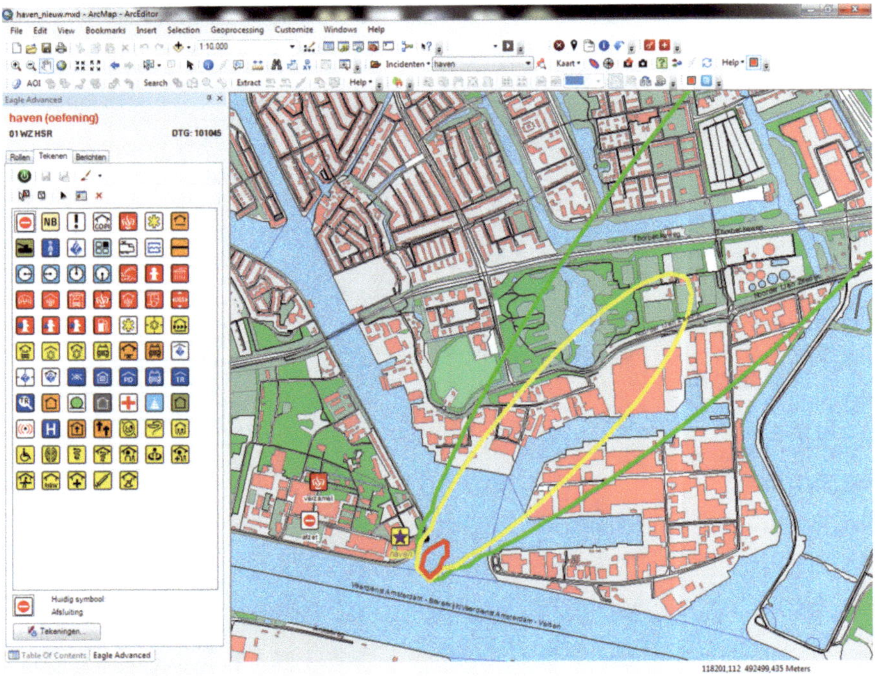

Fig. 7 The smoke plume contours displayed in the Eagle crisis management system

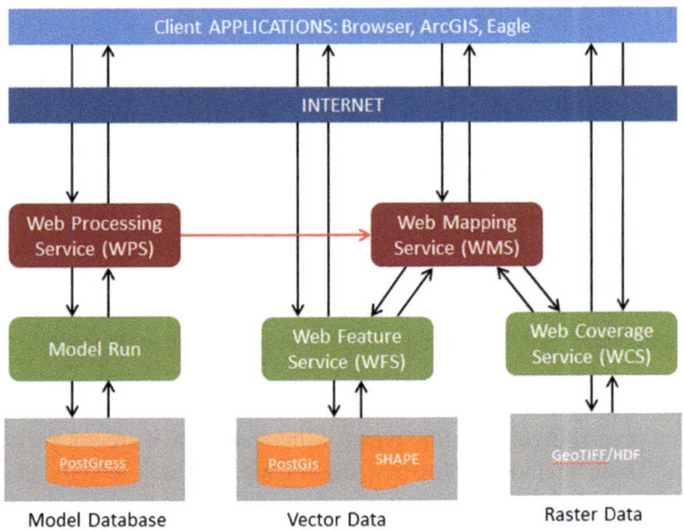

Fig. 8 Schematic of the system components. Web protocols are indicated between parentheses. *Red* and *green blocks* can be approached via the Internet. *Gray blocks* are implemented internally and cannot be accessed directly

The final, main challenge for this service is the market introduction. Although the added value of the service is obvious and the market (fire brigades, emergency services) is clearly present, it is complex to introduce such services. There are many parties involved (national institutes) with responsibilities during incidents. Each of these will have to judge how the service connects to their tasks and responsibilities. Results from complementary services (e.g. this service and a national service) should not contradict each other and the working methods should be the same. Discussions with the national institutes have been planned and are ongoing. We strive to see this service in full operation in the near future.

4 Summary and Conclusions

BMT ARGOSS and Geodan have developed an integrated system to provide emergency workers with smoke plume and gas cloud forecasts during incidents. The system is based on a coupled modeling system: a high resolution weather model drives a chemistry-transport model which computes the smoke plume transport. Model results are presented in Eagle, a state-of-the-art crisis management system. Eagleis a GIS-based tool, displaying all information in a common map. It allows transparent sharing of information between all parties involved in crisis management.

The integrated smoke plume modeling system has been tested by several fire departments in the Netherlands. Initial responses are very positive, and further tests during live exercises are planned for fall 2012.

This project has proven that it is possible to apply the state-of-the-art in weather forecasting and chemistry modeling toward real-world applications in crisis management and rescue work. We strive toward bringing this system to the market both in the Netherlands and internationally, so all emergency workers will have access to the best information possible during an incident.

Acknowledgments This project is sponsored by AgentschapNL.

References

1. CalPuff model, http://www.src.com/calpuff/calpuff1.htm
2. S. Fruijtier, E. Van der Zee, B. Gehrels, S. Zlatanova, M. Ryan, H.J. Scholten, Design, implementation and real-life evaluation of a disastermanagement architecture, in *GISWORX* Dubai, 2009
3. J.M.M. Neuvel, H.J. Scholten, A. van den Brink, *From Spatial Data to Synchronised Actions: The Network-centric Organisation of Spatial Decision Support for Risk and Emergency Management* (Springer, Applied Spatial Analysis, 2010)
4. H.J. Scholten, S. Fruijter, A. Dilo, E. Van Borkulo, in *Spatial Data Infrastructures for Emergency Response in the Netherlands*, ed. by S. Nayak, S. Zlatanova. Remote Sensing and

GIS Technologies for monitoring en Prediction of Disasters, (Springer, Berlin, 2008), pp. 179–197
5. W.C. Skamarock, J.B. Klemp, J. Dudhia, D.O. Gill, D.M. Barker, W. Wang, J.G. Powers, A Description of the Advanced Research WRF Version 2, UCAR Technical Report (2005)
6. WRF model, http://wrf-model.org
7. H. Zelle, P. Groenewoud, C. Calkoen, Forecasting Services for Wind Parks, Technical Report, BMT ARGOSS (2010)

Simulation System of Tsunami Evacuation Behavior During an Earthquake Around JR Osaka Station Area

Ryo Ishida, Tomoko Izumi and Yoshio Nakatani

Abstract In the case of a Tonankai earthquake or Nankai earthquake occurring, it is predicted that tsunami damage will occur at the JR Osaka Station area. The purpose of this study is to propose a computer simulation which can calculate the appropriate behavior of evacuees in order to avoid tsunami damage in that situation. There are many institutions, department stores and hotels, and an underground shopping center, around JR Osaka Station. Therefore, various situations must be considered, such as collisions between people trying to evacuate to buildings and people trying to get away from the underground shopping center and buildings to the outside, in order to plan safe evacuation. In this paper, we implement a simulation that assumes various situations, and collect data fundamental to devising a method for inducing evacuation.

Keywords Evacuation · Agent-based · Shortest path · Behavior · Refuge buildings

R. Ishida (✉) · T. Izumi · Y. Nakatani
Graduate School of Science and Engineering, Ritsumeikan University, Shiga, Kusatsu 5250058, Japan
e-mail: is003082@ed.ritsumei.ac.jp

T. Izumi
e-mail: izumi-t@fc.ritsumei.ac.jp

Y. Nakatani
e-mail: nakatani@is.ritsumei.ac.jp

S. Zlatanova et al. (eds.), *Intelligent Systems for Crisis Management*, Lecture Notes in Geoinformation and Cartography, DOI: 10.1007/978-3-642-33218-0_6,

1 Introduction

Earthquakes occur in Japan so often that is particularly well known as an earthquake-prone country. At present, there are four major earthquakes predicted to occur in Japan in the near future: a Tokai earthquake, Tonankai earthquake, Nankai earthquake, and a major earthquake in the Tokyo Metropolitan Area. In particular, the probability of occurrence of a Tokai earthquake in the next 30 years is 87 %, and that of Tonankai and Tokai earthquakes is also calculated to be greater than 60 % [1]. If these earthquakes actually occur, it has been predicted that there will be massive damage and many stranded persons unable to return home.

It has been predicted that there will be many stranded persons in Osaka City, which would be greatly affected by a Nankai earthquake or Tonakai earthquake. JR Osaka Station, Midosuji (a main street in the city), Osaka City Hall, etc. are located in Osaka City. These are specified as being within the flood region, and it has been predicted that massive damage to the municipal subway and underground shopping center will occur. There are many people, particularly tourists, shoppers and commuters, in the area surrounding JR Osaka Station, so there is the possibility of around 420,000 people being trapped and widespread confusion occurring as they take refuge from a tsunami during an earthquake disaster [2]. In addition, it has been predicted that secondary and tertiary disasters may result from the original disaster. They are the disasters that occur subsequent to when a disaster occurs. Therefore, there is a need to conduct quick and precise evacuation guidance in the context of reviewing prevention measures to reduce major damage in Osaka City. It is predicted that evacuation simulation by computer is effective because it is difficult to conduct an experiment of evacuation guidance on a real-life scale in the JR Osaka Station area, which is crowded even on weekdays.

In order to solve these problems, this research aims to construct a working environment that can simulate on a computer the behavior of disaster victims, targeting JR Osaka Station and Umeda Station areas.

2 Related Work

2.1 Physical Model Method

The physical model method assumes that a crowd will take action following specific physical laws.

This commonly used method regards crowds as fluid, and calculates the number of people per speed and space by design of the line of flow in buildings, which is called the fluid model. In cases when the crowd acts smoothly, this method is effective and highly reliable. It is capable of simulating situations where the flow of people is blocked, by setting the proper parameters using calculation, but it is also necessary to set parameters considering the structure of buildings.

An individual member of the crowd is regarded as a single rigid body, and the grains model has been used to model their movement in situations where they move while interacting with each other. Compared with the fluid model, this is capable of expressing detailed movement, but the number of calculations is higher.

Over the past few years, several studies have been conducted on simulation models using the potential model as a highly physical model [3]. It is possible to simulate various evacuation behaviors including blockades by setting appropriate potential to the spatial properties of route, individuality of evacuees, and the disaster situation, etc., but the reliability of results depends on the quality of the set potential. It is capable of simulating situations, but there are no principles or methods of determining potential objectively, so it is not suitable for predicting emergency evacuation behavior, where ensuring reliability is important.

2.2 Cellular Automaton Method

The cellular automaton method is used to calculate the movement of crowds as a state transition between cells, following a simple rule using probability by dealing with time and space discretely. Although the rule is simple, it is possible to form a complex transition pattern by splitting time and space between a large number of cells. In the case of evacuation behavior, if state transition between cells is modeled as the ensemble behavior of each pedestrian, it is possible to implement various patterns of behavior by including or ignoring various elements of the determined behavioral rules of each evacuee. As a consequence, it is capable of reproducing complex state transition. The degree of freedom in setting the form of walking space and positioning facilities is high, and it is possible to easily program and calculate complex boundary conditions, such as smoothly flowing evacuation patterns and blocked evacuation patterns.

This method is suitable for classifying qualitative patterns of state transition, but the quantitative prediction is limited. It is capable of predicting quantitatively by increasing the complexity of the rule, but this is not a strong point of this method.

2.3 Multi-Agent Simulation

The multi-agent method "constructs a system bottom-up from the behavior of many autonomous agents" [4]. It regards individual elements constituting the system as agents, makes them act autonomously and simulates the behavior of the system. This method is sometimes described as an advanced type of the cellular automaton method, but it is notable for modeling decision-making and actions of individual evacuees, and has been practiced in the field of artificial intelligence since the late 1980s.

At present, it is possible to obtain reliable results through designing suitable agents, so various multi-agent simulations have been researched. For example, multi-agent simulations of evacuation behavior in fire emergencies have been developed, and it is planned to diversify to other disaster situations. In social technology, multi-agent simulations of risk management in nuclear power plants have been developed. For earthquake disasters, multi-agent simulations have also been used to predict fire suppression and the actions of the police, etc.

2.4 Related Research

Various research and development of crowd evacuation behavior simulations have been conducted in relation to fire and tsunami situations. For example, Hori et al. conducted evacuation behavior simulations in underground spaces [3]. They experimented with multi-agent simulation on evacuation behavior from a subway station with five underground levels to the ground during an earthquake. Consequently, they were able to analyze evaluation findings such as that the evacuation time of individual evacuees is uneven and increases according to the number of people.

In addition, Watanabe et al. developed an evacuation simulation model for tsunami situations using multi-agent simulation. During planning of community development of disaster prevention for tsunami situations, this research can examine both maintenance of material aspects and policy [5]; for instance, questions such as what kind of maintenance will enable all evacuees to escape within a set period of time.

3 Summary of Proposed System

3.1 Current Problems

Until now, many studies have been made on simulations of human behavior using computer [6, 7]. In a conventional simulator, the principal purpose is to model the actions of evacuees simply as fluid and to simulate this movement, but simulators of not only physical interaction among evacuees but also social interaction have been developed recently. However, fundamental questions remain unanswered. As mentioned in Section II, there have been many simulations such as those targeting the damaged region, those conducted in internal spaces such as tower buildings, department stores, underground shopping centers, etc., but there are not many simulators that model evacuation in metropolises such as Tokyo and Osaka, in which there are many large buildings and various groups of people. In addition, large-scale and large-area simulations of evacuation behavior which consider evacuee inflow to buildings do not exist. For this reason, it is not possible to predict what situations will occur in Osaka City during an earthquake in terms of collisions between people trying to evacuate to buildings and people trying to get away from the underground shopping center, and from buildings to the outside.

Furthermore, there is an immediate need to reconsider planning of disaster prevention related to a Nankai earthquake affecting Osaka City in the wake of the 2011 Tohoku Earthquake and Tsunami. In particular, measures in relation to tsunami resulting from an earthquake are a main object to be reconsidered. As the Yodo River is located close to JR Osaka Station and Hankyu Umeda station, it is predicted that widespread damage and confusion will occur in the event of a tsunami. In order to reduce such damage, it is necessary to consider effective tsunami countermeasures. Recently, quick and precise evacuation guidance within the limited time period before a tsunami occurs is effective.

However, private companies and administrative powers in Osaka City do not share a specific control strategy for tsunami occurrence. For this reason, it can be assumed that firstly, the information needed to formulate a tsunami evacuation manual is lacking; and secondly, even if such information exists, it is not being shared. As a result, individual companies and institutions are making separate guidance manuals for evacuation. In the case of actually basing evacuation guidance on such manuals, it is not clear whether these constitute effective guidance for a particular region. For instance, even if tsunami refuge buildings are capable of accommodating stranded persons, evacuees have a psychological resistance to moving higher than the third floor in buildings in case of stopped elevators and escalators after an earthquake. In consequence, widespread confusion may occur in the case of many people remaining on lower floors while further evacuees enter there, and this may occur on both the ground and underground levels. In order to resolve such problems, it is necessary to make a system that is capable of inspecting how evacuees will behave considering various evacuation situations.

3.2 Proposed System

The purpose of this study is to develop a computing system that simulates evacuation behavior of people at the time of tsunami precaution in major cities, including in the ground and the underground, and buildings. Furthermore, it is our goal to provide fundamental data for specifying tsunami refuge buildings and selecting the evacuation guidance method and system, the guiding location, etc. This paper is the first step in this objective, and thus we have developed a simulator of evacuation behavior which targets people located only on the ground level. Using this, we conducted a simulation targeting the area around JR Osaka Station and Umeda Station in various situations.

The following functions are considered in the system.

- Functions
  - Target area: The target area is defined as the area within a kilometer radius of JR Osaka Station.
  - Evacuation route: The evacuation route used by evacuees is the shortest distance on the road network to JR Osaka Station. Further, it assumes that there are no obstacles such as cars on the road, and it is set so that the evacuees do not take any detours even if the walking speed decreases due to dense crowding. We plan to consider the influence of road traffic in future research.
  - Walking speed: The speed used as a standard is considered to be 66.6 m/min per one person/square meter from the relationship between crowd density and speed of the evacuees on the evacuation route. It is set so that the speed decreases as the congestion of people increases.
  - Tsunami refuge buildings: In order for evacuees to take refuge in places other than the goal, buildings in which large numbers of evacuees, such as Umeda Station, department stores, and tower buildings etc. are designated. As such, buildings specified as tsunami refuge buildings are designated as buildings for emergency evacuation.
  - Evacuee limit: A control panel can be created so that the number of evacuees who can be accommodated in those buildings can be adjusted, and it can be set up with various values.
  - Graphical representation: The X-axis indicates lapsed time and the Y-axis shows the number of evacuees who have taken refuge in each building on the time series graph of the evacuee number.

4 System Structure

4.1 Simulation Flow

The simulation flow of this system appears in Fig. 1. Each step is explained below.

①: The evacuation route calculation agent calculates the shortest route to the goal using Dijkstra's algorithm and saves the data file.

②: The evacuee agents located at each intersection take the shortest route from each position to the goal based on this data. When all evacuees have finished taking the route, they are indicated on the map of the intersections by a green circle.

③: As shown in Fig. 2, we set the evacuee limit for each emergency refuge building using a control panel. The slide bar is capable of adjusting the number of evacuees by units of 100 persons.

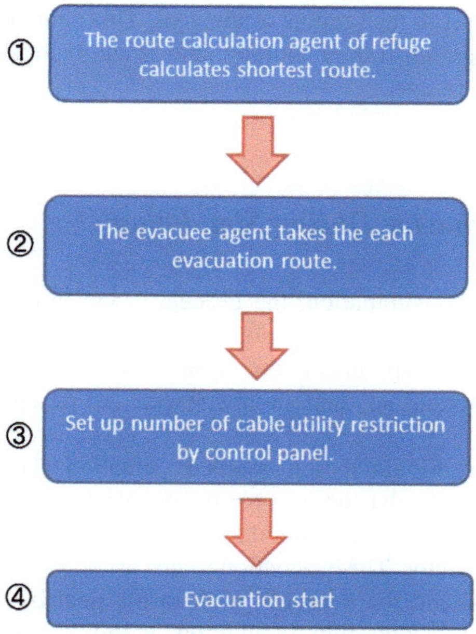

Fig. 1 Simulation flow

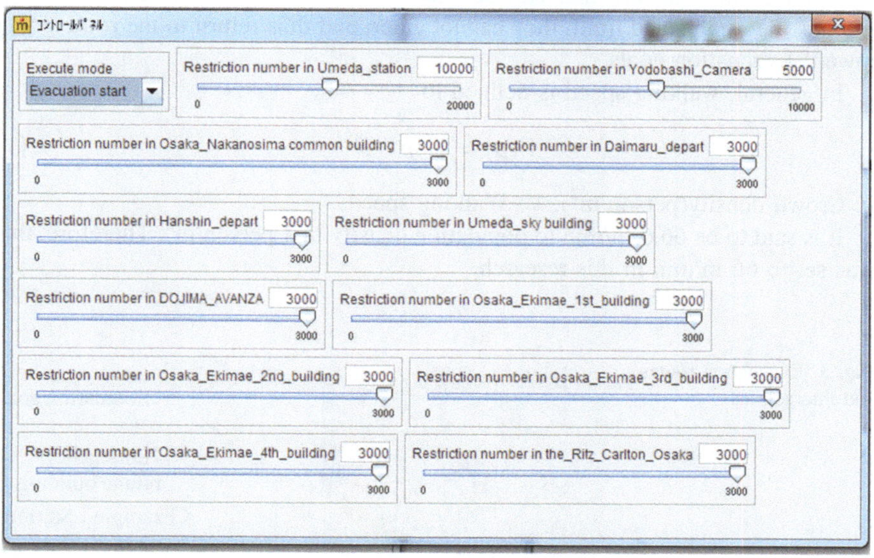

Fig. 2 Control panel

④: The evacuees start to take refuge at the goal. This simulation continues to run until all evacuees finish their journey, and it automatically finishes when all have completed refuge.

4.2 Evacuee Behavior During Simulation

When we execute this simulation, the evacuee agent is displayed with a green circle and positioned at each intersection. Each green circle represents a group of 100 evacuees. In this experiment, this system only simulated the behavior of evacuees on the ground, so their numbers were set at approximately 420,000 persons.

In addition to this, this simulation assumed a situation where all evacuees would go to three areas, JR Osaka Station, Northeast, Northwest, when an earthquake occurs, as shown in Fig. 3. Therefore, they are designated as the goal, which forms the basis of their behavior. The evacuation routes are shown in Fig. 4, where green lines indicate roads, and the evacuees walk to the goal by the shortest route as calculated by Dijkstra's algorithm. In addition, as shown in Fig. 5, when evacuees find emergency refuge buildings within three units surrounding themselves, they do not go to the goals but take refuge in these emergency buildings. However, if the number of evacuees that can be accommodated in each of these buildings reaches the established limit, they cannot enter, and thus return to the route going toward Evacuation goals.

In general, walking speed is defined to

$$V(\rho) = 1.1\rho^{-0.7954} \tag{1}$$

ρ: Crowd density(person/m^2); V: Walking speed

It is said to be 66.6 m/min to population density of 1 person/m^2. Therefore, this was set to 66 m/min in this research.

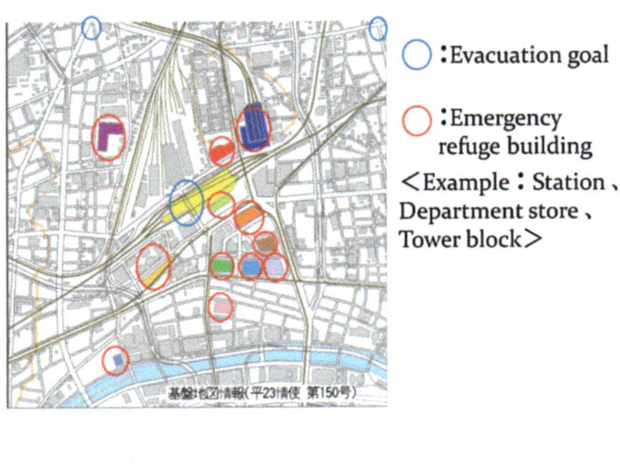

Fig. 3 Evacuation routes and interactions

Fig. 4 Evacuation routes and interactions

3	3	3	3	3	3	3
3	2	2	2	2	2	3
3	2	1	1	1	2	3
3	2	1	0	1	2	3
3	2	1	1	1	2	3
3	2	2	2	2	2	3
3	3	3	3	3	3	3

Fig. 5 Range of evacuee's view

Table 1 Walking speed

	Congestion factor		
	0–1	2–4	5–
Speed (m/min)	66	38	22

Furthermore, we set the speed to fluctuate according to the congestion situation on the evacuation routes. Table 1 summarizes the numerical values of these fluctuations.

4.3 Designation of Tsunami Refuge Buildings

In this simulation, the evacuees take refuge not only at JR Osaka Station but also in surrounding buildings. As shown in Fig. 2, a control panel can be created so that the number of evacuees that can be accommodated in those buildings can be adjusted, and it can be set up with various values. Table 2 summarizes the names of the tsunami refuge buildings and the number of evacuees that can be accommodated within.

5 Results

5.1 Case Study

In this simulation, we can set the maximum possible number of accommodated evacuees in each emergency refuge building using a control panel. In this research, the simulation was conducted using various proportions of evacuees accommodated in the refuge buildings, as described in the conditions listed below.

(1) When the proportion of evacuees who can be accommodated is 0 %.
(2) When the proportion of evacuees who can be accommodated is 20 %.
(3) When the proportion of evacuees who can be accommodated is 50 %.
(4) When the proportion of evacuees who can be accommodated is 80 %.
(5) When the proportion of evacuees who can be accommodated is 100 %.

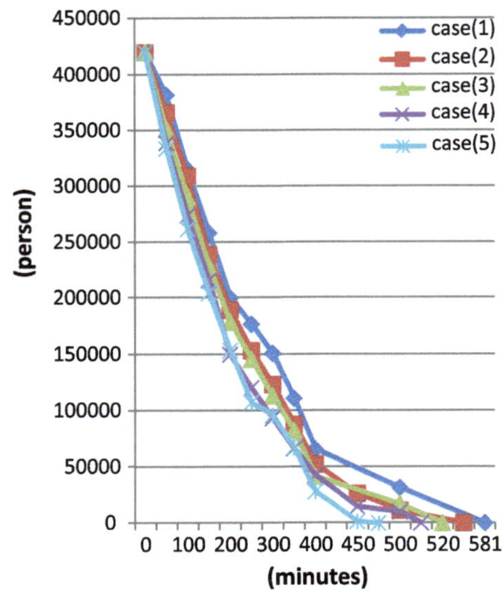

Fig. 6 Transition of evacuees unable to find refuge

Table 2 Building names and accommodated number of persons

	Color	Accommodated number (persons)
Umeda station	Blue	0–20000
Yodobashi_Camera	Red	0–5000
Osaka_Nakanoshima common building	Grayish blue	0–3000
Daimaru_Department store	Pea green	0–3000
Hanshin_Department store	Orange	0–3000
Umeda_Sky building	Purple	0–3000
DOJIMA_AVANZA	Pink	0–4000
Osaka_Ekimae_1st_building	Green	0–4000
Osaka_Ekimae_2nd_building	Sky blue	0–4000
Osaka_Ekimae_3rd_building	Light purple	0–4000
Osaka_Ekimae_4th_building	Brown	0–3000
Ritz_Carlton_Osaka	Bright yellow	0–2000

The data obtained by using the above proportions in shown in Fig. 6. The transition of evacuees unable to find refuge each time is shown by a time series graph.

5.2 Discussion

In the case of Condition 1, Fig. 6 shows that it takes a maximum time of 581 min until completion of refuge. Although evacuees go to JR Osaka Station and Kobe, it became apparent that it will take a considerable length of time for three groups of 420,000 people to converge. Furthermore, the simulation result clearly shows that the time taken to complete refuge becomes shorter as the proportion of evacuees accommodated in the refuge buildings becomes higher. As a result, in this paper it can be concluded that designating tsunami refuge buildings such as stations, department stores, tower buildings, etc. enables evacuees to take refuge effectively and quickly in cases where they converge on one location.

However, the predicted time period for a tsunami to hit Osaka City is about 140 min after an earthquake, so the time calculated by the simulation exceeds the tsunami warning time, even with a crowd of 420,000 people. In this simulation, although there is the time lag before evacuees reach to three goals, so further more people have to be considered in the simulation.

Further, evacuees can cross the road in front of the station in this simulation, but it is predicted that the street will be clogged with traffic and they will be difficult to do so freely. Although there are foot bridges, it is predicted that their utilization will be limited because of dangers due to use by large numbers of people.

In the future, it is necessary to make a system that is capable of considering factors such as what is the best way to accommodate crowds of evacuees, and how many buildings are required as evacuation base facilities, etc. on the assumption that a situation like the one described in this paper occurs.

6 Conclusion and Future Work

The purpose of this research is to develop an evacuation guidance method for the area around JR Osaka Station, and we have proposed a system that is able to simulate how evacuees will act in this area as the first step towards that purpose. So far, we have developed a simulator of evacuation behavior which targets only the people on the ground level. Using this, we conducted a simulation targeting the area around JR Osaka Station and Umeda Station in various situations. It is predicted that evacuation simulation using computer is effective because it is problematic to conduct an experiment of evacuation guidance on a real-life scale in the crowded JR Osaka Station area.

In this simulation, we did not deal with the interaction between people trying to evacuate from buildings, the subway and the underground shopping center to the outside and people who are already outside. In cases where large crowds stay on the ground, it is predicted that evacuees will not be able to go easily outside from buildings and the underground shopping area. Therefore, it is predicted that many will stay in the places around the station. At the same time, it is possible that the psychological pressure caused by the short time period until a tsunami hits will cause panic. Therefore, we plan to expand the functions of this system in order to be able to simulate cases where evacuees will go out from buildings to take refuge at ground level, as the next step in our research, and we consider the function of some disaster's phychological reaction in this simulation.

References

1. Cabinet Office, Government of Japan, White Paper on Disaster Management, Tokyo, 2010
2. Office of Emergency Management, What to do in an Emergency (Living Information), Osaka, 2011
3. M. Hori, Y. Inukai, K. Oguni, T. Ichimura, Study on developing simulation method for prediction of evacuation processes after earthquake. Socio Tchnica. **3**, 138–145 (2005)
4. A. Ouchi, H. Kawamura, and M. Yamamoto, The foundation and application of multi agent system, in *Calculation paradigm of the complex system engineering* Corona, Tokyo, Japan, 2002
5. K. Watanabe, A. Kondo, Development of Tsunami evacuation simulation model to support community planning for Tsunami disaster mitigation. J. Archit. Plann. AIJ **74**(637), 627-634 (2009)
6. B. Maury, J. Venel, Handling of contacts in crowd motion simulations, University of Paris, (in press), Orsay Cedex, France
7. A. Kirchner, A. Schadshneider, Simulation of evacuation processes using a bionics-inspired cellular automaton model for pedestrian dynamics. Physica A: Stat. Mech. Appl. **312**, 260–276 (2002)

Interactive Simulation and Visualisation of Realistic Flooding Scenarios

Christian Kehl and Gerwin de Haan

Abstract Floods are a permanent threat for urban environments and coastal regions. Due to the numerous environmental and climatological factors that cause floods, their prevention and prediction is complicated. Flood protection and prevention plans are assessed by computational models. The related risk analysis communication demands simulations of accurate inundation models and their interactive visualisation. In our new Dutch Knowledge for Climate project we work closely together with industrial partners with whom we develop a platform that supports this communication. Our research focuses on real-time flow simulations, their interactive visualisation and steering techniques for flooding scenarios. Our goal is an interactive, realistic problem-solving environment for flooding discussions amongst decision makers, water boards, hydrologists and the general public. Most important in this research are sophisticated algorithms that promote this goal. Related work in the field is done on small-scale examples and abstract computational models. We work on large-scale, high-resolution, realistic computations while maintaining interactivity. For this we use aerial terrain LiDAR point clouds of The Netherlands and most recent, complex Computational Fluid Dynamics (CFD) models. The rendering system will apply a combination of new point cloud compression algorithms and spatial Level-of-Detail data structures. Fast CFD simulations will be achieved by subgridding and parallel processing of

C. Kehl (✉) · G. de Haan
Department of Intelligent Systems, Delft University of Technology,
Mekelweg 4, 2628 CD Delft, The Netherlands
e-mail: c.kehl@tudelft.nl

G. de Haan
e-mail: g.dehaan@tudelft.nl

S. Zlatanova et al. (eds.), *Intelligent Systems for Crisis Management*, Lecture Notes in Geoinformation and Cartography, DOI: 10.1007/978-3-642-33218-0_7,

non-linear calculation models. Additionally, the integration of various geo-information (i.e. precipitation) is key to educated flooding decision-making. In this paper we describe in detail our project goals, our current progress and upcoming related research tracks.

1 Introduction

Floods are a permanent threat for urban environments and coastal regions worldwide. Therefore in Europe, The Netherlands are particularly threatened. Dikes at the coast are securing the inside of the country from floods while channels distribute water inside the country in a controlled way. Floods and high sea levels are a constant danger to this controlled ecosystem. The challenge for computational science is the construction of accurate simulations with models as close to reality as possible. The results of these simulations are flooding assessments of use cases, as a basis for flood protection and flood prevention concepts. The output of such simulations is large in size, therefore it needs to be visualized effectively to promote discussions of fore-mentioned concepts. Additionally, for applicable concepts, simulation and visualisation need to work on large-scale data and with an interactive speed.

Our dataset is a large-scale terrain- and bathymetry scan of The Netherlands, available as the AHN- and AHN-2 datasets. The size of the data sets demands new ways in visualisation, simulation and interaction. The main research question connected to the topic is as follows:

Which algorithms and data structures promote a multi-source, content-rich, interactive visualisation and simulation for flooding-aware environmental discussion?

Several topics and their related research questions, together with their available approaches and our initial sketches, are further presented. They form the basis of our future research. From the practical point of view, we are developing our own rendering system towards a multi-source decision support system (DSS) with two main purposes:

- enable water experts to explore better flood protection mechanisms on large scales;
- promote the discussions between water experts and decision makers in developing efficient worst-case-scenario and evacuation plans.

The following sections discuss in detail our 4 main research tracks. An overview of first results concludes the current progress of the presented topics. A final section will lay out the next steps of our research and development.

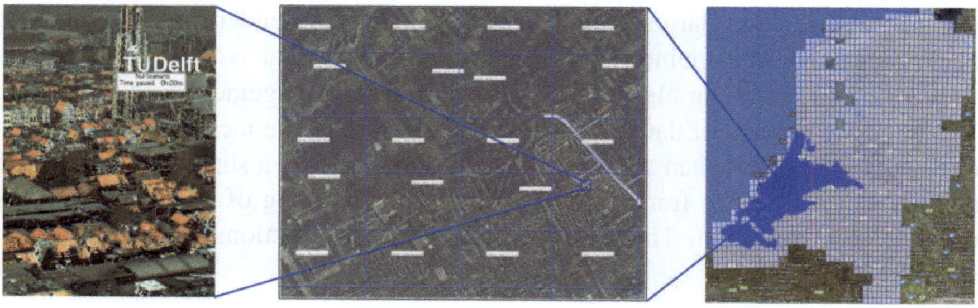

Fig. 1 Tile overview: overview of all AHN-2 point cloud tiles (*right*), in relation to the region of Delfland (*mid*) and the marketplace of Delft (*left*)

2 Large-Scale Point Cloud Rendering

The available terrain datasets used in our approach are represented as massive, unordered, roughly labelled 3D point clouds. Such point clouds receive growing attention in natural hazard management [20, 26]. These point clouds are part of the AHN-2 dataset[1] which is an aerial LiDAR[2] scan of the whole Netherlands. The scan consists of unordered point clouds which are geo-referenced as "Rijksdriehoekscoördinaten". Due to the data size the dataset is divided into around 44.000 separate tiles (Fig. 1), which consist of point clouds themselves. These point clouds have an average resolution of $30\frac{points}{m^2}$ and a precision of 0.05 m. Moreover each tile is separated into ground data (e.g. terrain) and additional data (i.e. buildings, trees). For the visualisation purpose a single or multiple tiles are linked to an aerial photograph. This colour information in return increases the memory consumption of each dataset.

Rendering concepts like Surfels [14] and QSplats [19], describe efficient ways of large-scale point visualization. We currently apply in our system circular splats as rendering primitives to compensate for locally-poor sampling densities. Alexa et al. [1] describe in their paper the generation of minimal samples for closed point set object perception via Moving-Least-Squares (MLS) surfaces, which is another way to efficiently render point cloud objects with varying sample densities. Gobbetti and Marton [7] prove the efficiency of large-scale point cloud rendering for static structures. Wand et al. [22] explore new ways of multi-resolution point cloud rendering by focussing on dynamic, efficient data structures for such data sets. We apply rendering techniques similar to Wand et al., with the difference that they store the full high-resolution base point set as bottom leaf layer. Lower resolution levels contain additional, simplified samples. Our approach, which is explained in further sections, distribute the points of the dataset over all levels of detail, therefore avoiding the generation of additional samples. Maeno et al. [11]

[1] AHN-2—Actueel Hoogtebestand Nederland 2nd version (http://www.ahn.nl).

[2] LiDAR—Light Detection and Ranging (http://www.lidar.com).

discuss in their paper large-scale data management and rendering techniques for heterogeneous terrain point clouds. Our focus in this field is on optimized data structures and rendering algorithms rather than data management.

Due to the amount of data, processing time as well as the memory consumption of each point, the data can not be managed in full scale by a single machine while maintaining reasonable framerates. Therefore, the handling of a dataset demands ways of data reduction. The corresponding research questions we hereby try to address are:

How do data reduction techniques contribute to interactive flooding visualisation and simulation?

Which techniques and designs are suitable for large-scale, interactive and realistic flooding simulations and visualisations?

2.1 Current Rendering and Visualisation

We created a prototypical 3D real-time rendering system to visualise AHN-2 datasets and the output of related flooding simulations. The system works with OpenSceneGraph and our internally-developed virtual reality framework "VRmeer". This program is extended during our research to apply new insights for subsequent user studies. While following paragraphs provide an overview of the used algorithms and techniques, more detailed information can be found in the paper of de Haan [6].

Our current algorithms are based on a spatial subdivision of points by employing a hierarchical, paged tree-data structure. The layout of the data structure can vary between octrees and quadtrees. The layout choice depends on the variation of the dimensional extent of the available tiles. For large, flat terrain the variation of x- and y-values is high while the variation of the z-axis is very low (for the used coordinate system we refer to Fig 2). Therefore, a quadtree is the preferred layout. For urban environments, the variation in all 3 dimensions is equal, thus an octree layout is the optimal choice. The tree depth depends on the overall number of points that needs to be shown.

This tree structure is used for Level-of-Detail (LoD) rendering. The necessary distance information for the LoD is stored for each point. When displaying the points only nodes (or leaves) of required resolution levels are shown. The arrangement of the points inside the tree is controlled via point cloud re-sampling. This re-sampling is done in a homogeneous manner, meaning that for N tree levels each N points are assigned to complementary nodes. This combination of spatial subdivision and Level-of-Detail can be seen in Fig. 3.

The water of the flooding scenarios is represented in one of 3 different ways. A first way is the conversion of a water depth map (one output of the simulation) to a colour texture. This is done by representing the integer-format water height by a saturation transfer function of blue, representing the water depth. The colour

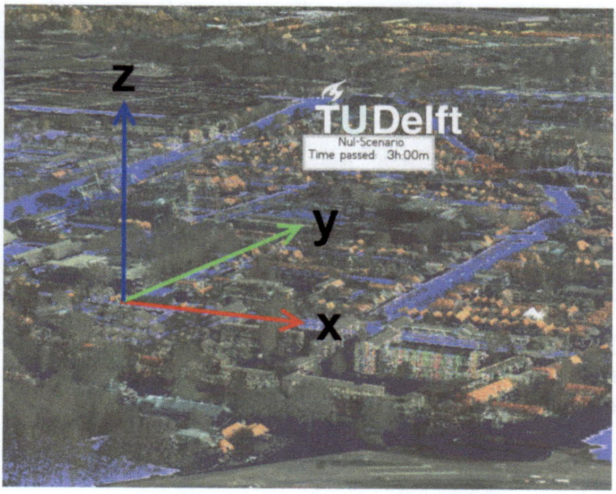

Fig. 2 Coordinate system: used global coordinate system with the z-axis orthogonal to terrain plane

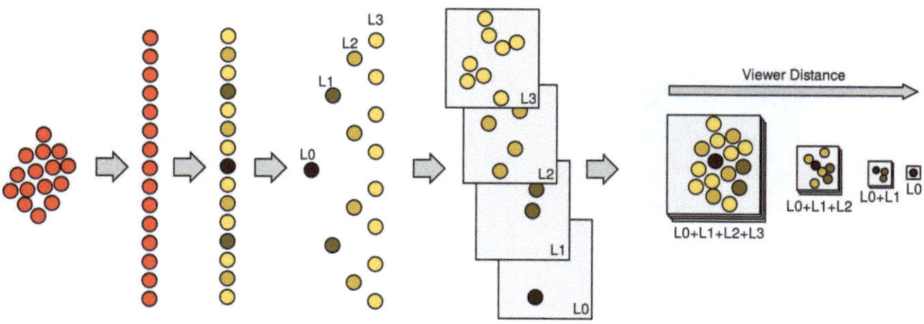

Fig. 3 Level-of-Detail: transformation from unordered points to tree-ordered LoD-cells containing points

values for each cell are then used in a separate overlay plane. This can be seen in Fig. 4a. The velocity vector information of the water is stored in addition to the water depth. This information is used in a second way of water visualisation. In this version, a shader program animates wave movements at each point (Fig. 4b). The drawback of these visualisation techniques is that the position of the overlay plane is set, leading to a visually homogeneous water depth. We therefore apply a third possibility of water visualisation. By transforming the depth map to a mesh that represents the actual 3D information of the water depth, it is easier to see the real flood impact. The mesh therefore fully- or partially-covers the terrain objects, depending on the simulation result and the height of each object (Fig. 4c). A comparative sketch of both methods, a fixed plane compared to a mesh, can be seen in Fig. 5.

Fig. 4 Water visualisation modes: our 3 modes of water visualisation. Water depth as *blue-colour* saturation transfer function on a static plane (**a**), realistic wave animation on a static plane (**b**) and a water mesh resembling the correct water depth (**c**)

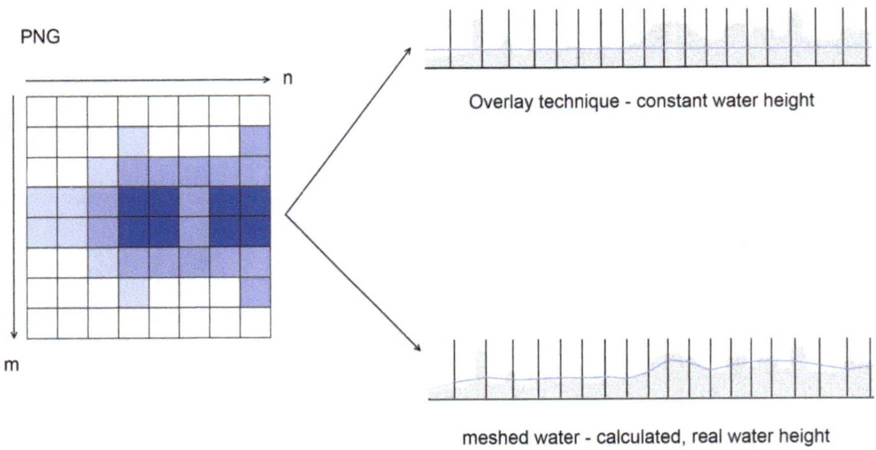

Fig. 5 Differences in water visualisation: sketch of the difference between the plane-based water visualisation and the meshed water

2.2 *Drawbacks and On-Going Work*

Although being a good starting point, the fore-mentioned algorithms of the system have some drawbacks. Generally, the visual quality of the visualised terrain heavily depends on the sampling-scheme of the point cloud. Issues connected to the point cloud sampling are:

- unnecessary high amount of points in overlapping regions of multiple tiles;
- high objects in urban environments are not well captured due to non-feature-adaptive sampling of the points;
- unnecessary high amount of points in low detail levels of homogeneous areas in the point clouds.

To overcome this sampling problem, we are working on more adaptive sampling schemes and feature-adaptive metrics for point clouds. Additionally, due to the fixed subdivision, there is a non-continuous transition between the different levels of detail. The reason for this behaviour are the discrete detail levels. This behaviour can irritate the user during the navigation in a dataset. We are currently working on a continuous Level-of-Detail technique, which improves the performance of the system and solves the transition issue.

In order to be able to render larger regions (i.e. whole provinces) or multiple data sets (i.e. historical changes in the landscape or multiple simulations at once) new ways of data reduction need to be followed. One of these ways is Labelling. We pursue a more precise labelling of the point cloud to efficiently re-sample the dataset and to chose the optimal tree container structure. We will apply new metrics during the re-sampling to obtain the optimal points for each level of detail.

3 Interactive and Adaptive Flooding Simulation

The American Geophysical Union (AGU) published an extensive summary on challenges in water monitoring on large scales in 2011 [25]. Here, a particular challenge is the need for accurate and fast water simulations that operate on large scales or global level. The available AHN-2 datasets are such large-scale regions on local level, due to the high resolution. The runtime of flooding simulations generally depends on the data size. Therefore, algorithms for data reduction and Level-of-Detail are beneficial to visualisation as well as simulation. Current systems (i.e. ArcGIS) rarely make use of data reduction techniques and, to our knowledge, no use of data compression. Consistent flooding simulation packages that combine modern parallel execution technologies with accurate computational models are rare. This is due to the non-linear nature of complex CFD models. Fast simulation execution times on complex CFD models are essential for our research to promote computational steering, as described by Liere et al. [21]. Our related research question is:

How can we achieve interactive, adaptive, accurate flooding simulations?

One prototypical sample project in the field is the Virtual Dike 3D water simulator [12]. This application shows the parallel potential of water simulations as well as the integration of counter-measure models in the simulation [16]. Chen et al. use the parallel OpenSees 3D simulation tool for calculation and steering of seismic wave propagation. The basis of our development is the realistic 1D/2D 3Di flooding simulation solver [8, 15]. The visualisation is created by first deriving precise bathymetry levels from the tiled AHN-2 data source. This bathymetry map is successively refined with a subgridding algorithm [3] to reduce the overall data size and therefore the calculation time. Afterwards, new water levels and velocity vectors are computed on grid cell level and saved in an appropriate exchange format. In the final step, the simulation is visualised according to Sect. 2. The whole process can be seen in Fig. 6.

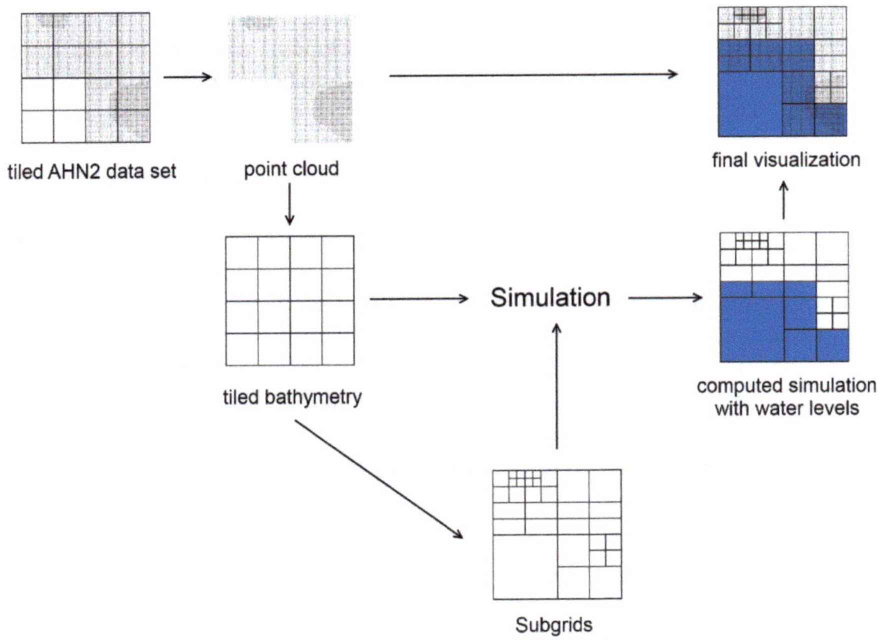

Fig. 6 Simulation pipeline: data flow- and transformation chart for our simulation

From our point of view, data reduction (i.e. information clustering with subgrid methods) and parallel processing (i.e. GPU-based CFD) are required for an accurate, full-scale, interactive simulation visualisation. Additionally, in-situ integration of additional, related data sources (i.e. precipitation) can be realized at the same time on modern, data-parallel architectures.

4 Geo-Information Integration

Flood occurrences, their progression and their impacts depend on numerous environmental factors. Consequently, flood simulation results should be reviewed depending on their environmental circumstances. Additionally, the discussion of flooding counter-measures depends on geographic context factors (i.e. landuse, polder extents) as well as the simulation result. Decisions are made based on disaster probability, effort and benefit. General geo-information is often available as image- or volume information. They can be easily integrated into modern GIS systems, that use meshes to visualise the environment. In contrast to these systems, our system is based on highly-detailed point clouds. There is only little algorithmic knowledge on the mapping and on-the-fly computation of mentioned geo-information on point clouds. Additionally, the efficient integration of meshes (i.e. counter-measures) into a point cloud-based simulation and visualisation system is a novelty in the field of GIS systems. Even standard GIS operations like colouring and clipping are hard to perform efficiently on large-scale point clouds. These points, the combination of map images, structural meshes and the point cloud-based terrain, are parts of our algorithmic research. We want to find new algorithms to enrich the highly-detailed point cloud with context data.

Another one of our research paths is the integration of environmental phenomena that are affecting flood occurrences. A specific environmental effect of major importance for floods in urban environments is precipitation. Our simulation system takes precipitation events into account. This is done by rising the local water depth on point basis. Currently, position and amount of precipitation is set by the user. Our work focuses on the integration of meteorological predictions as input to the simulation and the combination of visualisation and simulation of this effect. Research questions that are to be answered in this context are:

Which techniques are suitable to combine heterogeneous context information data structures and point cloud terrains?

How can we handle different sizes and boundary conditions of input data of environmental effects?

How can we visually combine environmental input- and output-data with a point-based terrain in a coherent way to promote the understanding of floods?

The main issue of the third question is the coherent visualisation of environmental effects. On the example of precipitation it can be seen that the visualisation of rainfall as particle systems, floods as meshes and terrains as point clouds is challenging. By choosing unsuitable visual representations of these data, users might get confused regarding the interplay of these processes instead of gaining insight from additional information.

While hereby specifically addressing issues regarding precipitation integration, challenges with the integration of other environmental effects (i.e. infiltration) are alike. We therefore expect that gained insights during the research (on basis of precipitation) are transferable to other environmental effects.

5 Multi-Scenario Comparative Simulation Visualisation

Available publications on flood assessment (i.e. precipitation affecting climate change [17], flood warning by taking rainfall and river discharge into account [13]) are use-case driven. They are based on pre-defined environmental conditions and pre-set flooding parameters. The interaction with the visualisation and simulation is limited. Changes in the simulation (e.g. adaptation of parameters) can not be done during calculation. Instead, the simulation needs to be stopped and re-calculated from the start with the new parameter set. Our goal is to develop an interactive, scenario-driven Decision Support System (DSS) based on formerly mentioned technologies, algorithms and concepts. This approach offers the possibility for peer-group users to collaboratively interact with the scene as well as with the simulation. The plan is to enrich the visualisation with new interaction possibilities and to create an on-the-fly adaptable simulation. This allows water experts to try out new prevention structures in-place and immediately see the planning result.

This concept is referred to as computational steering. We refer to this term by considering:

- a meaningful, multi-modal visualisation;
- a real-time, accurate CFD calculation;
- a multi-user interaction.

Our connected research question to this topic is:

How can multiple flooding scenarios be facilitated in a comparative visualisation and simulation?

An example in the field of computational steering of flooding scenarios is an application called "WorldLines" [23, 24]. This is a prototypical sample application that shows initial solutions to challenges connected with scenario-driven data exploration. The major differences between WorldLines and our DSS will be the spatial extent of each scenario as well as realism and complexity of the simulation calculations. As stated earlier, the large-scale, high-resolution AHN-2 point clouds as well as the complex, real-world-applicable CFD simulation demands new concepts. Thus, these concepts are aligned and embedded in a target-group-related steering system.

Another related example of computational steering is the 3D seismic wave visualisation by Chen et al. [4]. Here, frequency-band analysis and frequency displacement are used for information filtering. The results are visualized by direct volume rendering.

The final system has to deal with changing simulation parameters, water levels and terrain point clouds. These changes can be user- as well as time-dependent. A challenge closely linked to this approach is also the simulation visualisation of historical scenarios. These setups are often used in flooding research to understand the occurrences of floods as well as evaluation of new protection mechanisms. While historical flooding data is commonly taken to assess and evaluate new simulation methods, historical flooding events are rarely used for further studies. The challenges in historical flooding simulation and visualisation are:

- scanning, monitoring and storage of historical flood occurrences and causes;
- meaningful visualisation and animation of time-dependent data;
- time-dependent river discharge rates, demanding flexible, adaptive simulations;
- adaptive models, because floods and extreme precipitation change terrain and infrastructure (i.e. real estate damage, soil erosion).

The challenges are addressed partially by the following research publications. The Darthmouth Flood Observatory runs an exhaustive surface water data record of worldwide historical flooding events, visualised by highly accurate maps. Lienert

Fig. 7 Delfland: use case dataset "Delfland" in bird's eye perspective

et al. [10] researched the usage of historic flooding data as context information for current flooding simulations in a flood prediction system to support the understanding and decision-making process of current and future flooding events. This research led to a large-scale database of historic floods' measurements, visualised by 2D maps [9]. Brass and Blanchard [2] visualised 5 historic, precipitation-based floods in Texas for assessing the influence of social factors, like personal experience and profession, in the human ability to recognise and rate flood risks. Ribarsky et al. [18] lay out in their paper the needs for the visualisation of dynamic, time-dependent, modifiable terrains, buildings and other geoinformation for the purpose of weather research. Clevis et al. [5] used geoarchaeological simulations to re-track the motion of water to subsequently gain insight into former terrain shapes and archaeological sites that are destroyed by soil erosion. As a conclusion it can be seen that each above-mentioned challenge is approached separately in different approaches. Due to our demands, we will address multiple above-listed challenges during our research. Of particular interest for us are the last 2 listed bullet points.

6 Recent Results on Use Case "Delfland"

This section presents our recent results on a part of the dutch municipality Delfland. The dataset consists of 55 tiles, measuring a physical extent of 67.26 km^2 and being 8.2 GB big. It includes point cloud information, aerial photos and 1 simulation dataset.

The computer for the runtime measurements is an Intel Quad-Core processor and an NVIDIA Quadro FX 3700 graphics card. The dataset in bird's eye perspective (Fig. 7), where all tiles are rendered exclusively in the coarsest LoD,

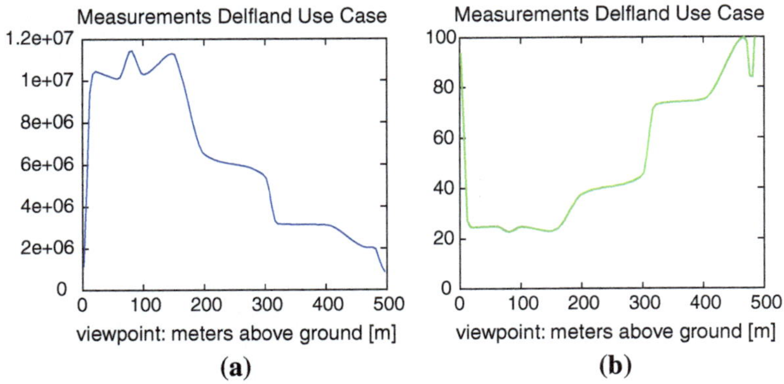

Fig. 8 Measurement plots: plots of the number of vertices in main memory (**a**) and the frames per second (**b**) in respect to the view distance, approximated in meters above ground level. This is because the Level-of-Detail choice depends of the distance "view point—vertex position". The measurements were taken in bird's-eye perspective

Fig. 9 Level-of-Detail in point clouds: a typical "wide-angle" view. The colouration shows the maximum loaded level of detail, from *blue* (level 0—farthest) to *red* (level 4—closest). It can be seen that a large amount of level-0-points are loaded, even in distances that can not be seen. That results a slow rendering in this viewpoint

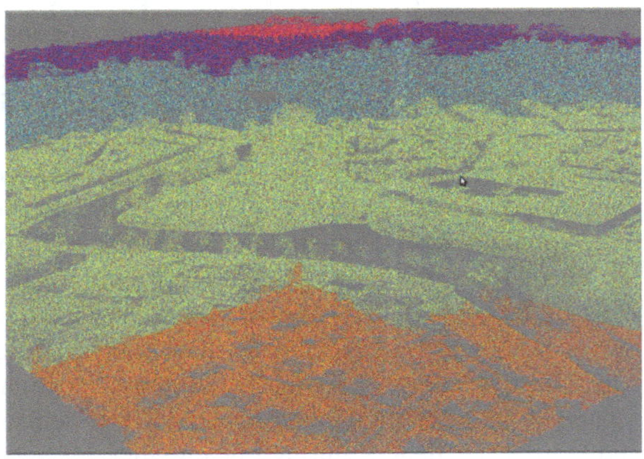

is running with an average speed of 90 fps, showing 772.000 vertices. While approaching ground level with the viewpoint, the frame rate drops consecutively while loading new levels of detail, from 90 to 20 fps. This happens because more points in higher detail levels are loaded. This behaviour can be seen in Fig. 8. The frame rate drops sharply when choosing a wide-angle view at near-ground view positions (around 11 fps with more than 20 million points in the view). An example of a wide-angle view is shown in Fig. 9. This is because all tiles in viewing direction need to be loaded, in their respective LoD. Nevertheless, the wide-angle view is important because it provides the best insight on the inundation situation in urban environments. Initial user studies have shown that most experts choose the wide-angle view for the assessment of flooding situations. Because of this regular usage of a wide-angle view, the explained performance behaviour is a point for improvement in the future.

(a) (b) (c)

Fig. 10 Non-continuous transition problem: the image series show the abrupt change from a start viewpoint (**a**) to an end viewpoint (**c**) while loading the tiles in (**b**)

Another unappreciated behaviour is the abrupt appearance of new tiles or tiled LoD-nodes, as described in Sect. 2. This can also be seen in Fig. 10.

7 Future Research and Assessment

Due to the presented state of each research track it can be seen that there are open questions which demand experiments and new algorithmic approaches. We will focus our future efforts on the improvement of the simulation in terms of speed by experimenting with GPU Computing technology in combination with our detailed CFD models. Our future progress can be expected in the concepts of:

- interactive, realistic, large-scale flooding simulations;
- point cloud-based terrain visualisation, derived from LiDAR data;
- point-based GIS operations;
- computational steering of large-scale simulations based on complex computational fluid dynamics (CFD) models;
- integration of precipitation and other environmental factors into a coherent fluid dynamics simulation visualisation.

Currently, water experts assess and use the system on professional workshops. They discuss pre-calculated flooding scenarios of chosen areas. The system has therefore, at the moment, an educational purpose to improve the understanding of floods.

The system is also used in museum installations to educate the public on flooding, based on large-scale, historic, pre-computed flooding scenarios. In this context, hydrologists at the TU Delft use the system to assess their CFD calculation models via comparison with real, historic floods.

Our vision is to use the software in the future as an immediate-response system of occurring floods to coordinate and control public safety. This is reflected in our focus on large-scale, interactive Simulation Visualization algorithms.

Acknowledgments This work is supported by the federal Dutch project for water- and climate knowledge "Kennis voor Klimaat", grant IGF-IGF26G. The work is embedded in the water planning project of "3Di". The work is carried out in close collaboration with Deltares; Nelen & Schuurmans; BiG Grid, advanced ICT research infrastructure for e-Science; SARA Computing and Networking Services; the TU Delft CiGT department, particularly Dr. Ir. Guus Stelling, and the TU Delft OTB department.

References

1. M. Alexa, J. Behr, D. Cohen-Or, S. Fleishman, D. Levin, C.T.Silva, Computing and rendering point set surfaces. IEEE Trans. Vis. Comput. Graph. **9**(1), 3–15 (2003)
2. M.W. Bass, R. Denise Blanchard, Examining geographic visualization as a technique for individual risk assessment. Appl. Geogr. **31**(1), 53–63 (2011)
3. V. Casulli, G.S. Stelling, Semi-implicit subgrid modelling of three-dimensional free-surface flows. Int. J. Numer. Methods Fluids **67**(4), 441–449 (2011)
4. C.-K. Chen, C. Ho, C. Correa, K.-L. Ma, A. Elgamal, Visualizing 3d earthquake simulation data. Comput. Sci. Eng. **13**(6), 52–63 (2011)
5. Q. Clevis, G.E. Tucker, G. Lock, S.T. Lancaster, N. Gasparini, A. Desitter, R.L. Bras, Geoarchaeological simulation of meandering river deposits and settlement distributions: a three-dimensional approach. Geoarchaeology **21**(8), 843–874 (2006)
6. G. de Haan, Scalable visualization of massive point clouds. Nederlandse Commissie voor Geodesie KNAW **49**, 59 (2009)
7. E. Gobbetti, F. Marton, Layered point clouds: a simple and efficient multiresolution structure for distributing and rendering gigantic point-sampled models. Comput. Graph. **28**(6), 815–826 (2004)
8. A. Leskens, Watermanagement is informatiemanagement. H2O **44**, 14–19 (2011)
9. C. Lienert, R. Weingartner, L. Hurni, Post-event flood documentation and communication using a hydrological map information system, in *Geospatial Data and Geovisualization: Environment, Security and Society*, vol. XXXVIII, ISPRS, 2010
10. C. Lienert, R. Weingartner, L. Hurni, Real-time visualization in operational hydrology through web-based cartography. Cartogr. Geograph. Inf. Sci. **36**(1), 45–58 (2009)
11. T. Maeno, H. Date, S. Kanai, in *A Data Management Method for Efficient Search and Rendering of Multiple Large Scale Point Clouds*, 2011
12. N.B. Melnikova, G.S. Shirshov, V.V. Krzhizhanovskaya, Virtual dike: multiscale simulation of dike stability. Procedia CS **4**, 791–800 (2011)
13. F. Pappenberger, J. Bartholmes, J. Thielen, H.L. Cloke, R. Buizza, A. de Roo, New dimensions in early flood warning across the globe using grand-ensemble weather predictions. Geophys. Res. Lett. **35**(10), L10404– (2008)
14. H. Pfister, M. Zwicker, J. van Baar, M. Gross, Surfels: surface elements as rendering primitives, in *Proceedings of the 27th Annual Conference on Computer Graphics and Interactive Techniques, SIGGRAPH '00*, New York, NY, USA. (ACM Press/Addison-Wesley Publishing Co., New York, 2000), pp. 335–342
15. O. Pleumeekers, J.-M. Verbree, Overstromingsmodellen met een hoge resolutie. H2O **43**, 20–21 (2010)
16. A.L.A. Pyayt, I.I.A. Mokhov, A.A. Kozionov, V.A. Kusherbaeva, N.B.B. Melnikova, V.V.B. Krzhizhanovskaya, R.J.C. Meijer, in *Artificial Intelligence and Finite Element Modelling for Monitoring Flood Defence Structures*, pp. 34–40, 2011
17. J. Räisänen, R. Joelsson, Changes in average and extreme precipitation in two regional climate model experiments. Tellus A **53**(5), 547–566 (2001)
18. W. Ribarsky, N.L. Faust, Z.J. Wartell, C.D. Shaw, J. Jang, in *Visual Query of Time-Dependent 3D Weather in a Global Geospatial Environment*, 2002
19. S. Rusinkiewicz, M. Levoy, Qsplat: a multiresolution point rendering system for large meshes, in *Proceedings of the 27th Annual Conference on Computer Graphics and Interactive Techniques, SIGGRAPH '00*, New York, NY, USA. (ACM Press/Addison-Wesley Publishing Co., New York, 2000), pp. 343–352

20. S. Slob, R. Hack, 3d terrestrial laser scanning as a new field measurement and monitoring technique, in *Engineering Geology for Infrastructure Planning in Europe*, ed. by R. Hack, R. Azzam, R. Charlier. Lecture Notes in Earth Sciences, vol. 104 (Springer, Berlin, 2004), pp. 179–189
21. R. van Liere, J.D. Mulder, J.J. van Wijk, in *Computational Steering*, 1997
22. M. Wand, A. Berner, M. Bokeloh, P. Jenke, A. Fleck, M. Hoffmann, B. Maier, D. Staneker, A. Schilling, H.-P. Seidel, Processing and interactive editing of huge point clouds from 3d scanners. Comput. Graph. **32**(2), 204–220 (2008)
23. J. Waser, H. Ribicic, R. Fuchs, C. Hirsch, B. Schindler, G. Blöschl, M.E. Groller, Nodes on ropes: a comprehensive data and control flow for steering ensemble simulations. IEEE Trans. Vis. Comput. Graph. **17**(12), 1872–1881 (2011)
24. J. Waser, R. Fuchs, H. Ribicic, B. Schindler, G. Blöschl, E. Groller, World lines. IEEE Trans. Vis. Comput. Graph. **16**(6), 1458–1467 (2010)
25. E.F. Wood, J.K. Roundry, T.J. Troy, L.P.H. van Beek, M.F.P.Bierkens, E. Blyth, A.A. de Roo, P. Doll, M. Ek, J. Famiglietti, D. Gochis, N. van de Giesen, P. Houser, P.R. Jaffe, S. Kollet, B. Lehner, D.P. Lettenmaier, C. Peters-Liedard, M. Sivapalan, J. Sheffield, A. Wade, P. Whitehead. Hyperresolution global land surface modelling: meeting a grand challenge for monitoring earth's terrestrial water, in *Water Resources Research*, vol. 47 (American Geophysical Union, Washington, DC, 2011)
26. M. Zwicker, M. Pauly, O. Knoll, M. Gross, Pointshop 3d: an interactive system for point-based surface editing, in *Proceedings of the 29th Annual Conference on Computer Graphics and Interactive Techniques, SIGGRAPH '02*, New York, NY, USA. (ACM, New York, 2002), pp. 322–329

Part II
Sensors and Data Integration

Identification of Earthquake Disaster Hot Spots with Crowd Sourced Data

Reza Hassanzadeh and Zorica Nedovic-Budic

Abstract This paper explores the value of the application of Crowd Sourced (CS) data in identification of areas damaged in the aftermath of an earthquake. A survey was conducted to collect CS data based on two stage cluster sampling method from people who experienced the earthquake in Bam city, Iran in 2003. The CS data submission time was considered for data analysis, including continuous, discrete and complete data submission. The CS data reporting on the level of building destruction, the number of fatalities and the number of injuries was used to identify hot spot areas for dispatching response operation teams. To test the value of CS data in identification of hot spots, the results were compared with the Actual Earthquake (AE) data by using of Fuzzy Kappa index, Fuzzy Inference System, and cross tabulation to calculate similarity and dissimilarity, quality and allocation disagreement between them. The similarity and dissimilarity measures indicate that there is a low to moderate similarity between hot spot maps based on the application of CS and those based on the AE data. They suggest that CS data has a moderate potential role in identifying highly damaged areas (hot spots) and low damaged areas (cold spots). The results of this study show that the CS data is better

R. Hassanzadeh (✉) · Z. Nedovic-Budic
Geography, Planning and Environmental Policy School, University College Dublin, Dublin, Ireland
e-mail: reza.hassanzadeh@ucdconnect.ie

R. Hassanzadeh
GPEP Research Unit, Blackwell Room, University College Dublin, Richview, Dublin 4, Ireland

Z. Nedovic-Budic
Geography, Planning and Environmental Policy School, Planning Building, University College Dublin, Richview, Dublin 4, Ireland
e-mail: zorica.nedovic-budic@ucd.ie

S. Zlatanova et al. (eds.), *Intelligent Systems for Crisis Management*, Lecture Notes in Geoinformation and Cartography, DOI: 10.1007/978-3-642-33218-0_8,

suited for more general determination of hot and cold spot areas than to provide exact locations where the resources could be dispatched. Consequently, we conclude that CS data is useful for decision making process by disaster managers if combined with the other sources of information to allocate the limited resources in affected areas.

Keywords Crowd sourced data · Earthquake disaster management · Real time data · Hot spot analysis · Fuzzy kappa · Cross tabulation analysis · Similarity and dissimilarity

1 Introduction

Crowd Sourced data has rapidly become an essential source of data in disaster response especially after the successful deployment of the Ushahidi Crisis Map during the 2010 Haiti earthquake [22]. The value of this data has been enhanced by the creation of crisis maps based on location data extracted from social media communications platform [20, 21]. Harvard Humanitarian Initiative [18] mentioned that "extracting, categorizing, visualizing, and evaluating such information presents serious research challenges, including the problem of managing and extracting meaningful information from the large volume of contributions, applying the information to decision support workflows, and the development of formal information sharing protocols". MacEachren et al. [21] also stated that the research on "the integration of Geographic Information Systems (GIS) and Crowd Sourced information from social media has focused more on the challenges of extracting action items and location information from social media feeds" and less on the utility of the extracted information and the effectiveness of associated crisis maps to support emergency response and also to test the effects of CS data in the disaster responses aftermath of an event specifically an earthquake.

Therefore, this research examines the application of CS data in disaster response operation based on spatial analysis in GIS environment and also tests the value of CS data in this process. We begin with background information, including an overview of Crowd Sourced data and their role in disaster response, followed by geospatial analysis methods in GIS such as hot spot analysis. Next, we test the value of CS data in disaster response by comparing it to AE data using Fuzzy Kappa index, Fuzzy Inference System and cross tabulation for similarity, dissimilarity and quality and allocation disagreement analysis.

2 The Role of CS in Disaster Response: Concepts and Previous Work

Crowd Sourced data are the community generated data which are contributed by people who are not trained and not supervised by an authority [5, 23, 30]. Goodchild [11] referred to this type of data as volunteered geographic information (VGI) which refers more specifically to the geographic content contributed via the web. The advances in technologies have created potential environment for users to contribute in CS data collection. These technologies include web mapping applications (Google Maps, Ushahidi platform, and OpenStreetMap), GeoRSS (standard data representation environment), web application Person Finder, social network (Facebook and Tweeter), and online video websites.

Data gathering methods which are used for creating Crowd Sourced data include: identifying the location by using Global Positioning System (GPS), Internet or mobile phone; locating an address or a place name on a map (Geo-coding); drawing a layer on a computer-based map; commenting on a feature already existing on a map; uploading photographs and videos to a website and describing it; and texting the information to the specific data gathering centres by using smart phone and iPhone.

Agrios [2] stated that Crowd Sourced data and social networks facilitate the process of data gathering about what is happening in areas affected by the disaster. The information is delivered to disaster managers and teams before official responders arrive on the scene. Therefore, the community involvement plays a significant role in data collection during a disaster because they provide complementary information sources [2]. Furthermore, the integration of this data as local knowledge with the other data has the potential to improve the accuracy and completeness of the required data in supporting decision making process in response phase. However, there are many challenges for using Crowd Source data that include the availability of data, data quality, bias in reporting events, localisation of received data, data collection methods and applications, and data verification [28].

Few examples of the application of Crowd Sourced data are illustrated in the following:

- Do you feel it project: The web interface that called "do you feel it?" this web interface lunched by the U.S. Geological Survey to produce intensity maps by integrating Crowd Sourced data [35].
- Applying Crowd Sourced data in flood damaged modeling: Poser and Dransch [28] carried out a preliminary study in Eilenburg city which was heavily affected by the 2002 flood of the Elbe River to assess the possibility of applying Crowd Sourced data in assessing of the flood damages. They collected the inundation depth data from affected people by telephone interview 6 months after the event. They concluded that CS data can be a great opportunity to support and improve disaster management process.

- Open Street Map databases after the Haiti earthquake: public participation for data gathering specifically for complementing of Open Street map databases increasingly enhanced after Haiti earthquake in 2010 [25].
- Wildfires Crowd Sourced data: Santa-Barbara 2007–2009 wildfires Crowd Sourced data aided the process of disaster management [12].
- Ushahidi platform in the Haiti earthquake: Meier and Munro stated that in the Haiti earthquake (2010), people were able to send information by using mobile phones and they applied the Ushahidi platform which is a free and open source mapping tool to create a live map of events. Accordingly, people will participate in data gathering after an earthquake. Goodchild and Glennon [12] stated that Crowd Sourced data could be relevant for disaster management.
- Ushahidi platform in snowstorms in Chicago, a non-profit technical company, set up a platform that allowed Chicago residents to post both needs and solutions during snowstorms in Chicago, Illinois [2].

The ability to aggregate data, evaluate the situation, and plan for response via logistical back support is a fundamental part of any response. As the case of Haiti has shown, crowd sourcing can play a key role in these logistics [36].

Research on the Crowd Sourced data analysis is still scarce. Empowering the Public with Information in a Crisis (EPIC) is a project on Tweet standards. Its purpose is to facilitate processing of CS data for machines and people [4, 13]. Also working with Tweet, Earle and Guy (2010) developed a prototype software application that collects Tweets containing the word "earthquake" and uses spatial data mining techniques to map and summarize short personal accounts that are reported within seconds after an earthquake strikes [15]. They suggested that this type of Crowd Sourced geographic information can be "a useful supplement to instrument-based estimates of quake location and magnitude" [8, p. 221]. A study was conducted by European Commission's Joint Research Center on the spatial analysis of the pattern of aggregated text messages via Ushahidi. It showed "a strong attraction between the patterns exhibited by SMS messages and building damages" [7]. However, Fruchterman [10] criticized the result of this research and stated that "one of the concerns about Crowd Sourced crisis information is that the data is not representative of the affected population and also there is a low possibility to have short messages from all parts of the affected area" (p. 1). These examples emphasize the significance of Crowd Sourced data in disaster management process, but there has been less research on the utility of the extracted information and the effectiveness of associated crisis maps to support emergency response. To contribute further knowledge, we examine the value of CS data in identifying the areas for response in the aftermath of an earthquake.

3 Methodology

This research was based on simulating the submission of CS data by conducting a survey, integrating the response data into GIS database, using the data to identify hot spots, and comparing them with actual earthquake data. The assumption in this research is that the affected people would be able to use communication and web application technologies to submit real-time CS data from any location in damaged area to the disaster management centre's database. These technologies include mobile phone, the Facebook, the Person Finder, and Ushahidi platform. Although, the communication systems' failure may happen either on the whole area or locally, it affects community participation in data submission, violates the sample distribution in entire damaged area and affects the reliability of the analysis in looking for hot spots.

3.1 Survey

We administered a questionnaire [3, 9] with people who experienced the earthquake in Bam city, Iran, in 2003. The sampling unit was a family unit, as 19,087 families [31] were living in Bam city before the earthquake. Based on two stages cluster sampling method [9] 396 people were sampled and surveyed. First, 80 % or 39 out of 49 administrative zones were randomly selected; second, the sample respondents in each zone have been selected, with the numbers proportional to the number of families in each zone. The sample included 79 % males; 21 % females; the average age was 47 years; 44 % of respondents had high school; 21 % had university degree (Figs. 1, 2; Table 1). The questionnaire was administered face to face and inquired about population loss and property damage (Fig. 2).

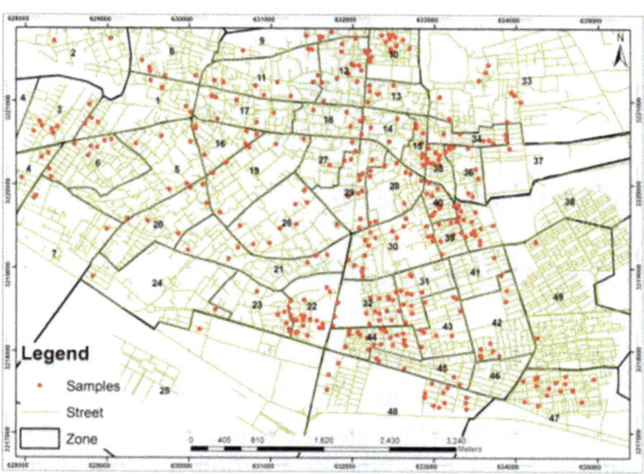

Fig. 1 Distribution of respondents in the study area

Post Earthquake Crowd Sourced Data Collection

A) Basic Information:

1. Could you please record some information about yourself?

Sex: Male☐ Female☐

Age: ....

Level of Education:

Primary☐ Lower Secondary☐ Higher Secondary☐

Diploma☐ Degree☐ Postgraduate☐

B) Household Demographic Information

2. Were you a resident of Bam city, when the earthquake occurred in 2003?

Yes☐ No☐

3. How long have you lived in Bam city? .... years

4. Do you still live in the same residence? Yes☐

No☐, why did you move: ....

5. How many people were living in the building? .... person/persons.

6. What ages were they?

..... Under 18 .... 19-34 ...... 35- 65 .... Over 65

7. Do you know what material the building was?

Adobe Building☐ Unreinforced masonry Building☐

Unreinforced masonry building with Reinforced floor☐ Steel building☐

Reinforced masonry building☐ Reinforced concrete building☐

Others☐

8. How many storeys had the building at the time of the earthquake?

9. What was the impact of the earthquake on the building?

No destruction☐ Minor cracks in the walls☐ Major cracks in the walls (1cm)☐

One wall was collapsed☐ one wall and some part of roof were collapsed☐

Completely collapsed ☐

10. Were any people in the building when earthquake happened?

Yes☐, How many?........................... No☐, Why?...............................

11. Were there any people trapped under the debris?

Yes☐, How many?........................... No☐, Why?...............................

12. Were there any people injured?

Yes☐, How many?........................... No☐, Why?...............................

13. Were there any fatalities?

Yes☐, How many?........................... No☐, Why?...............................

14. After the earthquake what happen for you? When were you able to submit this data to the disaster management centre? (We will explain the assumptions)

Only office use:

-Survey code:

- Building's coordinate: X: Y:

-Address: No................... Alley.................................. Street...........................

-Extra explanation:

Fig. 2 The questionnaire form

3.2 Data Preparation

The CS data would be available after an earthquake occurrence through community's contributions. The data gathered in the survey included: the number of people trapped in buildings, the number of injured people trapped in buildings, the number of fatalities, the destruction level of each building, the fire status of a

Table 1 The sample of respondents by zone

No	Zone	Buildings (number)	Families (number)	Population (number)	Respondents (number)	No	Zone	Buildings (number)	Families (number)	Population (number)	Respondent (number)
1	1	642	432	2163	10	21	28	646	381	1570	10
2	3	567	490	2427	10	22	29	707	351	1467	11
3	5	1040	624	2696	6	23	30	747	511	2118	18
4	6	866	500	2461	8	24	31	485	360	1699	10
5	8	528	381	1965	8	25	32	516	389	1820	17
6	9	428	249	1210	9	26	33	875	456	2119	9
7	10	747	573	2663	20	27	34	456	450	2080	7
8	11	632	326	1525	10	28	35	355	373	1562	9
9	12	850	440	2027	14	29	36	492	476	2012	10
10	13	549	256	1081	5	30	38	546	498	2443	15
11	14	1097	278	1189	5	31	39	423	422	2145	7
12	15	434	275	1240	7	32	40	415	327	1528	10
13	16	577	364	1580	6	33	42	521	447	2189	10
14	17	735	419	1901	9	34	43	634	508	2537	9
15	18	938	300	1208	7	35	44	489	366	1741	19
16	20	623	387	1665	5	36	45	613	548	2698	5
17	21	694	357	1467	8	37	47	385	286	1323	16
18	22	444	342	1545	20	38	48	656	526	2630	14
19	23	416	363	1541	7	39	49	228	161	751	3
20	26	755	478	2004	13		Sum	23751	15670	71990	396

building and its surrounding, power outage status, food and water status, road blockages, and security issues.

The CS data was stored in a database from multi-sources. This would require conversion of different formats into a common GIS format for data processing. We selected GIS format because of the capability of GIS software to analyze the data spatially and temporally. To record the CS data submissions within the GIS database, a point shape file for each data entry was created, based on the coordinate of location (X, Y) of the data submission device and by geocoding addresses of people in the damaged area. For example, the data from mobile phones uses coordinates of the points (X and Y); the data from Facebook and Person Finder applies geocoding extensions. On the other hand, the collected data from the survey was geo-coded based on their addresses in GIS to be utilized in data processing analysis.

The other assumption is about the validity of CS data. We considered the collected data as accurate data to concentrate on their usage and usefulness in identifying hot spot damage areas. The methods used to control for data accuracy include the comparison to existing databases, the triangulation among the responders in one specific area, and reliance on experts.

The control data that were used for comparison with CS data were the AE data. AE data was gathered in a street survey after Bam earthquake by Statistical Center of Iran in 2004. These data include the building destruction level, the number of fatalities and the number of injuries in each building block [31].

3.3 Hot Spot Analysis

We utilized the simulated real-time CS data to examine their value in disaster response operation. This data was considered based on their submission time as continuous, discrete, and complete submission. They were analyzed with application of geospatial statistics Getis-Ord (Gi*) [26] to identify hot spots in GIS environment. This method was used to identify the spots with high values in each specific parameter. The output was a Z score and *p* value for each feature. These values represented the statistical significance of the spatial clustering of values, given the spatial relationships and the scale of analysis (distance parameter). A high Z score and small *p* value (probability) for a feature indicates a spatial clustering of high values. A low negative Z score and small *p* value indicates a spatial clustering of low values. The higher (or lower) the Z score, the more intense the clustering.

To test the value of CS data in identification of hot spot areas we undertook a comparative analysis based on Fuzzy Kappa index [17, 34], Fuzzy Inference System [29] and cross tabulation [27]. Therefore, the hot-spot end results with the application of CS data were compared to the same end results with application of AE data to determine how similar (percentage of agreement) the CS data were to the actual earthquake data. The three analytical procedures are explained below.

3.3.1 Fuzzy Kappa Index

The Fuzzy Kappa statistic shows "the agreement between two categorical raster maps. The statistic goes beyond cell-by-cell comparison and gives partial credit to cells based on the categories found in the neighborhood" [16, p. 4]. When matching categories are found at shorter distances, the agreement is higher. In this method two concepts of the 'fuzziness of the categories' (quantity) and the 'fuzziness of the location' will be measured [17]. With support of 'Map Comparison Kit' software, the fuzzy comparison index produces a map for each cell with the degree of similarity on a scale of 0–1. In addition of this spatial assessment of similarity, an overall value for similarity is also derived [17].

3.3.2 Fuzzy Inference System

The Fuzzy Inference System comparison algorithm [29] compares "the characteristics of polygons rather than cells found in both maps". The characteristics that are taken into account in this evaluation are area of intersection, area of disagreement and size of polygon. This analysis also determines how similar (percentage of agreement) the CS data are to the actual earthquake data. In order to apply this analysis method, the hot spot maps are converted to raster files and then compared with the "Map Comparison Kit" software.

3.3.3 Cross-Tabulation Analysis

Another statistical technique for accuracy assessment and map comparison is the cross-tabulation matrix which is based on two summary parameters—quantity disagreement and allocation disagreement. Quantity disagreement is defined as the difference between two maps due to an "imperfect match in overall proportions of all mapped categories". Allocation disagreement is defined as the difference between two maps due to an "imperfect match between the spatial allocations of all mapped categories" [27, p. 4409]. The two measures are used to evaluate both the validity of a modeled map (comparing to a reference map) and the differences between two scenarios [27].

4 Identification of Hot Spots

We focused on the analysis of three measures derived from CS data: the level of destruction of each building, the number of fatalities in buildings, and the number of injured people in buildings. We identified areas with high values of destruction level based on European Macrocosmic Scale (EMS-98) [14] high values of fatalities and high values of injured people. We then completed the analysis of AE data for the same parameters.

4.1 Hot Spot Analysis with Crowd Sourced Data

The CS data were analyzed according to three different time periods of submitted data. These are continuous, discrete and complete data submission.

4.1.1 Continuous Submission

The CS data were sent gradually to the main database of the disaster management centre. Disaster manager would take the data into account immediately after receiving them. The submission time for this data would be based on submission time of collected data. For instance, based on "Twitcident: Fighting Fire with Information from Social Web Streams" study, Abel et al. [1] suggested that "all incidents reach their maximum peak of CS data's submission within the first 4 h after the incident occurred". To identify the proportion of submitted data in each time period, the 396 surveys were released over the period of 1440 min (24 h) (Table 2). As shown in Fig. 3, after 5 min only one person participated in data gathering; after 20 min there were 68 participants; and it continued until 240 min by when the cumulative number of participants was 368. After this time the participation decreased gradually, thus indicating that the most crucial time would be the first 4 h after earthquake occurrence (Fig. 4).

The first step in the hot spot data analysis was to determine the spatial scale (distance band). Therefore, for the fixed distance option, the distance band used for analysis should be based on the understanding of the spatial interaction among the features. Alternatively, features may be evaluated for a range of distance values or at the specific distance where spatial autocorrelation is maximized. Thus, spatial autocorrelation based on the Global Moran I [19, 24] was run with 15 distance bands to find out the distance in which the spatial clustering was mostly pronounced while the value of Z score was the highest. The distance bands for 20, 60, 120, 180, 360, and 1440 min of CS data submission time were calculated, which resulted in 1700, 1800, and 2200 m respectively as seen in Fig. 5. These were the appropriate distance bands for hot spot analysis in each specific time. In the process of continuous submission of CS, the data was analyzed for only one parameter—the level of building destruction (Fig. 6).

As per Fig. 6, after 20 min, the high damage areas (hot spots) were located in south-east and low damage areas (cold spots) mostly in the north-west and west

Table 2 Real-time continuous data submission after earthquake occurrence

Submission time (min)	5	10	15	20	25	30	40	45	50	60	90	120	150
Count of CS data in each time	1	11	8	68	3	98	2	40	1	53	18	24	6
Submission time (min)	180	210	240	270	300	360	390	420	450	480	660	720	1440
Count of CS data in each time	21	2	12	1	7	9	1	2	1	1	1	2	3

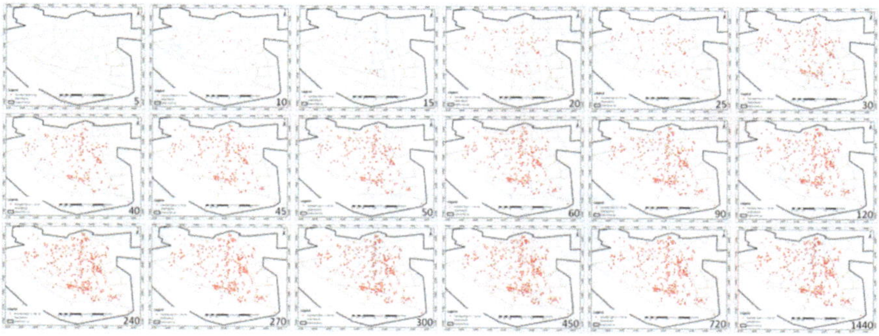

Fig. 3 CS data entry after earthquake in Bam city based on each submission time

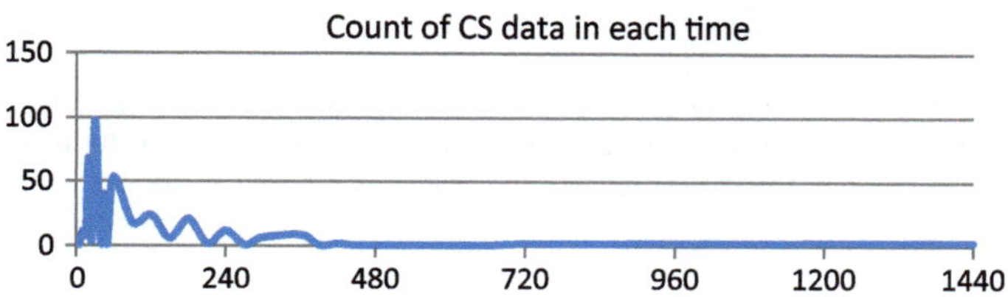

Fig. 4 Continuous CS data submission after earthquake occurrence (X: time (minute); Y: Count of CS data)

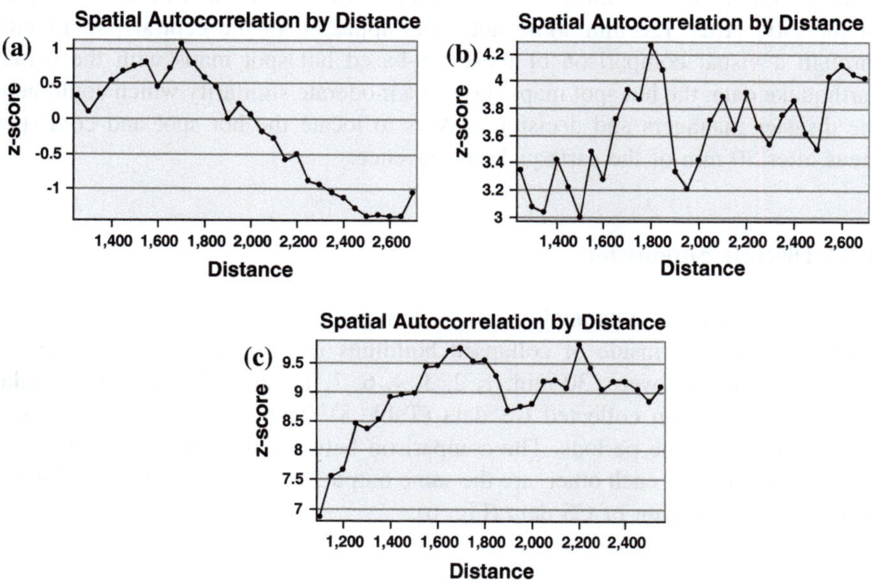

Fig. 5 Distance band for: **a** submission time of less than 20 min: 1700 m; **b** submission time of less than 60 min: 1800 m; **c** submission time of less than 120, 180, 360, 1440 min: 2200 m

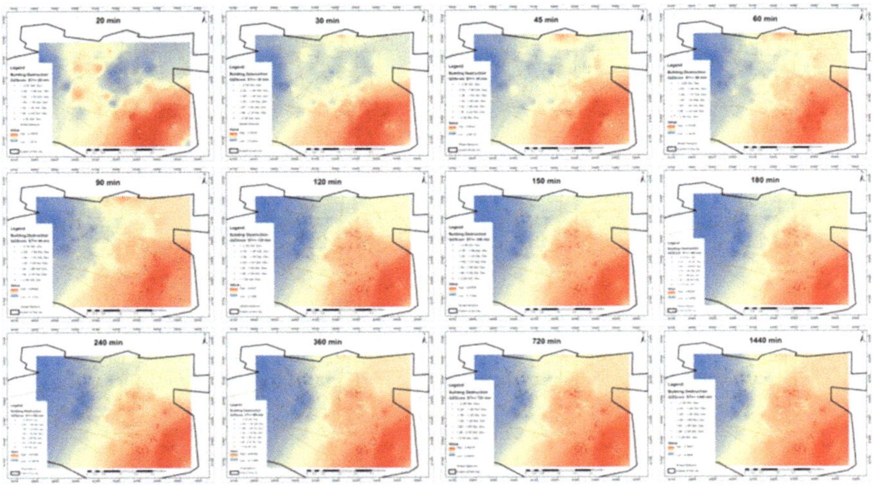

Fig. 6 Hot spot maps for the level of building destruction based on simulated continuous submission of CS data after Bam earthquake (if it is in *black* and *white* color; for interpretation of the references to color in this figure legend, the reader is referred to the web version of this article)

part of the city. As the time goes by, the pattern was changing as a result of changes in the value of building destruction level and increase in the CS data submissions. For example, after 60 min the highly damaged areas were located in south-east, but the low damaged areas concentrated in the north-west and west part of the city. After 120 min some hot spots appeared in the central part of city through a visual comparison of these CS-based hot spot maps with the actual earthquake data, the hot spot maps showed a moderate similarity which could help the disaster managers and decision makers to locate the hot spot and cold spot areas after 30 min of the earthquake occurrence.

4.1.2 Discrete Submission

The discrete time for data analysis was calculated based on critical survival time for trapped people inside of collapsed buildings in an earthquake [6] (Fig. 7). These time periods were: 30 min, 1, 2, 3, 4, 6, 7, 12 and 24 h after earthquake occurrence based on collected CS data (Table 3). Thus, we produced the end-results for eight time periods. The comparison between the end-results for these time periods against each other, are the same outputs which were generated for the continuous submission of CS data (Fig. 6).

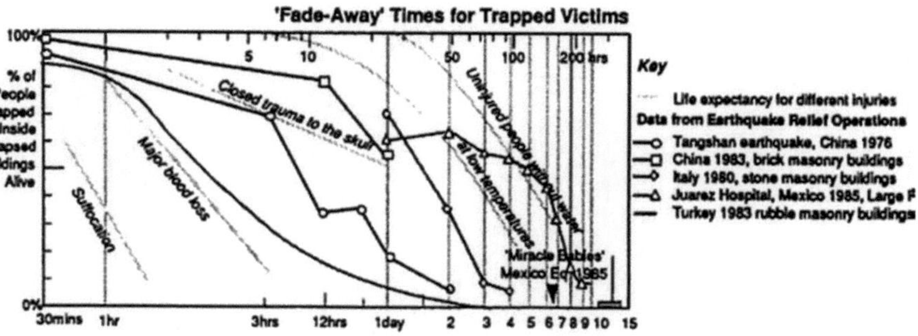

Fig. 7 The survival time for trapped people [6]

Table 3 Discrete CS data submission after the earthquake occurrence

Class	1	2	3	4	5	6	7	8
Survival percentage of trapped people	90	80	75	70	65	60	40	20
Submission time (Minute)	30	60	120	180	240	360	720	1440
Count of the CS data submissions in each time period	189	96	42	27	14	17	3	8

Fig. 8 Spatial autocorrelation by distance for complete submission of CS data: 2200 m

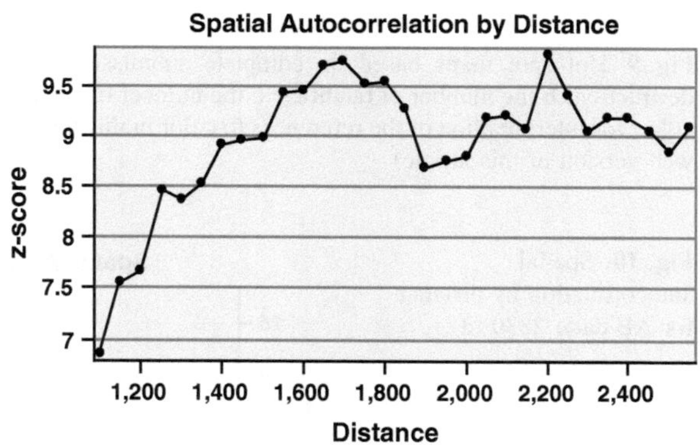

4.1.3 Complete Submission

Under this scenario, the CS data were submitted to the disaster management center's database in one specific time after the earthquake occurrence. We calculated this time as the average submission time of the collected CS data, which was in the 92nd min for 396 sample responses. The hot spot analysis was run to create hot spot maps for all 396 points of CS data and also actual earthquake data. Again, the distance bands were calculated to take into consideration the most pronounced spatial processes that promoted clustering. So, the distance band for CS data was 2200 m (Fig. 8). In this process three parameters were taken into account: the level of destruction of each building, the number of fatalities of each

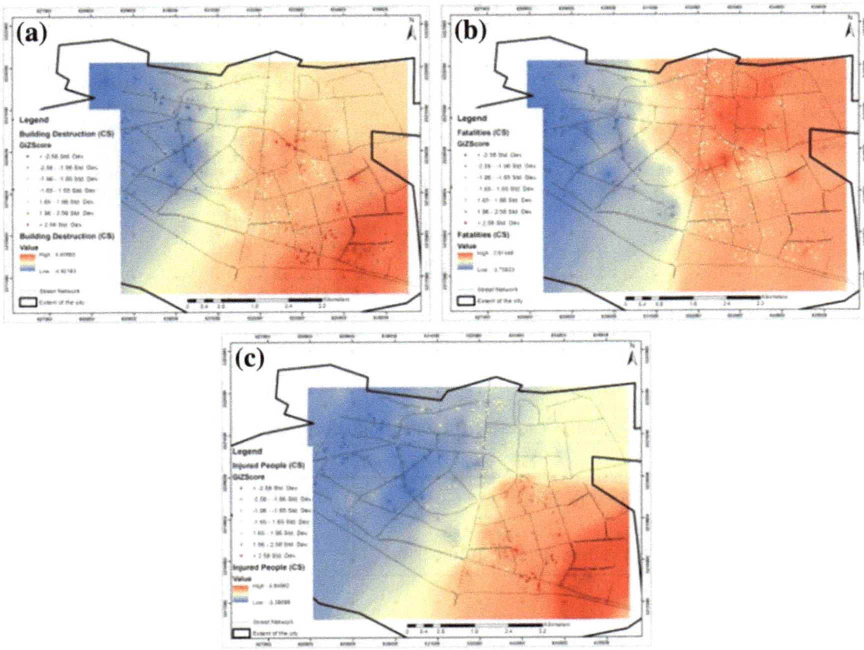

Fig. 9 Hot spot maps based on complete submission of CS data for: **a** level of building destruction; **b** the number of fatalities; **c** the number of injured people (if it is in *black* and *white* color; for interpretation of the references to color in this figure legend, the reader is referred to the web version of this article)

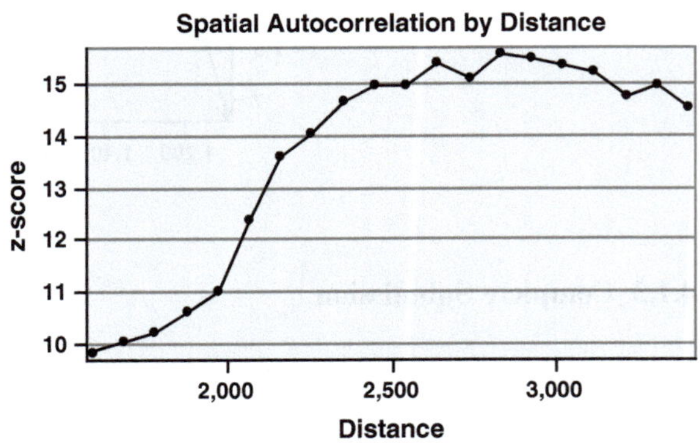

Fig. 10 Spatial autocorrelation by distance for AE data: 2830 m

building, and the number of injured people of each building. The related hot spot maps were produced based on CS data (Fig. 9a–c).

A visual comparison [33] identified that the hot spots areas for all three parameters, the level of building destruction, the number of injuries and the number of fatalities, were located in south-east and east part of the city and cold

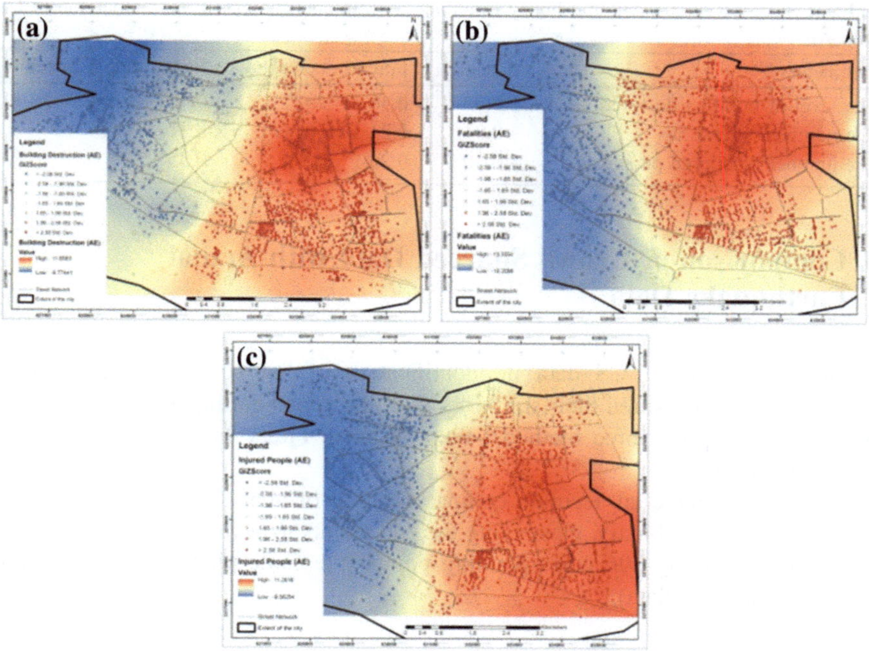

Fig. 11 Hot spot maps based on Actual Earthquake data for **a** level of building destruction; **b** the number of fatalities; **c** the number of injured people (if it is in *black* and *white* color; for interpretation of the references to color in this figure legend, the reader is referred to the web version of this article)

spots were in the west part of the city. Thus, this result would give an idea to disaster managers and decision makers where would be the possible places to start the allocation of limited resources even after 60–90 min after an earthquake occurrence. The more detailed account of damages would be confirmed with the application of other sources of information.

4.2 Hot Spot Analysis with Actual Earthquake Data

The hot spot analysis with AE data was based on 1415 building blocks in Bam city. The distance band for AE data was 2830 m (Fig. 10). Hot spot maps for the same three parameters were produced (Fig. 11a–c). These maps were used for the comparison of hot spots generated by CS and AE data, presented in the next section.

Table 4 Fuzzy Kappa Index and Fuzzy Inference System analysis based on comparison of CS with AE

Parameters	K_{fuzzy}	Fraction correct	Fuzzy global matching
The level of building destruction	−0.59	0.46	0.46
The number of fatalities	−0.04	0.21	0.32
The number of injured people	−0.30	0.26	0.35

5 Comparison of Crowd Sourced Data with Actual Earthquake Data

To test the value of CS data in identification of hot spots after earthquake occurrence, we measured their similarity with the actual earthquake data. For this purpose, we applied three methods: Fuzzy Kappa, Fuzzy Inference System and cross tabulation.

In order to create an accurate comparable unit of analysis, two aspects were taken into account. Firstly, classifying the end results to seven similar classes and secondly, specifying 396 corresponding points (the same place) in both hotspot maps (CS and AE data).

5.1 Fuzzy Kappa and Fuzzy Inference System

We employed Fuzzy Kappa comparison approach to assess the degree of agreement between hot spot maps generated with CS data and those generated with the actual earthquake data, used for ground truth.

To allow for numerical comparison, all maps should be displayed with the same categories. Although all maps that we produced had seven classes, the values of the upper and the lower class were different. Therefore, we applied categorical map comparison by first converting all numerical classes to categorical classes based on the Z-Score of hot spot analysis. In Fuzzy Kappa algorithm, the radius of neighborhood was set at 4 km and an exponential decay distance function was used to assess the accuracy of the end-results of the hot spots based on complete submission of CS data. They were compared with the end-results of hot spots generated with the actual earthquake data for three parameters: the level of building destruction, the number of injured people, and the fatalities after the earthquake.

The calculated K_{fuzzy} and fraction correct similarity value for the hot spots in terms of the level of building destruction, fatalities and injured people showed in Table 4. The results confirmed a low to moderate overall spatial similarity between the hot spots which resulted from CS data and AE data (Fig. 12a–c and Table 4). We would note here that the negative Kappa value implied similarities between maps which were worse than randomly produced maps containing the

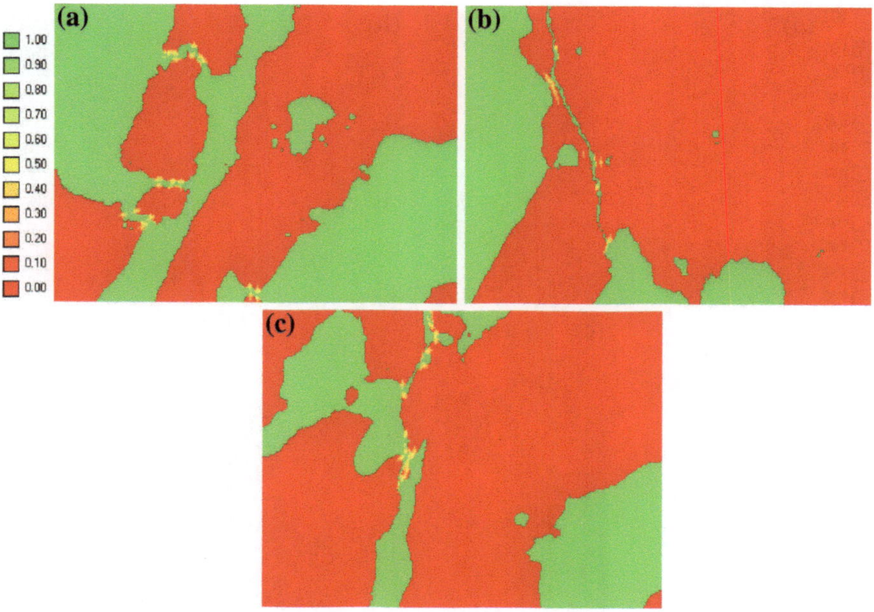

Fig. 12 Map comparison based on Fuzzy Kappa Index: **a** the level of building destruction, **b** fatalities, **c** injured people; *Red color*: dissimilar = 0 and *green color*: similar = 1 (if it is in *black* and *white color*; for interpretation of the references to color in this figure legend, the reader is referred to the web version of this article)

same number of cells for each specific category. The similarity index of 0.80 (light green area) was obtained in west part of the city in the area of cold spots; the similarity index of 0.10 (light red area) was achieved in the center of the city where a hot spot area is located.

The comparison based on the Fuzzy Inference System method showed higher similarities between two sets of hot spot map. The fuzzy global matching for the comparison hot spots maps for the level of building destruction, fatalities and injured people were calculated (Fig. 13a–c and Table 4). Again the results showed low to moderate similarity between hot spot maps. The similarity index of 0.60 (yellow area) was obtained in west part of the city in the area of cold spots; the similarity index of 0.30 (light brown area) was achieved in the center of the city where a hot spot area is located.

5.2 *Cross Tabulation*

To compare the similarity of CS data to the actual earthquake data, 396 corresponding points (the same place) in both hotspot maps (CS and AE data) were considered. The cross-tabulation analysis [27] of these data was undertaken for the

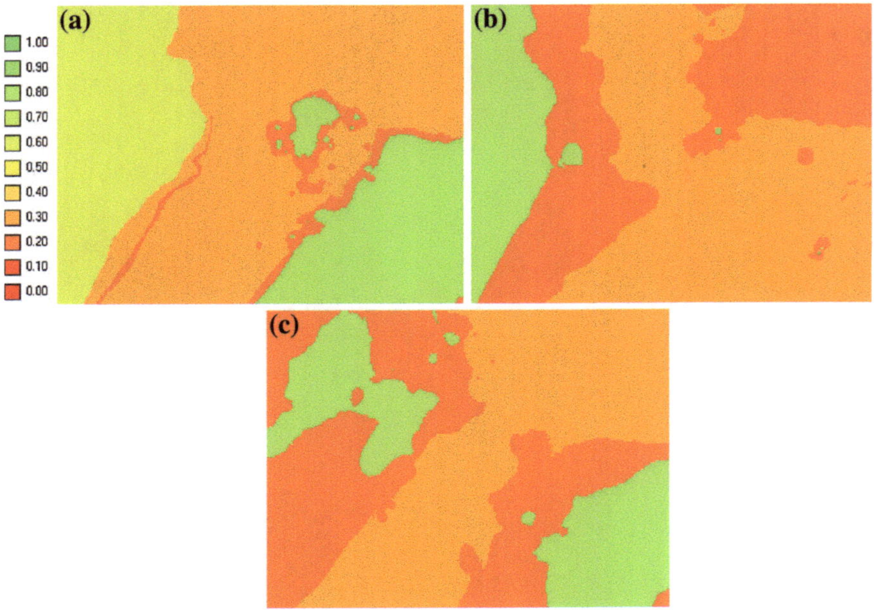

Fig. 13 Map comparison based on Fuzzy Inference System for **a** the level of building destruction, **b** fatalities, **c** injured people. *Red color* : dissimilar = 0 and *green color*: similar = 1 (if it is in *black* and *white color*; for interpretation of the references to color in this figure legend, the reader is referred to the web version of this article)

three mentioned parameters. The output includes the kappa and disagreement indices (Tables 5a–c and 6a–c).

In Table 6a–c. we provided quantity disagreement and allocation disagreement for each paired set of CS and AE data. The lowest total disagreement was for the level of building destruction, while the greatest disagreement was for the fatalities. The total disagreement in building destruction level was 66 %. It was demonstrated the agreement of 44 % between the CS and AE hot spots regarding building destruction. And also total disagreement of 77 % regarding injured people showed the low similarities between these parameters.

Finally, to summarize the results obtained for each of the hot spot maps based on CS and AE data and for three parameters, we calculated the agreement, omission disagreement and commission disagreement by object category (Fig. 14a–c). There were seven categories in each map with each category indicating the level of damages: 1: completely damaged, 2: very heavily damaged, 3: heavily damaged, 4: moderately damaged, 5: lightly damaged, 6: very lightly damaged and 7: no damages. For example, agreement for category X was where both the reference information and the map indicated category X. Omission disagreement for category X was where the reference information indicated category X, while the map showed a different category. Commission disagreement for category X was where the map showed category X, but the reference information indicated a different category.

Table 5 Cross tabulation based on hot spot analysis of CS data and AE data

	CS data							
AE data	1	2	3	4	5	6	7	Total
Count of building destruction level								
1	43	34	34	144				255
2				9				9
3				5				5
4			1	51		4	17	73
5				3				3
6				5		3	5	13
7				2		1	35	38
Total	43	34	35	219	0	8	57	396
Count of Fatalities								
1	3	43	24	244	11	3		308
2		2	2	12		2		18
3				6				6
4				6	1	8	5	20
5				1				1
6					2	1		3
7					2	6	32	40
Total	3	45	26	249	16	20	37	396
Count of Injured People								
1	33	55	28	145	4	7		272
2				15	1			16
3				10				10
4				16	5	5	3	29
5						1		1
6						1	1	2
7				1‘	6	20	39	66
Total	33	55	28	187	16	34	43	396

(a) the level of building destruction, (b) fatalities, and (c) injured people

Table 6 Kappa and disagreement indices

Indices	(a) Building destruction level	(b) Fatalities	(c) Injured people
Kno	0.12	−0.19	−0.03
Kallocation	0.58	0.34	0.48
Kquantity	−0.36	−0.51	−0.51
Khisto	0.33	0.18	0.25
Kstandard	0.19	0.06	0.12
Chance agreement	19	5	12
Quantity agreement	0	0	0
Allocation agreement	15	6	11
Allocation disagreement	11	11	11
Quantity disagreement	55	78	66

(a) the level of building destruction, (b) fatalities, (c) injured people

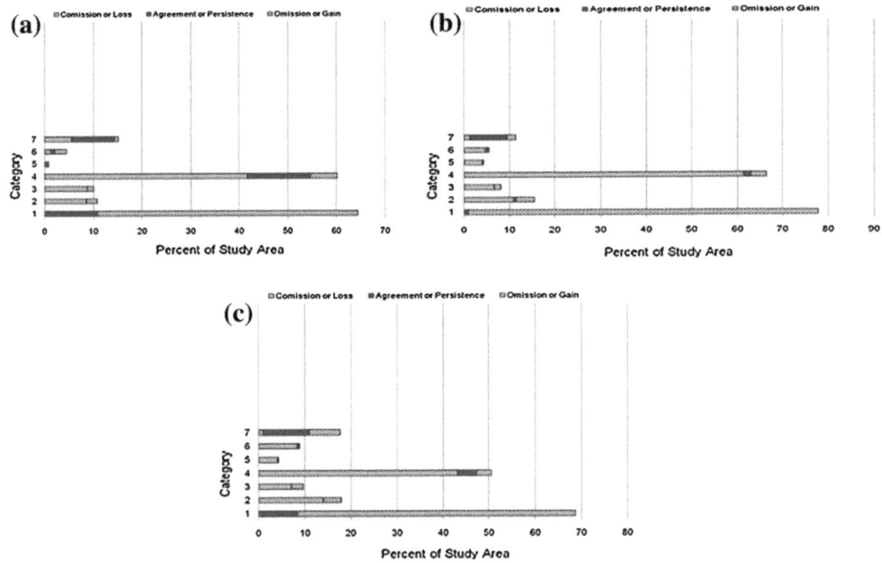

Fig. 14 Comission and omission in study area: **a** the level of building destruction, **b** fatalities, and **c** injured people

Consequently, omission for one category would commission for a different category. "If the commission disagreement is greater than the omission disagreement for a particular category, the mapped points show more of that category than the reference points'' [32, p.14]. In our analysis, for all maps the quantity of category 2 and category 4 were overestimated. Category 1 and category 7 showed the best results among the classes which these two categories were the most crucial in terms of hot spot analysis. The hot spots and cold spots in two maps were, therefore, covering each other with the agreement of 15 % in the level of building destruction, 6 % in fatalities and 11 % in injured people. The results showed low degree of agreement between CS and AE.

6 Conclusion

The previous research emphasized on the significance of Crowd Sourced data in disaster management process, but there has been less research on the utility of the extracted information to support emergency response. Therefore, we examined the value of CS data in identifying the damaged areas for response in the aftermath of an earthquake. This has been done based on hot spot (geospatial statistics Getis-Ord; Gi*) and comparative analysis (Fuzzy Kappa index, Fuzzy Inference System, and cross tabulation) of CS and AE data. In order to create an accurate comparable unit of analysis, two aspects were taken into account. Firstly, classifying the end results to seven similar classes and secondly, specifying 396 corresponding points (the same place) in both hotspot maps (CS and AE data).

Based on Fuzzy Kappa the similarity between hot spot maps of building destruction, fatalities, and injured people was 46, 21, and 26 % respectively. Fuzzy Inference System showed a similarity of fuzzy global matching of 46, 32, and 35 % for the same measures. Cross tabulation analysis pointed to a disagreement between hot spot maps which had the allocation dissimilarity of 11 % for all three measures. These similarity and dissimilarity measures clarified that there was a low to moderate similarity between hot spot maps based on CS and AE.

Consequently, CS data could have a moderate potential role in identifying highly damaged areas (hot spots) and low damaged areas (cold spots). Specifically, in this study the CS data were more indicative of in the determination of cold spot areas and divided the city into two general areas—east (hot spots with highly damaged areas) and west (cold spots with less damaged areas).

The results would suggest that the direct use of CS data could give a general idea about the high and low damages in affected area. However, for effective disaster response decisions, the CS data should be only used as a complementary data with combination of existing data such as various existing datasets (spatial data infrastructure—SDI) and remotely sensed data. This study provides the initial ideas for future research regarding the integration of CS data with SDIs to facilitate earthquake disaster response.

Acknowledgments The authors greatly appreciate the support provided by the Disaster Management Center of Kerman Municipality (KDMC) in Iran.

References

1. F. Abel, C. Hauff et al., Twitcident: fighting fire with information from social web streams. in *Proceedings of the 21st International Conference Companion on World Wide Web* (ACM, Lyon, 2012), pp. 305–308
2. B. Agrios, *Thinking Spatially About Crowd Sourcing* (ArcWatch e-Magazine, Esri, 2011)
3. American Educational Research Association (AERA), *Standards for Educational and Psychological Testing* (American Educational Research Association, Washington, 1999)
4. G. Barbier, R. Zafarani et al., Maximizing benefits from crowdsourced data. Comput. Math. Organ. Theory. **18**(3) ,257–279 (2012)
5. N. Budhathoki, B. Bruce et al., Reconceptualizing the role of the user of spatial data infrastructure. GeoJournal **72**(3), 149–160 (2008)
6. A. Coburn, R. Spence, *Earthquake Protection* (Wiley, Hoboken, 2002)
7. C. Corbane, G. Lemoine et al., Relationship between the spatial distribution of SMS messages reporting needs and building damage in 2010 Haiti disaster. Nat. Hazards Earth Syst. Sci. **12**, 255–265 (2012)
8. P. Earle, Earthquake twitter. Nat. Geosci. **3**(4), 221–222 (2010)
9. A. Fink, *The Survey Kit: How to Sample in Surveys* (Sage Publications, Thousand Oaks, 2003)
10. J. Fruchterman, Issues with crowdsourced data (2011), http://benetech.blogspot.ie/2011/03/issues-with-crowdsourced-data-part-2.html

11. M. Goodchild, Citizens as sensors: the world of volunteered geography. GeoJournal **69**(4), 211–221 (2007)
12. M.F. Goodchild, J.A. Glennon, Crowdsourcing geographic information for disaster response: a research frontier. Int. J. Digit. Earth **3**(3), 231–241 (2010)
13. R. Goolsby, Social media as crisis platform: the future of community maps/crisis maps. ACM Trans. Intell. Syst. Technol. **1**(1), 1–11 (2010)
14. G. Grünthal, European Macroseismic Scale 1998 (EMS-98). (Cahiers du Centre Européen de Géodynamique et de Séismologie 15, Centre Européen de Géodynamique et de Séismologie, Luxembourg, 1998), p. 99
15. M. Guy, P. Earle et al., in *Integration and Dissemination of Citizen Reported and Seismically Derived Earthquake Information via Social Network Technologies Advances in Intelligent Data Analysis IX*, vol. 6065, ed. by P. Cohen, N. Adams, M. Berthold (Springer, Berlin, 2010), pp. 42–53
16. A. Hagen Zanker, An improved Fuzzy Kappa statistic that accounts for spatial autocorrelation. Int. J. Geog. Inform. Sci. **23**(1), 61–73 (2009)
17. A. Hagen-Zanker, B. Straatman et al., Further developments of a fuzzy set map comparison approach. Int. J. Geog. Inform. Sci. **19**(7), 769–785 (2005)
18. Harvard Humanitarian Initiative, Disaster Relief 2.0: The Future of Information Sharing in Humanitarian Emergencies. (Foundation and Vodafone Foundation Technology Partnership, Washington, 2011)
19. P.F. Kuo, X. Zeng, et al., Guidelines for Choosing Hot-Spot Analysis Tools Based on Characteristics, Network Restrictions, and Time Distributions. in *The 91st Annual Meeting of the Transportation Research Board*, Washington, 2011)
20. S.B. Liu, L. Palen, The new cartographers: crisis map mashups and the emergence of neogeographic practice. Cartography Geog. Inform. Sci. **37**, 69–90 (2010)
21. A.M. MacEachren, A. Jaiswal, et al., SensePlace2: Geotwitter analytics support for situation awareness. in *IEEE Conference on Visual Analytics Science and Technology*, Providence, 2011
22. P. Meier, R. Munro, The unprecedented role of SMS in disaster response: learning from Haiti. SAIS Rev **30**(2), 91–103 (Johns Hopkins university press, 2010)
23. T. Milo, Crowd-based data sourcing. in *7th International Workshop of Databases in Networked Information Systems,* Aiuz Wakamatsu, Japan, ed. by S. Kikuchi, A. Madaan, S. Sachdeva, S. Bhalla (Spring, Heidelberg, 2011)
24. P.A.P. Moran, Notes on continuous stochastic phenomena. Biometrika **37**, 17–23 (1950)
25. Open Street Map, Haiti earthquake and OSM (2010), http://www.openstreetmap.org/. Accessed 23 May 2011
26. J.K. Ord, A. Getis, Local spatial autocorrelation statistics: distributional issues and an application. Geog. Anal. **27**, 286–306 (1995)
27. J.R.G. Pontius, M. Millones, Death to kappa: birth of a quantity disagreement and allocation disagreement for accuracy assessment. Int. J. Remote. Sens. **32**(15), 4407–4429 (2011)
28. K. Poser, D. Dransch, Volunteered geographic information for disaster management with application to rapid flood damage estimation. Geomatica **64**(1), 89–98 (2010)
29. C. Power, A. Simms et al., Hierarchical fuzzy pattern matching for the regional comparison of land use maps. Int. J. Geog. Inform. Sci **15**, 77–100 (2001)
30. K.-F. Richter, S. Winter, in *Citizens as Database: Conscious Ubiquity in Data Collection Advances in Spatial and Temporal Databases*, vol. 6849, ed. by. D. Pfoser, Y. Tao, K. Mouratidis et al. (Springer, Berlin, 2011), pp. 445–448
31. Statistical Centre of Iran (SCI),*A Report of Bam Earthquake Impact on the Population and Building*. (Information and Publication Office of Statistical Centre of Iran (SCI), Programming and Planning Organization of Iran, Iran, 2004), p. 251
32. E. Tarantino, B. Figorito, Mapping rural areas with widespread plastic covered vineyards using true color aerial data. Remote. Sens. **4**(7), 1913–1928 (2012)
33. H. Visser, *The Map Comparison Kit: Methods, Software and Applications* (Research Institute for Knowledge Systems, Bilthoven, 2004)

34. H. Visser, T.D. Nijs, The map comparison kit. Environ. Model. Softw. **21**(3), 346–358 (2006)
35. D.J. Wald, V. Quitoriano et al., Utilization of the internet for rapid community intensity maps. Seismol. Res. Lett. **70**(6), 680–697 (1999)
36. M. Zook, G. Mark et al., Volunteered geographic information and crowdsourcing disaster relief: a case study of the Haitian earthquake. World Med. Health Policy **2**(2), 7−33 (2010)

Remote Sensing Based Post-Disaster Damage Mapping with Collaborative Methods

Norman Kerle

Abstract Remote sensing has become an essential tool in post-disaster response, including damage assessment, traditionally done by professional analysts. However, geodata and—tools have become ubiquitous, which has allowed other organizations and laypersons to play a more prominent role in post disaster response and assistance. This chapter addresses the prospects and challenges of collaborative damage mapping. It discusses limitations and problems of traditional damage mapping that may be overcome by modern Web 2.0-based methods. The response to the 2010 Haiti earthquake, in particular the collaborative damage mapping by the GEO-CAN initiative and a number of problems associated with this activity, are discussed in detail. These center on the analysis problem inherent in image-based damage assessment, but also the limited mutual understanding among the mapping organizers, map user and volunteer mappers. Cognitive task analysis (CTA) is recommended to address and overcome the cognitive challenges and demands in collaborative mapping.

Keywords Image-mapping · Volunteer mapping · Damage mapping

1 Introduction

Rapid and accurate assessment of structural damage is essential after disaster events, especially in densely built-up urban areas. The results provide guidance for rescue forces and other immediate relief efforts, as well as subsequent

N. Kerle (✉)
Department of Earth System Analysis, University of Twente, Hengelosestraat 99,
P.O. Box 6, 7500 AA Enschede, The Netherlands
e-mail: kerle@itc.nl

S. Zlatanova et al. (eds.), *Intelligent Systems for Crisis Management*, Lecture Notes in Geoinformation and Cartography, DOI: 10.1007/978-3-642-33218-0_9,

rehabilitation and reconstruction. Especially for spatially extensive events, ground-based mapping is too slow, potentially dangerous for rescue forces, and typically hindered by disaster-related site access difficulties. Remote sensing has long been seen as a potential solution. The utility of airborne and spaceborne remote sensing in emergency response was reviewed by Kerle et al. [16] and Zhang and Kerle [27], respectively.

Many studies have investigated the utility of imagery for damage mapping, covering the entire spectrum from low-cost and uncalibrated ground- or airborne data (still or video imagery), deployed on balloons, kites, unmanned aerial vehicles or piloted aircraft, to sophisticated multi- or hyperspectral, lidar, thermal or radar devices mounted on air- or spaceborne platforms. To respond rapidly to a disaster anywhere in the world, often only satellites offer a solution. To realize this potential, and to create a dependable disaster support instrument, the International Charter "Space and Major Disasters" was established in 2000 (see [2]), and to date has been activated more than 330 times. At the same time there has been a democratization of geodata and—tool access, which, together with growing spatial literacy outside the professional geoinformatics domain, also offers the potential for more collaborative damage mapping. The suitability and potential challenges of such approaches are addressed in this chapter.

2 Traditional Image-Based Damage Mapping

Damage map generation based on Charter data has long been a routine activity, carried out by professional analysts at *UNOSAT*, the *German Aerospace Center' Center for Crisis Information (DLR-ZKI)* and *SERTIT* (based at the University of Strasbourg, France). It is generally seen as a successful example of international cooperation [24], and also the resulting damage mapping has been repeatedly praised (e.g., [25]). However, recent disaster events have highlighted (i) limitations in the process, (ii) newly emerging and potentially competitive methods, and (iii) a far wider field of stakeholders (both map producers and users) than ever before. In particular the increasing time pressure of Charter activations that have reached an average of 40 per year, the growing number of maps produced per disaster, the typical lack of ground information, and the largely Europe-based map production, mean that map accuracies are rarely determined [15]. While the variable map accuracy can be attributed to a number of factors, such as image type and cloud contamination, it is part of a broader set of problems:

- universally accepted damage map nomenclature and style are lacking;
- instead, damage is depicted on various scales and in different categories using point or line signatures, damage clusters, grid-based damage averages, damage per city block using color ramps, damage aggregated per neighborhood, or as continuous damage density maps (Fig. 1);

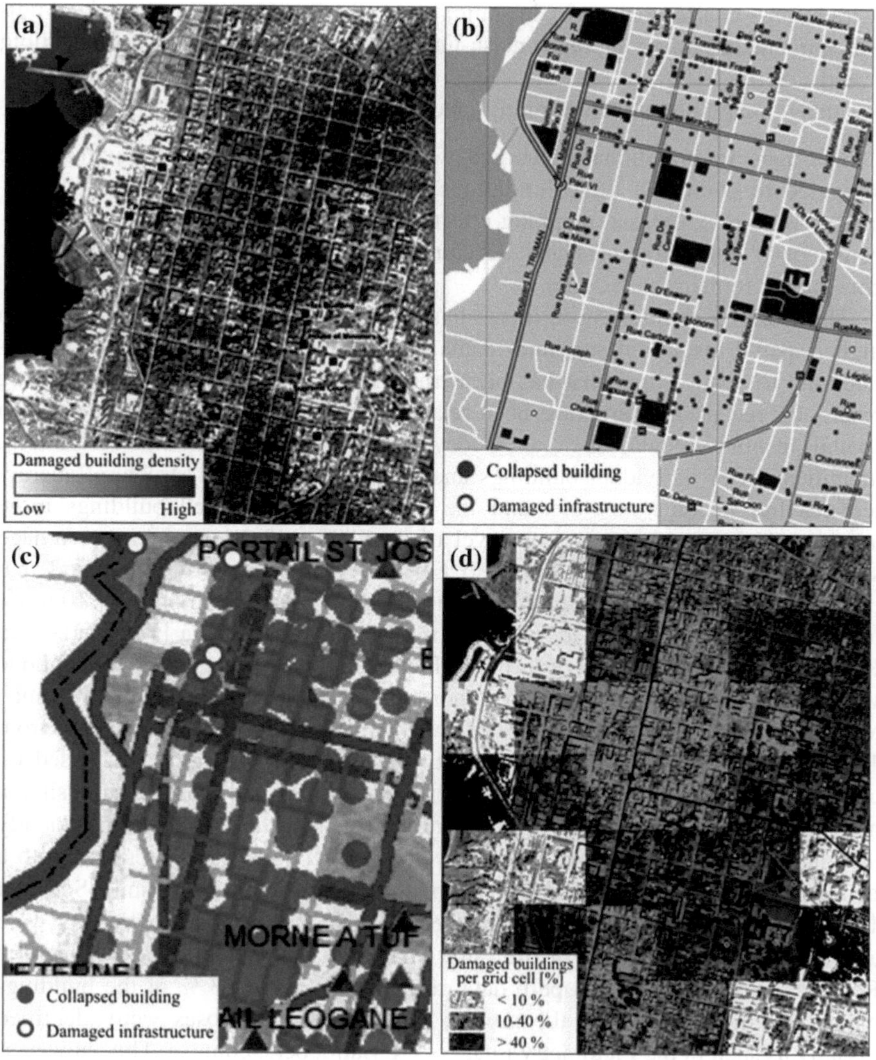

Fig. 1 Sample damage maps of parts of Port-au-Prince (Haiti) following the 2010 earthquake; **a** SERTIT, **b** ITHACA, **c** iMMAP, **d** DLR-ZKI; legend adapted from original sources to show only elements related to structural damage. Map scales are variable

- the decisions for a given mapping style do not appear to be based on what users have identified as useful or understandable, nor to reflect the needs of specific user groups;
- a growing number of organizations produces damage maps, including non-experts, leading to duplication of mapping efforts and potential disagreement;
- the number of damage map users, and their information needs, have been growing rapidly;

- traditional charter maps remain static, being distributed as print-optimized pdf-documents, not allowing ready mash-ups with other data and map customization; and
- damage mapping validation rarely takes places.

The agencies traditionally charged with image-based damage mapping have struggled to overcome these limitations. Whether the fresh approaches of collaborative Web 2.0-based mapping efforts will allow these problems to be overcome, or if flat hierarchies and decentralized stakeholders simply add further constraints, remains to be seen.

The working assumption in image-based damage mapping has been that with higher spatial resolution structural damage mapping becomes increasingly accurate, and to some extent that has been true. Overall, however, damage mapping results for recent disasters have been relatively sobering. With modern very high spatial resolution data, such as Geoeye-1 (0.41 m panchromatic resolution), identification of individual buildings and some of its external structural characteristics has become possible, and the detection of collapsed buildings more accurate. However, even at such level of detail identification of lower damage scales remains difficult ([5]; see also Fig. 2). Also in Haiti it was found that building damage mapped in the 15 cm aerial imagery was approximately 10 times higher than what had been identified in the 50 cm satellite data [18]. Gerke and Kerle [7] studied the utility of airborne Pictometry data (imagery with 4 oblique views and a vertical image) acquired shortly after the Haiti earthquake for (semi-) automatic damage-assessment. The data ranged in spatial resolution from 10–16 cm (fore- to background, respectively), and the approach was based on classification of façades, intact and destroyed roofs, and vegetation, using a number of image-derived spectral, textural and geometric features. Subsequently the classification results from different viewing directions were integrated into a per-building damage score adapted from the European Macroseismic Scale of 1998 (EMS 98; [9]), with accuracies reaching 63 %. Comparable results were achieved by visual analysis of the same image data by Cambridge Architectural Research Ltd. Despite the superb data quality, only about 63 % of the buildings mapped as D4 and D5 on the ground for validation purposes were identified visually as such in the Pictometry imagery (K. Saito, pers. comm.). This, however, also reflects that image- and ground-based damage scores are not always directly comparable. Even the EMS 98 scale that is commonly used in image-based damage mapping was actually designed for in situ damage assessment by structural engineers, not laypersons or remote sensing experts. This limits its use for image-based damage assessment.

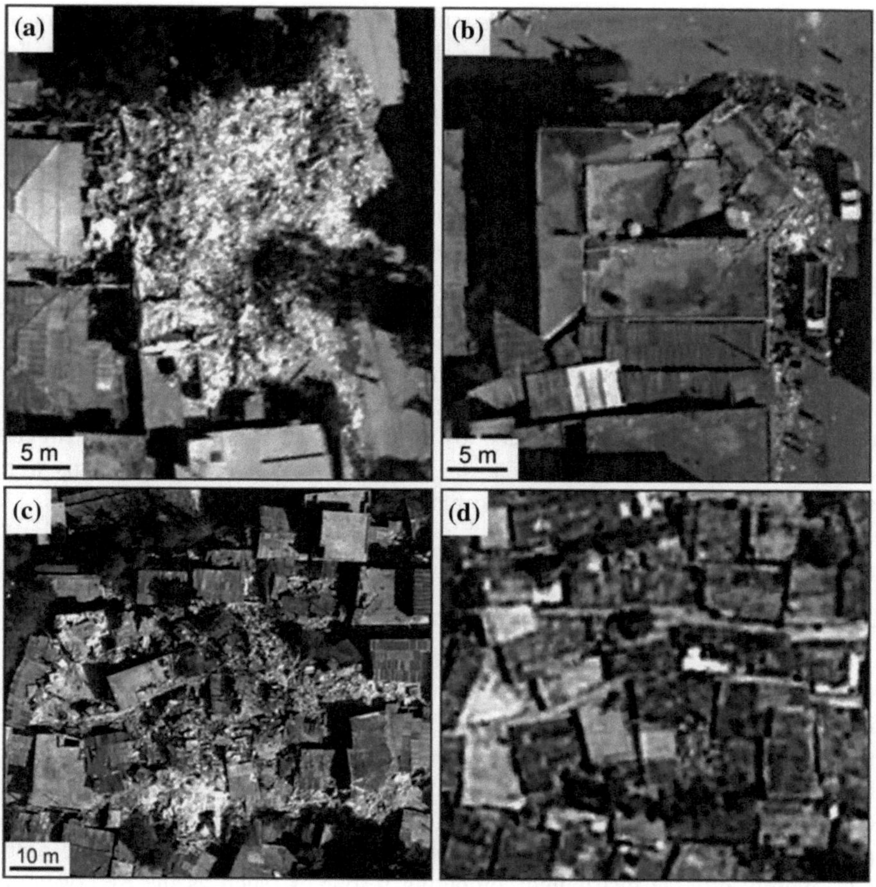

Fig. 2 Vertical aerial imagery of D5 damage (total destruction) of an adobe (**a**) and a reinforced concrete building (**b**) in Port-au-Prince resulting from the 2010 Haiti earthquake. Image-based damage mapping in Haiti was particularly challenging in densely built-up areas of small and partly overlapping buildings (**c**), where also pre-disaster satellite data as reference (**d**) were of limited help

3 Recent Trends in Disaster Damage Mapping

3.1 Maps by and for the Masses

With the digital age cartography has moved beyond its traditional boundaries, developing from a highly accurate and tightly defined craft into a core component of a growing number of enabling technologies [21]. This essentially allows anyone to be a map maker (though not necessarily a cartographer), and makes anyone with a digital media device a likely map user, a trend also evident in disaster damage mapping. On the “official” damage mapping side, the Charter-data processing agencies used to be the only major damage mappers [11]. Recent disasters, in particular the 2010 Haiti earthquake, have demonstrated how much that has changed, with maps also being made by *Information Technology for Humanitarian Assistance, Cooperation and Action* (ITHACA, based at Torino University, Italy), the *United Nations Cartographic Section* (UNCS), the *European Union Satellite Center* (EUSC), *Information Management & Mine Action Programs* (iMMAP), the *Joint Research Centers* (JRC), and an alliance of the Italian Space Agency and Telespazio that produces damage maps under its own *e-GEOS* label, as well as that of its GMES service *G-MOSAIC* (see Fig. 1). In principle such an extensive mapping is good, given the growing number of disaster response agencies (at the time of the 2010 Haiti disaster an estimated 10,000 non-governmental organizations were active in the country; [14]) who require ever more detailed information for more effective and coordinated work. In reality, however, the more than 2,000 damage maps for Haiti catalogued by Reliefweb (www.reliefweb.int) suggest considerable duplication and a lack of coordination. Which of those maps were actually used (and found useful) also remains to be assessed.

3.2 Neogeography and Collaboration Methods

Interaction and collaboration are key concepts in the Web 2.0 philosophy that is increasingly entering scientific territory. However, citizen science, i.e., the contribution of the wider population to answer scientific questions, already has a long history (an Audubon Society's bird counting program that started in 1900 is often cited as the origin). It can comprise contributors who may or may not be lay persons, and who either actively contribute to answering a given scientific problem, merely provide distributed computing resources (such as the SETI@home project), or passively provide spatial information (e.g., mobile phone users leaving spatial profiles based on the location of their handsets).

In the disaster arena it is a more recent phenomenon, but one that is quickly establishing itself in a number of different forms. People serve as distributed sensors in the USGS's *"Did you feel it"* to map seismic intensities, or neogeography tools [6] are used to aggregate and visualize community damage reports, such as after the 2011 Queensland, Australia, floods (http://queenslandfloods.crowdmap.com/main), or other crisis field reports as in Ushahidi [23]. At its simplest, geotagged photos provide a rough overview of a disaster site. However, the geotools by Google have had the most profound impact on damage mapping. Google Earth has clearly led to hugely enlarged spatial literacy and wider interest in geodata, led to a democratization of geodata access, and has served as an effective platform for distribution or image data and derivatives, as well as a collaboration platform. This interest, as well as the growing strength of the citizen mapper, is well reflected in the *Crisismappers* community (http://www.crisismappers.net) that now has more than 4,000 members, and that largely took charge of coordinating the unofficial Haiti damage mapping.

3.3 Collaborative Damage Mapping

Following the Haiti earthquake there were two prominent approaches to post-disaster mapping using Google tools. The most visible was Google Map Maker (GMM; and its open-source equivalent OpenStreetMap, OSM), whose rapid mapping of Port-au-Prince has been well documented (e.g., [11]). Hundreds of people with local knowledge created a comprehensive base map of the disaster area within a few days. Following the Wiki principle, anyone was allowed to map new or edit existing map elements, with a moderation component aiming at maximizing accuracy and error correction. This approach may not meet all traditional cartographic benchmarks, but results from Haiti have indicated high coverage and geometric accuracy [11]. The potential of such base data mapping also became clear quickly, with many other services being built around it (see http://wiki.openstreetmap.org/wiki/WikiProject_Haiti/Earthquake_map_resources).

The second form of collaborative mapping is best illustrated by the Global Earth Observation-Catastrophe Assessment Network (GEO-CAN; [4]) initiative led by the World Bank and the Global Facility for Disaster Reduction and Recovery (GFDRR). More than 600 registered individuals mapped damage visually on air- and spaceborne imagery, identifying D4 and D5 buildings. This followed experiences from the 2008 Wenchuan, China, earthquake, where 85 volunteers had mapped damage using the proprietary Virtual Disaster Viewer developed by ImageCat (VDV; vdv.mceer.buffalo.edu). For Haiti Google Earth was used as the platform instead and, unlike for base data mapping in GMM and OSM, the mapping was done only by remote sensing experts. Initially, 50 cm Geoeye imagery (down-sampled from the original 41 cm due to US government regulations) was used to derive point locations of collapsed buildings. Subsequently a more detailed and extensive mapping on 15 cm aerial imagery, using polygons to outline D4 and D5 buildings. The volunteers were provided with instructions comprising text-based descriptions of damage to be mapped, and supported by example illustrations. The work was distributed as 500 × 500 m grid cells, for which mappers submitted kmz (Google's Keyhole Markup Language) files of damage outlines that were later integrated at ImageCat. Those outlines were post- processed as described by Ghosh et al. [8] and became part of the Joint Remote Sensing Damage Assessment Database of UNOSAT, World Bank and the JRC.

4 Is Collaborative Damage Mapping the Way Forward?

For a number of reasons image-based damage assessment can be considered more challenging than base data mapping (roads, bridges, etc.), which independent assessments have found to be very accurate (e.g., [10]). Damage mapping requires that a single ordinal scale label is given to a structure in a complex state, in which individual damage elements (e.g., roofs or facades) do not add linearly to a given damage scale, and where damage classifications vary for different building types, such as masonry or reinforced concrete. That essentially makes "damage" a concept rather than a physical state. Furthermore, for the detection of damage indicators direct perception is often impossible; instead, a variety of proxies needs to be used ([7]; Kerle and Hoffman, in review). This includes the use of shadow to estimate buildings heights or changes compared to the pre-event situation, or blow-out debris near a building as evidence of pancake collapse. This explains why CEO-CAN only relied on expert contributions, but also raises further questions.

4.1 The Volunteer and the Role of Instructions

Kerle and Hoffman (in review) noted how expert mapping becomes a series of ad hoc decisions that reflect (i) how the volunteer understood and mentally adjusted the instructions and examples provided, (ii) the image analysis expertise of the mapper, (iii) the learning process during the mapping task, and (iv) the time allocated to the mapping. This can lead to variations in mapping accuracy (damage level) and precision (building outline), compounded when during the mapping new image data or information from the disaster scene become available, or additional mapping tasks are added. The study identified a number of additional factors that may compromise the accuracy and, ultimately, value of the analysis results: (i) the collaborative method assumes that all volunteers have the perceptual and damage recognition skills needed, (ii) it is unclear how general remote sensing expertise (GEO-CAN required a MSc degree and 4–5 years experience) relates to the ability to map damage accurately, (iii) the specific purpose of the damage data was unknown to the volunteer during past collaborative mapping exercises, and (iv) volunteers received no corrective feedback during the mapping, even though it is considered essential in voluntary collaborative work [12]. Further, (v) the effect of mapping complexity on accuracy is unknown. In the Wenchuan case, in addition to structural building damage, also damage to infrastructure, location of shelter sites, and landslides triggered by the earthquake had to be mapped.

Even a basic level of consistency can only be achieved if both mapping goal and strategy are clearly explained [22]. For both Wenchuan and Haiti detailed instructions were provided that explained the damage to be mapped and gave example illustrations. Volunteers were provided with text descriptions and example imagery meant to be comprehensive and unambiguous. However, ambiguity is not a function of the number or choice of illustrations, but rather results from the interaction of instructions with the knowledge and experience of the mapper and the context of the mapping task. Therefore, this ambiguity cannot be reduced simply by expanding the instruction manual, but only through corrective feedback provided to the mapper during the analysis (Kerle and Hoffman, in review).

4.2 The Need for Cognitive Task Analysis

Collaborative damage mapping is a complex interplay of three actors: the mapping organizers, the volunteers, and the damage information users. While ideally each has a good understanding about the other two, the reality is different and characterized by many assumptions. Those include (i) the organizers assuming that all volunteers unambiguously understand the mapping instructions and have the ability to translate them into accurate analysis results, (ii) the volunteers assuming any damage information *per se* will be useful to the map user, and any map entry understandable, and (iii) the users assuming that the mapping results are accurate, i.e. not appreciating the uncertainty associated with image-based damage mapping, which is typically not explained in terms understandable by non-experts. Collaborative mapping and emergency response can thus be seen as complex cognitive systems. Cognitive systems engineering, specifically its primary method, Cognitive task analysis (CTA), appears to be a suitable tool to identify and analyze cognitive challenges and demands in collaborative mapping. Kerle and Hoffman (in review) discuss how CTA can be used to address the following issues: (i) how mapping instructions can be conveyed to the volunteer in the most comprehensive and least ambiguous manner, (ii) to assess whether the instructions actually result in valid results, (iii) how it can be ensured that the mapper interprets the instructions in the way they were meant, and (iv) how the collaborative process, including feedback, can best be organized and the mapping results by many volunteers validated and merged. The latter includes practical issues, such as whether several volunteers should map the same tile, or a professional should moderate the work of several volunteers and facilitate communication among them. It can (v) also be used to assess how map users interpret mapping results. CTA thus allows the development of information systems that support and enhance human performance (see [13]).

4.3 Open Access or Closed Systems?

Web 2.0 technologies allow a number of ways to support collaborative mapping, with suitability dictated by the mapping aims, the qualifications and spatial literacy of the mapper, and confidentiality and data access restrictions. Early tools for post-disaster collaboration tended to be of the sophisticated GIS and web map server technology, including both proprietary and open-source systems. Many are meant to support multi-stakeholder decision making, such as the Distributed Virtual Geographic Environment (DVGE; [26]) or the User-Defined Operational Picture

(UDOP; [3]), built for the US Southern Command to facilitate Google Earth-based data integration and decision making for the Haiti disaster response. Others developed more rigid Geo Web Services to share damage mapping results and allow controlled interaction with registered users, e.g. to resolve mapping ambiguities [20]. Both system types can serve a specific purpose well; however, with system transparency and ease of use being critical components of successful collaborative mapping [22], tools based on familiar technology of the Google type can more easily reach a large number of contributors. GEO-CAN worked with a hybrid approach of an essentially open platform used by a closed community.

4.4 Design of a More Robust Collaborative Mapping Strategy

The collaborative mapping employed in the GEO-CAN effort is attractive, since it can facilitate an efficient and coordinated parallel image analysis by hundreds of experts. The controlled access conditions also allow tighter control over the instructions issued to the group and over the integration of submitted results than is possible with GMM or OSM, and every damage element mapped is clearly linked to a specific contributor. However, a number of technicalities need to be resolved: (i) how should mapping instructions best be crafted and presented to avoid ambiguities and ensure consistent mapping?, (ii) how can an optimal balance between more mapping detail (e.g., annotated polygons vs simple damage points) and accuracy and consistency declining with increasing complexity best be achieved?, (iii) should the same tile be mapped by several experts to provide a consensus and highlight mapping errors?, and (iv) how can the accuracy of the results be objectively assessed, and remaining errors and ambiguities identified? Whatever the strategy, it is critical that the mappers are provided with feedback on their work and the results [19]. As proposed by Kerle and Hoffman (in review), CTA can be used to identify and resolve cognitive challenges that result from the interaction of three distinct actors in collaborative mapping. This also implies that new collaborative approaches aiming at opening damage mapping also to interested lay volunteers [1], such as Tomnod (tomnod.com/geocan), may fail to realise a number of basic limitations in collaborative mapping.

5 Summary

What do the three distinct damage mapping approaches—static maps based on Charter data, collaborative grid-based damage mapping by many experts, and the mapping largely done by lay person in GMM and OSM—teach us? The rapid and flexible mapping response in the aftermath of the Haiti Earthquake carried out outside the official image-based damage mapping domain has not only created competition for traditional Charter-based map products. It has also shown that the technology to support collaborative mapping is mature, with the apparently most robust and accessible platform having emerged from open access virtual globes such as Google Earth, rather than from more traditional web map services previously explored by disaster response researchers. The methods developed or refined for Haiti have shown that (i) considerable potential in post-disaster information gathering, including of ground-based damage evidence, also lies with individuals other than traditional mapping professionals, (ii) there is a great willingness by non-mandated people to contribute to such efforts, and (iii) there is a great need for flexible and customized map products and that those can be readily provided. The traditional Charter-type mapping process needs to move away from static, one-directional mapping, and strive for a better understanding of map user needs, as well as damage maps with standardized and agreed-upon nomenclature.

However, actual image-based structural damage by lay persons is not advisable. Here an expert-based collaborative approach is more useful, though an optimal methodology both for the mapping and the subsequent data integration and processing has yet to be developed. This process needs to be transparent, and feedback must be given to the mappers. It must also be underpinned by the realization that damage assessment based on imagery, even very high resolution and multi-perspective oblique airborne data, has limits. This must be clear to the mapping organizers and enter the way volunteer mappers are instructed, and it also needs to be clearly communicated to potential users that map accuracies will be modest only even for high resolution data. Finally, mapping organizers, map users and the volunteers are all engaged in complex cognitive work. Cognitive task analysis (CTA) can be a useful way to understand the information and decision requirements of the map users, and how the volunteers can be optimally instructed and their mapping contributions integrated into suitable map products. Here an iterative approach involving map users and remote sensing specialists, but also cognitive systems engineers and instructional designers, is recommendable.

References

1. L. Barrington, S. Ghosh, M. Greene, S. Har-Noy, J. Berger, S. Gill, A.Y.M. Lin, C. Huyck, Crowdsourcing earthquake damage assessment using remote sensing imagery. Ann. Geophys. **54**, 680–687 (2011)
2. J.L. Bessis, J. Bequignon, A. Mahmood, Three typical examples of activation of the international charter “space and major disasters”. Adv. Space. Res. **33**, 244–248 (2004)
3. A.J. Clark, P. Holliday, R. Chau, H. Eisenberg, M. Chau, Collaborative geospatial data as applied to disaster relief: Haiti. In: ed. by T.-H. Kim, W.-C. Fang, M.K. Khan Security Technology, Disaster Recovery and Business Continuity, (Springer Berlin Heidelberg, 2010), pp. 250–258
4. C. Corbane, K. Saito, L. Dell’Oro, S.P.D. Gill, B.E. Piard, C.K. Huyck, T. Kemper, G. Lemoine, R.J.S. Spence, R. Shankar, O. Senegas, F. Ghesquiere, D. Lallemant, G.B. Evans, R.A. Gartley Joaquin Toro, A. Ghosh, W.D. Svekla, B.J. Adams, R.T. Eguchi, A comprehensive analysis of building damage in the 12 January 2010 Mw7 Haiti earthquake

using high resolution satellite and aerial imagery. Photogramm. Eng. Remote Sens. **77**, 997–1009 (2011)
5. D. Ehrlich, H.D. Guo, K. Molch, J.W. Ma, M. Pesaresi, Identifying damage caused by the 2008 Wenchuan earthquake from VHR remote sensing data. Int. J. Digit. Earth **2**, 309–326 (2009)
6. S. Elwood, Geographic information science: new geovisualization technologies—emerging questions and linkages with GIScience research. Prog. Hum. Geogr. **33**, 256–263 (2009)
7. M. Gerke, N. Kerle, Automatic structural seismic damage assessment with airborne oblique pictometry imagery. Photogramm. Eng. Remote Sens. **77**, 885–898 (2011)
8. S. Ghosh, C.K. Huyck, M. Greene, S.P. Gill, J. Bevington, W. Svekla, R. DesRoches, R.T. Eguchi, Crowdsourcing for rapid damage assessment: the global earth observation catastrophe assessment network (GEO-CAN). Earthq. Spectr. **27**, S179–S198 (2011)
9. G. Grünthal, European Macroseismic Scale 1998 (EMS-98). Cahiers du Centre Européen de Géodynamique et de Séismologie 15. Centre Européen de Géodynamique et de Séismologie, p. 99 (1998)
10. M. Haklay, How good is volunteered geographical information? A comparative study of open street map and ordnance survey datasets. Environ. Plan. B-Plan. Des. **37**, 682–703 (2010)
11. C. Heipke, Crowdsourcing geospatial data. ISPRS J. Photogramm. Remote. Sens. **65**, 550–557 (2010)
12. R.R. Hoffman, Remote perceiving: a step toward a unified science of remote sensing. Geocarto Int. **5**, 3–13 (1990)
13. R.R. Hoffman, L.G. Militello, *Perspectives on Cognitive Task Analysis: Historical Origins and Modern Communities of Practice* (CRC Press/Taylor and Francis, Boca Raton, 2008)
14. K. Jobe, Disaster relief in post-earthquake Haiti: unintended consequences of humanitarian volunteerism. Travel Med. Infect. Dis. **9**, 1–5 (2011)
15. N. Kerle, Satellite-based damage mapping following the 2006 Indonesia earthquake—how accurate was it? Int. J. Appl. Earth Obs. Geoinform. **12**, 466–476 (2010)
16. N. Kerle, S. Heuel, N. Pfeifer, Real-time data collection and information generation using airborne sensors, in *Geospatial Information Technology for Emergency Response*, ed. by S. Zlatanova, J. Li (Taylor and Francis, London, 2008), pp. 43–74
17. N. Kerle, R.R. Hoffman, Collaborative damage mapping for emergency response: the role of cognitive systems engineering. Nat. Hazards Earth Syst. Sci. (in review)
18. G. Lemoine, *Validation of Building Damage Assessments Based on Post-Haiti 2010 Earthquake Imagery Using Multi-Source Reference Data, Valgeo 2010–2nd International Workshop on Validation of Geo-Information Products for Crisis Management* (Ispra, Italy, 2010)
19. A.M. MacEachren, G.R. Cai, Supporting group work in crisis management: visually mediated human-GIS-human dialogue. Environ. Plan. B-Plan. Des. **33**, 435–456 (2006)
20. L. Maiyo, N. Kerle, B. Kobben, Collaborative post-disaster damage mapping via Geo Web Services, in *Cartography and Geoinformatics for Early Warning and Emergency Management: Towards Better Solutions*, ed. by M. Konecny, S. Zlatanova, T. Bandrova, L. Friedmannova (Springer Berlin, 2010), pp. 221–231
21. M. Monmonier, Cartography: uncertainty, interventions, and dynamic display. Prog. Hum. Geogr. **30**, 373–381 (2006)
22. G. Newman, D. Zimmerman, A. Crall, M. Laituri, J. Graham, L. Stapel, User-friendly web mapping: lessons from a citizen science website. Int. J. Geog. Inform. Sci. **24**, 1851–1869 (2010)
23. S. Roche, E. Propeck-Zimmermann, B. Mericskay, GeoWeb and crisis management: issues and perspectives of volunteered geographic information. GeoJournal 1–20 (2011)
24. T. Stryker, B. Jones, Disaster response and the international charter program. Photogramm. Eng. Remote. Sens. **75**, 1342–1344 (2009)
25. S. Voigt, T. Kemper, T. Riedlinger, R. Kiefl, K. Scholte, H. Mehl, Satellite image analysis for disaster and crisis-management support. IEEE Trans. Geosci. Remote Sens. **45**, 1520–1528 (2007)

26. J.Q. Zhang, J.H. Gong, H. Lin, G. Wang, J.L. Huang, J. Zhu, B.L. Xu, J. Teng, Design and development of distributed virtual geographic environment system based on web services. Inf. Sci. **177**, 3968–3980 (2007)
27. Y. Zhang, N. Kerle, Satellite remote sensing for near-real time data collection, in *Geospatial Information Technology for Emergency Response*, ed. by S. Zlatanova, J. Li (Taylor and Francis, London, 2008), pp. 75–102

Automatic Determination of Optimal Regularization Parameter in Rational Polynomial Coefficients Derivation

Junhee Youn, Tae-Hoon Kim, Changhee Hong and Jung-Rae Hwang

Abstract Recently, massive archives of ground information are provided by imagery from new sensors. To establish the functional relationship between image and ground space, sensor models are required. The rational functional model (RFM), which is used as an alternative to rigorous sensor model, becomes attractive due to its generality and simplicity. To get a rational polynomial coefficients (RPC) in RFM, we encounter the problem for obtaining a stable solution, because design matrix for solutions usually ill-conditioned in the experiments. To solve such an unstable problem, regularization techniques are generally used. In this paper, we describe the determination of optimal regularization parameter in the regularization technique during RPC derivation. A brief mathematical background of RFM is presented, followed by numerical approaches for automatic determination of optimal regularization parameter with the Euler Method. Experiments are carried out; assuming tilted aerial image is taken with known rigorous sensor model.

Keywords Sensor model · Rational functional model · Rational polynomial coefficients · Regularization parameter

1 Introduction

Increasing demand for accurate spatial information, massive archives of ground information are provided by imagery from new sensors (e.g. IKONOS satellite, GeoEye satellite, QuickBird satellite etc.). To establish the functional relationship

J. Youn (✉) · T.-H. Kim · C. Hong · J.-R. Hwang
Korea Institute of Construction Technology, 283 Goyangdae-ro,
Ilsanseo-gu, Goyang-si, Gyeonggi-do, Korea
e-mail: younj@kict.re.kr

S. Zlatanova et al. (eds.), *Intelligent Systems for Crisis Management*, Lecture Notes in Geoinformation and Cartography, DOI: 10.1007/978-3-642-33218-0_10,

between the image space and the ground space, sensor models are required. There are two categories for sensor models. One is a rigorous sensor model and the other is a generalized sensor model.

A rigorous sensor model produces high accuracy, however it has several disadvantages. A rigorous sensor model, which is based on the collinearity equations, presents the rigorous imaging geometric relationship between image point and the homologous ground point, with parameters of physical meanings [1]. The advantages of rigorous sensor model are that it is very suitable for adjustment by analytical triangulation, and normally yields high accuracy. However, it is complicated and the parameters used in the rigorous sensor model are usually kept confidential in some sensors. Some commercial satellite vendors have adopted the generalized sensor model instead of rigorous sensor model to the end users [2, 3].

The rational functional model (RFM), which is one of the most popular generalized sensor model [1, 4], has drawn special attention in the photogrammetry and remote sensing fields. One of the advantages of RFM is its sensor independence. The transformation between the image point and ground point is represented as some rational functions without modelling the rigorous imaging process. Although rigorous sensor model produces more accurate results, the difference is negligible [5]. Grodecki [6] reported that IKONOS rational model differs by no more than 0.04 pixel from the rigorous sensor model, with the RMS error below 0.01 pixel. Also, the information of sensors is effectively kept confidential by providing RFM, due to the difficulty of transforming from RFM to the rigorous sensor model.

The RFM expresses image coordinates as a ratio of two polynomials with variables of ground coordinates. The polynomial coefficients in RFM are called rational polynomial coefficients (RPC). Comparing with the rigorous sensor model and RFM, the RFM would be over parameterized [7]. The ground to image transformation can be achieved with 8-coefficient in affine model. On the other hand, the usual number of RPC is 80. That may cause the design matrix to become almost rank deficient due to the complex correlation among RPC [8]. Therefore, it may result in numerical instability in the least square adjustment [8]. To solve such an instability problem, regularization technique with regularization parameter (RP) is generally used.

Determination of optimal RP has still been a challenging topic. There are several methods for the determination of RP, including L-curve method, U-curve method, ordinary cross validation (OCV) method, ridge tracing method, and so on. The L-curve [9] is a log-log plot of the norm of a regularized solution versus the norm of the corresponding residual (fitting error) as the RP is varied [10]. In Hansen [9], the point on the L-curve, that has maximum curvature, should be chosen as the corner of the curve and consequently be the optimal RP. This can be automated without plotting the curve. An improvement of this approach using the so-called "U-curve" is reported in Krawczy-Stando and Rudnicki [11]. In this approach, the authors proposed the function which is expressed as a summation of inverse number of solution norm and inverse number of residual norm, and chose the optimal RP as producing minimum value in the function.

Thite and Thompson [12] used OCV for choosing an optimal RP. Choi et al. [10] compared the performance of the L-curve method and OCV. In this paper, they concluded that the L-curve method is most effective when the problem is relatively well-conditioned (low condition number) with high noise level whereas OCV provides a more reasonable amount of regularization for ill-conditioned (high condition number) problem where noise levels are relatively high. In ridge tracing method, root mean square errors are computed for a large number of different RP, selecting the best one by suitable heuristics (e.g. trial method). Applying ridge tracing methods to obtain RP in RPC derivation is tried by Tao and Hu [4] and Zhan et al. [13]. They chose the optimal RP within the range extracted by experiments, and proved that RMSE is not sensitive to the particular RP as long as RP is within the specific range. The advantage of Ridge tracing method is its simplicity and generality. That means it is independent of condition number and noise. In this paper, we introduce a numerical approach for automatic determination of RP in ridge tracing method by using the Euler root finding method. Our approach is independent of ill-condition or well-condition unlike L-curve and U-curve. RP can be determined at the target accuracy. Also, RP can be automatically and simply calculated.

This paper is organized in the following three steps. Next section presents a mathematical background of RPC derivation in RFM with the terrain independent model. This is followed by numerical approach for automatic detection of regularization parameter in regularization process. Our algorithm is exercised with terrain independent scenarios.

2 General Solution for RFM

The RFM relates ground coordinates (X, Y, Z) to image coordinates (r, c) in the form of rational functions that are ratio of polynomials. For the ground to image transformation, the defined ratios of polynomials have following form for each section [14]:

$$r_n = \frac{p_1(X_n, Y_n, Z_n)}{p_2(X_n, Y_n, Z_n)} = \frac{\sum_{i=0}^{m1}\sum_{j=0}^{m2}\sum_{k=0}^{m3} a_{ijk} X_n^i Y_n^j Z_n^k}{\sum_{i=0}^{n1}\sum_{j=0}^{n2}\sum_{k=0}^{n3} b_{ijk} X_n^i Y_n^j Z_n^k} \tag{1}$$

$$c_n = \frac{p_3(X_n, Y_n, Z_n)}{p_4(X_n, Y_n, Z_n)} = \frac{\sum_{i=0}^{m1}\sum_{j=0}^{m2}\sum_{k=0}^{m3} c_{ijk} X_n^i Y_n^j Z_n^k}{\sum_{i=0}^{n1}\sum_{j=0}^{n2}\sum_{k=0}^{n3} d_{ijk} X_n^i Y_n^j Z_n^k}, \tag{2}$$

where

$$P_1 = a_0 + a_1X + a_2Y + a_3Z + a_4XY + a_5XZ + a_6YZ + a_7X^2 + a_8Y^2 + a_9Z^2 + a_{10}XYZ + a_{11}X^3 + a_{12}XY^2 + a_{13}XZ^2 + a_{14}X^2Y + a_{15}Y^3 + a_{16}YZ^2 + a_{17}X^2Z + a_{18}Y^2Z + a_{19}Z^3 \tag{3}$$

$$P_2 = b_0 + b_1X + b_2Y + b_3Z + b_4XY + b_5XZ + b_6YZ + b_7X^2 + b_8Y^2 + b_9Z^2 + b_{10}XYZ + b_{11}X^3 + b_{12}XY^2 + b_{13}XZ^2 + b_{14}X^2Y + b_{15}Y^3 + b_{16}YZ^2 + b_{17}X^2Z + b_{18}Y^2Z + b_{19}Z^3 \tag{4}$$

$$P_3 = c_0 + c_1X + c_2Y + c_3Z + c_4XY + c_5XZ + c_6YZ + c_7X^2 + c_8Y^2 + c_9Z^2 + c_{10}XYZ + c_{11}X^3 + c_{12}XY^2 + c_{13}XZ^2 + c_{14}X^2Y + c_{15}Y^3 + c_{16}YZ^2 + c_{17}X^2Z + c_{18}Y^2Z + c_{19}Z^3 \tag{5}$$

$$P_4 = c_0 + c_1X + c_2Y + c_3Z + c_4XY + c_5XZ + c_6YZ + c_7X^2 + c_8Y^2 + c_9Z^2 + c_{10}XYZ + c_{11}X^3 + c_{12}XY^2 + c_{13}XZ^2 + c_{14}X^2Y + c_{15}Y^3 + c_{16}YZ^2 + c_{17}X^2Z + c_{18}Y^2Z + c_{19}Z^3, \tag{6}$$

where r_n and c_n are the normalized row and column indices of pixels in image space, and X_n, Y_n, and Z_n represent normalized coordinate value of object point in ground space. Here, a_{ijk}, b_{ijk}, c_{ijk}, and d_{ijk} are polynomial coefficients called RPC. The total number of a_{ijk}, b_{ijk}, c_{ijk}, and d_{ijk} are the 80. And b_1, and d_1 are usually set to 1. Then the number of RPC is 78. The normalization of the coordinates is calculated as,

$$r_n = \frac{r - r_0}{r_s}, c_n = \frac{c - c_0}{c_s}, X_n = \frac{X - X_0}{X_s}, Y_n = \frac{Y - Y_0}{Y_s}, Z_n = \frac{Z - Z_0}{Z_s}, \tag{7}$$

where r_O and c_O are row offset value and column offset value for the image coordinates. Also, r_s and c_s are row scale value and column scale value for the image coordinates. Similarly, X_O, Y_O and Z_O are offset values for ground coordinates. Also, X_s, Y_s and Z_s are scale values for the ground coordinates. The offsets and scales normalized the coordinates to [−1. 1].

To obtain the a_{ijk}, b_{ijk}, c_{ijk}, and d_{ijk}, least square solution is used. Equations (1) and (2) are rewritten as nonlinear condition equation,

$$F(l, x) = 0, \tag{8}$$

where l represents the observation vector (r_n, c_n, X_n, Y_n, and Z_n) and x represents the parameter vector (a_{ijk}, b_{ijk}, c_{ijk}, and d_{ijk}). For the least square adjustment, the linearization of Eq. (8) follows

$$v + B\Delta = f, \tag{9}$$

in which f is given by

$$f = -[F(l, x^0)], \quad (10)$$

where l is the approximation vector for the observation, and x^0 is the approximation vector for the parameters. Δ is the vector of corrections to the approximations for the parameters. v is the vector of observational residuals, and B is the design matrix of the partial derivation of F with respect to the parameters like as follows,

$$B = [F(l, x^0)] \quad (11)$$

The normal equation matrix and the vector of correction matrix are represented respectively as

$$N = B^T W B \quad (12)$$

$$\Delta = N^{-1}(B^T W f - f_x), \quad (13)$$

in which f_x is given by

$$f_x = x^0 - l_x, \quad (14)$$

where l_x is a vector of parameter observations, and W is weight matrix. x^0 is updated x^0 plus Δ. Iteration step will be stopped when Δ is less than threshold which is insignificantly small. The final least square estimates of parameter $\hat{x}$ is

$$\hat{x} = x^0 + \Delta \quad (15)$$

3 Regularization

The RPC in RFM are highly correlated between coefficients. As a result, the matrix B in Eq. (11) is usually ill-conditioned and matrix N in Eq. (12) could become singular. It happens often when the high order (i.e., more than second-order) polynomials in the FRM are used [4]. The negative impact of this is that the iterative solution cannot be converged [4].

In order to tackle the possible ill-conditioned problem during the least square adjustment, the regularization technique, in which a small multiplication (i.e., RP) of the identity matrix is added, is applied. With this modification, the normal equation matrix in Eq. (12), correction vector in Eq. (13), and final estimates of parameter matrix in Eq. (15) are rewritten as

$$N' = B^T W B + \lambda^2 I \quad (16)$$

$$\Delta' = N'^{-1}(B^T W f - f_x) \quad (17)$$

$$\hat{x}' = x^0 + \Delta', \quad (18)$$

where λ is RP.

Determination of RP λ is non trivial. We propose root mean square error function as,

$$F_{RMSE}(\lambda) = \frac{\sum_{i=1}^{N_{CK}} \sqrt{\left[(r_{\lambda i} - r_{rigi})^2 + (c_{\lambda i} - c_{rigi})^2\right]}}{N_{CK}} \tag{19}$$

Here, N_{CK} is the number of check points. $r_{\lambda i}$ and $c_{\lambda i}$ are image coordinates obtained by using the Eqs. (1) and (2) with the known ground coordinates and adjusted parameters, which are from Eq. (18) at specific λ. r_{rigi} and c_{rigi} are image coordinates obtained by using the rigorous sensor model with the known ground coordinates.

From the Euler method [15], the first derivatives of root mean square error function in Eq. (19) can be numerically written as

$$\frac{F_{RMSE}(\lambda)}{\partial \lambda} = \frac{F_{RMSE}(\lambda + \Delta\lambda) - F_{RMSE}(\lambda)}{\Delta\lambda} \tag{20}$$

Euler step for finding optimum λ is derived as

$$\lambda_{t+1} = \lambda_t - h\frac{F_{RMSE}(\lambda_t)}{\partial \lambda} \tag{21}$$

where h is step size. This equation is iteratively solved and iteration step will be stopped when the value of first derivative of λ function is insignificantly small. Theoretically, small $\Delta\lambda$ and h make more accurate solution while iteration time is increased.

4 Experiment and Results

The RPC in the RFM can be solved for with or without knowing the rigorous sensor model [4]. The rigorous sensor model being available, the terrain independent solution is able to be developed. In the terrain independent scenarios, RFM can be solved using a ground grid with its grid-point coordinates determined using a rigorous sensor model. We modify the terrain independent scenarios proposed by Tao and Hu [4].

First, a 3D object grid in ground space is established. The maximum and minimum easting-ground UTM coordinates of grid are 506,600 and 507,460 m respectively. The maximum and minimum northing-ground UTM coordinates of grid are 4,474,600 and 4,475,630 m respectively. The relief range is from 0 to 400 m. The intervals of grid of easting, northing, and height are 100, 100, and 50 m respectively. Second, image grid in the image space is determined. By using the available rigorous

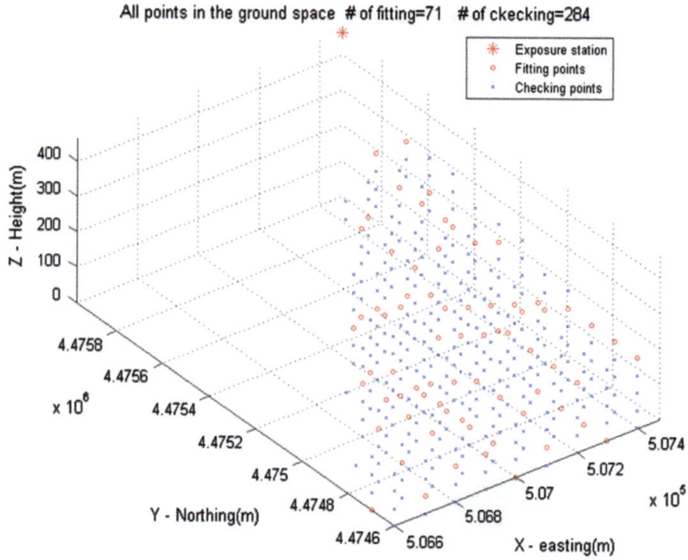

Fig. 1 Fitting points, checking points, and exposure station in the ground space

sensor model, corresponding image coordinates of ground coordinates are calculated. Figure 1 shows the used fitting points and checking points of the object grid. The location of exposure station in the ground space is also presented in Figs. 1, 2 and 3. In Fig. 4, we present fitting points and checking points in the image space. The number of fitting points and checking points are 71 and 284 respectively. The used parameters for rigorous sensor model are shown in Table 1. For the details of the notation in the Table 1, see the Wolf and Dewitt [16].

The unknown RPC can be calculated by using the corresponding image and object grid points with the Eqs. (1–18) presented in previous section. To show the RP effects to RMSE, we calculate the RMSE of check points with increasing the RP. Figures 5 and 6 show the results of RP versus RMSE. In Figs. 5 and 6, x-axis denotes the RP, and y-axis denotes the RMSE calculated by Eq. (19). Figure 5 shows the RMSE of check points with the RP range from 4×10^{-7} to 3.2×10^{-5}. The step size of RP is 1×10^{-7}. RMSE of check points with the RP range from 4×10^{-5} to 1×10^{-1} are presented in Fig. 6. Figures 5 and 6 shows that RMSE curve is decreased when RP increasing from near 0 RP, and RMSE curve continues near flat with increasing RP. The RMSE curve is increased from the specific RP. Therefore, we come to conclusion that optimal RP can be determined with the Euler root finding method.

Optimal RP is calculated by using Eqs. (19)–(21). Initial RP is set to 1×10^{-1}, and the step size h is set to 1×10^{-3} in Eq. (21). 1×10^{-5} for $\Delta\lambda$ is used in

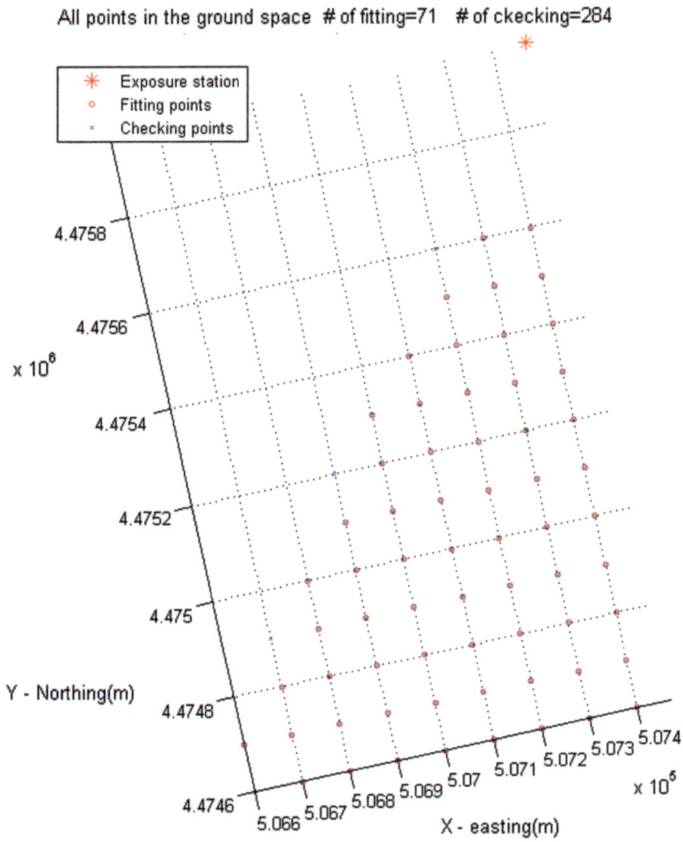

Fig. 2 Fitting points, checking points, and exposure station in the ground space (X–Y plane)

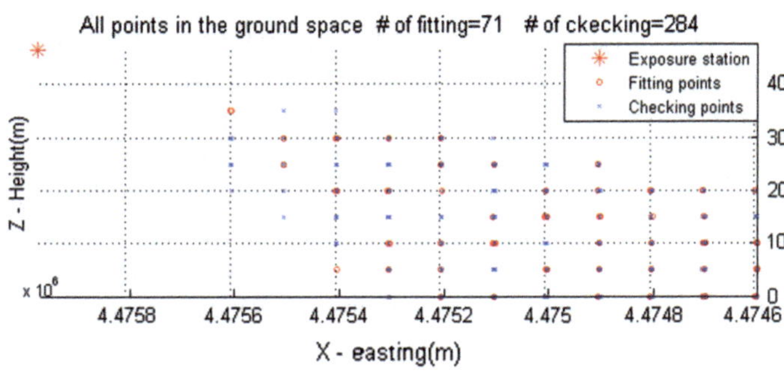

Fig. 3 Fitting points, checking points, and exposure station in the ground space (X–Z plane)

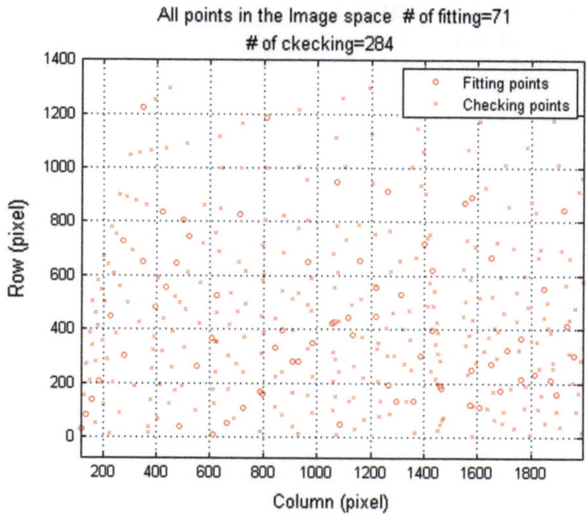

Fig. 4 Fitting points and checking points in the image space

Table 1 Used parameters for rigorous sensor model

Rotation angle ω	−1.18445382 rad
Rotation angle φ	0.30607712 rad
Rotation angle κ	3.03356237 rad
Focal length f	3310.0 pixel
Coordinates of principal point (x_a, y_a)	(45.0, 50.0) pixel
Exposure station X_L	507,471.474 m
Exposure station Y_L	4,475,970.091 m
Exposure station Z_L	463.380 m

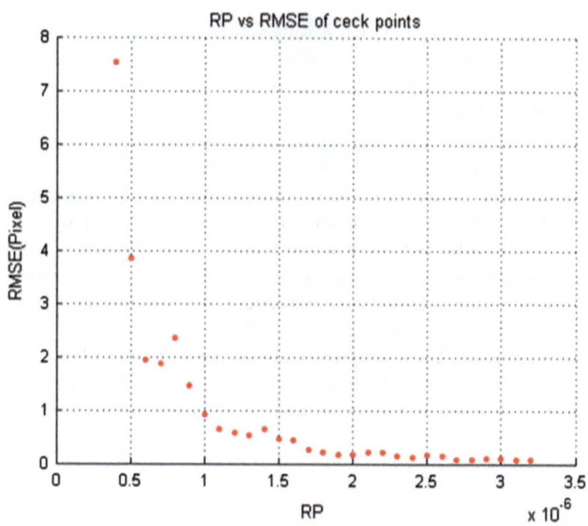

Fig. 5 RP versus RMSE, a range of RPs is from 4×10^{-7} to 3.2×10^{-5}

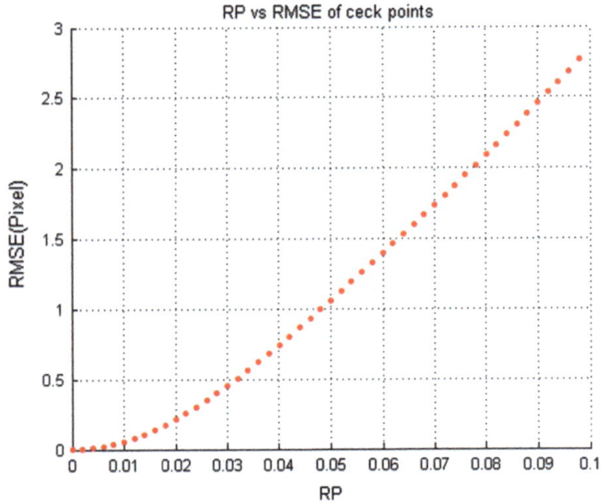

Fig. 6 RP versus RMSE, a range of RPs is from 4×10^{-5} to 1×10^{-1}

Eq. (20). With the experiment, determined RP is 2.024×10^{-4}, and RMSE for RP is 0.1612 pixel.

Acknowledgments This work was supported by a grant from a Strategic Research Project (Development of Site Investigation and Monitoring System for Extreme Cold Region) funded by the Korea Institute of Construction Technology.

References

1. X. Tong, S. Liu, Q. Weng, Bias-corrected rational polynomial coefficients for high accuracy geo-positioning of QuickBird stereo imagery. ISPRS J. Photogramm. Remote Sens. **65**, 218–226 (2010)
2. K. Di, R. Ma, R.X. Li, Rational functions and potential for rigorous sensor model recovery. PE&RS **69**(1), 33–41 (2003)
3. C.V. Tao, Y. Hu, 3D reconstruction methods based on the rational function model. PE&RS **68**(7), 705–714 (2002)
4. C.V. Tao, Y. Hu, A comprehensive study of the rational function model for photogrammetric processing. PE&RS **67**(12), 1347–1357 (2001)
5. G. Dial, J. Grodecki, RPC Replacement camera models, in *Proceedings of the 2005 ASPRS Annual Conference*, Baltimore, Maryland, unpaginated CD-ROM, 7–11 March 2005
6. J. Grodecki, IKONOS stereo feature extraction—RPC approach, in *Proceedings of the 2001 ASPRS Annual Conference*, Louis, Missouri, unpaginated CD-ROM, 23–27 April 2001
7. C.S. Fraser, G. Dial, J. Grodecki, Sensor orientation via RPCs. ISPRS J. Photogramm. Remote Sens. **60**, 182–194 (2006)
8. X. Lin, X. Yuan, Improvement of the stability solving rational polynomial coefficients. Int. Arch. Photogramm. Remote Sens. Spat. Sci. **XXXVII**(B1), 711–716 (2008)
9. P.C. Hansen, Analysis of discrete Ill-posed problem by means of L-curve. SIAM Rev. **34**(4), 561–580 (1992)

10. H.G. Choi, A.N. Thite, D.J. Thompson, Comparison of methods for parameter selection in Tikhonov regularization with application to inverse force determination. J. Sound Vib. **304**(3–5), 894–917 (2007)
11. D. Krawczy-Stando, M. Rudnicki, Regularization parameter selection in discrete Ill-posed problems—the use of the U-curve. Int. J. Appl. Comput. Sci. **17**(2), 157–164 (2007)
12. A.N. Thite, D.J. Thompson, The quantification of structure-borne transmission paths by inverse method. Part 2: use of regularization techniques. J. Sound Vib. **264**(2), 433–451 (2003)
13. Y. Zhan, C. Liu, G. Qiao, Accuracy evaluation of rational polynomial coefficients solution for QuickBird imagery based on auxiliary ground control points. Int. Arch. Photogramm. Remote Sens. Spat. Sci. **XXXVII**(B7), 1287–1294 (2008)
14. OGC (OpenGIS Consortium), The OpenGIS abstract specification-topic 7: the earth imagery case (1999), http://www.google.co.kr/url?sa=t&rct=j&q=the%20opengis%20abstract%20specification-topic%207%3A%20the%20earth%20imagery%20case&source=web&cd=1&ved=0CFIQFjAA&url=http%3A%2F%2Fportal.opengeospatial.org%2Ffiles%2F%3Fartifact_id%3D892&ei=urPtT4PyHJSfiQfP7KWsDQ&usg=AFQjCNFWbvJSPJ4wnAMgxMLobpDr2gLTYA&cad=rjt
15. C.F. Gerald, P.O. Wheatley, *Applied Numerical Analysis*, 5th edn. (Addison-Wesley Publishing Company, New York, 1994), pp. 398–403
16. P.R. Wolf, B.A. Dewitt, *Elements of Photogrammetry*, 3rd edn. (McGraw-Hill Companies, New York, 2000), pp. 552–553

Granular Computing and Dempster–Shafer Integration in Seismic Vulnerability Assessment

Fateme Khamespanah, Mahmoud Reza Delavar, Hadis Samadi Alinia and Mehdi Zare

Abstract Iran is one of the seismically active areas of the world due to its position in the Alpine-Himalayan mountain system. Tehran has several faults hence huge earthquakes will permeates human settlement there. Production of seismic vulnerability map could help disaster management organizations to develop and implement a plan to promote awareness of earthquake vulnerability and implementation of seismic vulnerability reduction measures in Tehran. The process of seismic vulnerability assessment is a supervised classification problem which undertaken by implementation of classification rules obtained from relationships between classes defined by a set of attributes and a unified decision of a group of experts. Therefore, seismic vulnerability assessment is a multidisciplinary problem which needs a multi criteria decision making. The influencing factors make the problem and the process of decision making a complicated disaster management

F. Khamespanah (✉)
Department of Surveying and Geomatics Engineering, College of Engineering, Tehran University, Tehran, Iran
e-mail: Fa_khamespanah@ut.ac.ir

M. R. Delavar
Center of Excellence in Geomatics Engineering in Disaster Management, Department of Surveying and Geomatics Engineering, College of Engineering, Tehran University, Tehran, Iran
e-mail: mdelavar@ut.ac.ir

H. S. Alinia
GIS Division, Department of Surveying and Geomatics Engineering, College of Engineering, University of Tehran, Tehran, Iran
e-mail: alinia@ut.ac.ir

M. Zare
Seismology Research Center, International Institute of Earthquake Engineering and Seismology (IIEES), Tehran, Iran
e-mail: mzare@iiees.ac.ir

S. Zlatanova et al. (eds.), *Intelligent Systems for Crisis Management*, Lecture Notes in Geoinformation and Cartography, DOI: 10.1007/978-3-642-33218-0_11,

problem. To overcome this problem, this paper proposes an integrated model based upon the granular computing and Dempster–Shafer to extract classification rules for classification of urban areas regarding seismic vulnerability. One of the significant properties of granular computing is induction of more compatible rules having no inconsistency. In this paper, Dempster–Shafer theory is used to integrate and model the conflict among different experts' viewpoints to get an informed decision regarding the measure of seismic vulnerability in each statistical unit in the study area.

Keywords Granular computing model · Dempster–Shafer theory · Geospatial information system (GIS) · Seismic vulnerability assessment · Decision-making

1 Introduction

Earthquake is a natural disaster that is unpredictable in short periods and is known as one of the main hazards in human beings lives. Tehran, capital of Iran, is located on a number of known and unknown faults which make this city exposed to huge earthquakes. Hundreds of buildings and infrastructure can be flattened in seconds and number of killed and injured inhabitants will be remained as a result.

The shattering impacts of earthquakes in municipal areas have provoked an increasing concern about potential seismic hazard and the corresponding vulnerability of the built areas. Accordingly, the necessity of genuine attempts allocated to rational estimates, predictions and mitigation is more evident.

The effective mitigation efforts and successful post-disaster decision-making processes are not at our disposal without assessing the expected damage and the associated loss caused in the urban areas of Tehran.

The main objective of this paper is the extraction of association rules using the integrated model of granular computing and Dempster–Shafer Theory for determining locations and intensity of seismic vulnerability of urban areas of the north of Tehran by considering various effective criteria including number of building floors, age and quality of buildings, slope and intensity of earthquake.

In this study, the result of north of Tehran fault hazard analysis [1] is applied to the vulnerability assessment process and activation of other faults has been ignored. It is assumed that the northern fault of Tehran is activated and then a seismic physical vulnerability gained. A pilot area of Tehran Metropolitan Area located in the north of Iran was selected for the purpose of this study. The study area is located between 51° 23′ N, 51° 33′ N Latitude and 34° 46′ E, 35° 49′ E Longitude [2].

To achieve this, different point of views of five experts on seismic vulnerability of 25 sample areas are combined by employing Dempster–Shafer Theory. Then granule decision tree is used as a data mining approach to induct more consistent seismic classification rules. In this way, all processes of managing spatial data layers are done using geospatial information systems (GIS).

Dempster–Shafer Theory is an extension of Bayesian theory that is also used for dealing with incomplete data [3]. Dempster–Shafer Theory is the result of Dempster and Shafer's works [4, 5].

The scheme of this work starts with an overview of Dempster–Shafer Theory. In the next section, the basic ideas of granular computing, construction of a granular decision tree, its ability to evaluate the classification performance level and finally extraction of more consistent classification rules are presented. An overview of the study area and data employed as the input dataset of the Dempster–Shafer and granular decision tree is discussed in the next section. Finally, association rules with their reliabilities are presented.

2 Dempster–Shafer Theory

Dempster–Shafer Theory is an alternative of traditional probabilistic theory and used in problem solving and uncertainty handling. The idea of this theory is that it assigns a probability mass to sets or intervals. Important aspect of Dempster–Shafer Theory is the capability of this theory for integrating multiple sources (evidences) and modeling their conflicts [3].

Principles of Dempster–Shafer Theory are followed.

Let $\Theta = \{h_1, h_2,\ldots,h_n\}$ be a finite set of n hypotheses (frame of discernment), basic probability number (BPN) is defined as m: $2^{\Theta} \rightarrow [0, 1]$ considering [6]:

$$\begin{aligned} &1.\ 0 \leq m(x) \leq \ x \in 2^{\Theta} \\ &2.\ \mathrm{m}(\phi) = 0(\phi - \text{empty set}) \\ &3.\ \sum\nolimits_{x \in 2^{\Theta}} m(x) = 1 \end{aligned} \tag{1}$$

where 2^{Θ} is the power set of Θ. m (x) is called a focal element which shows the exact belief in the proposition depicted by x [6].

Based on BPN, for any subset A of the frame of discernment Θ, belief and plausibility function can be defined.

Belief function, Bel: $2^{\Theta} \rightarrow [0, 1]$ is defined as the sum of BPN of the subsets of A, presented in Eq. 2 [6]:

$$\mathrm{Bel}\ (\mathrm{A}) = \sum\nolimits_{\mathrm{B} \subseteq \mathrm{A}} \mathrm{m(B)}\ \text{for all}\ \mathrm{A} \subseteq \Theta \tag{2}$$

Plausibility measure is a function, Pls: $2^{\Theta} \rightarrow [0, 1]$ which is defined in Eq. 3 [6]:

$$\mathrm{Pls}\ (\mathrm{A}) = \sum\nolimits_{\mathrm{B} \cap \mathrm{A} \neq \Theta} \mathrm{m(B)}\ \text{for all}\ \mathrm{A} \subseteq \Theta \tag{3}$$

The Pls (A) represents the extent to which it is failed to disbelieve A.

Dempster–Shafer Theory is used to combine the measures of evidence (BPN) from different sources. It is assumed that the evidence sources are independent. The function, m1 $\oplus$ m2: $2^{\Theta} \rightarrow [0, 1]$ defined in Eq. 4 [6]:

$$[\mathrm{m1} \oplus \mathrm{m2}](\mathrm{y}) = \begin{cases} 0 & \mathrm{y} = \Theta \\ \frac{\sum_{A \cap B = y} m_1(A) m_2(B)}{1 - \sum_{A \cap B = \Theta} m_1(A) m_2(B)} & \mathrm{y} \neq \Theta \end{cases} \quad (4)$$

3 Granular Computing Model

The basic idea of granular computing model is problem solving with different granularities [7, 8]. In granular computing, model granulation is defined as grouping of individual elements of universe into classes, based on available information [9].

In this model, information tables are base knowledge which provide a convenient way to describe a finite set of objects called a universe by a finite set of attributes [10; 9; 11].

To construct a granular tree, at first the universe is divided into grouping or partitions of the same class with a set of atomic formula of attribute-values. A rule can be expressed in the form, $\Phi \Rightarrow \Psi$, where Φ and Ψ are intensions of two concepts. In machine learning and data mining, a rule is usually paraphrased by an if–then statement, "if an object satisfies Φ, then the object satisfies Ψ".

This section provides an overview of the granular computing concepts based on information table model.

3.1 Information Table

Definition of an information table is presented in Eq. 5 [10]:

$$s = (U, At, L, \{V_a | a \in At\}, \{I_a | a \in At\}) \quad (5)$$

where,

U Finite non-empty set of objects,

At A finite non-empty set of attributes

L A language defined by using attributes in

V_a A non-empty set of values of a $\in$ At

I_a U $\rightarrow$ Va is an information function that maps objects of U to exactly one possible value of attribute V_a

Language L is used for describing objects of the universe in an information table. In this research, the decision logic language (DL-language) studied by Pawlak [10] is defined to provide formal descriptions of various notions. For this, some measures for single granule, relationship between two granules and relationship between a granule and a family of granules are applied automatically by the granule network algorithm. These measurements are discussed in Sect. 3.2.

3.2 Generality

Generality is the measuring of a single granule and for a formula Φ is defined as [12, 13]:

$$G(\Phi) = \frac{|m(\Phi)|}{|U|} \tag{6}$$

where $|m(\Phi)|$ is the size of constructive granule of concept Φ and $|U|$ is the size of universe.

This quantity identifies relative size of the granule $m(\Phi)$. A concept Φ is more general if it covers more instances of the universe [12, 13].

3.3 Confidence

For two formulas Φ and Ψ, the confidence or absolute support of Ψ provided by Φ is presented in Eq. 7 [12, 13]:

$$AS = \frac{|m(\Phi \wedge \Psi)|}{|m(\Phi)|} = \frac{|m(\Phi \cap \Psi)|}{|m(\Phi)|} \tag{7}$$

where $|m(\Phi \wedge \Psi)|$ is the size of constructive granule of concept Φ and ψ and $|m(\Phi)|$ is the size of constructive granule of concept Φ.

This quantity displays the conditional probability of a randomly selected object satisfying both Ψ and Φ.

3.4 Coverage

Coverage is a measure of the applicability or recall of the inference. The coverage Ψ provided by Φ is defined in Eq. 8 [12, 13]:

$$CV = \frac{|m(\Phi \wedge \Psi)|}{|m(\Psi)|} = \frac{|m(\Phi \cap \Psi)|}{|m(\Psi)|} \tag{8}$$

where $|m(\Phi \wedge \Psi)|$ is the size of constructive granule of concept Φ and Ψ and $|m(\Psi)|$ is the size of constructive granule of concept Ψ.

This quantity displays the conditional probability of a randomly selected object satisfying Φ and Ψ. In the other word, it indicates fraction of data in a class correctly classified by the rule [14].

3.5 Conditional Entropy

It provides a measure that is inversely related to the strength of the inference. This measure which depends on the confidence, is the most commonly used measure for selecting attribute-value in the construction of decision tree for classification [15].

The conditional entropy, H (Ψ |Φ), is defined using Eq. 9 [12, 13]:

$$\mathrm{H}(\Psi|\Phi) = -\sum\nolimits_{i=1}^{n} p(\Psi i|\Phi) \log p(\Psi i|\Phi) \tag{9}$$

where, $p(\Psi \mathrm{i}|\Phi) = \frac{|\mathrm{m}(\phi \cap \Psi \mathrm{i})|}{|\mathrm{m}(\Phi)|}$ is defined as conditional probability of a randomly selected object satisfying Φ and Ψi.

Entropy demonstrates the concept of inconsistency. For certain concepts, entropy reaches minimum value, 0 [13].

4 Methodology

In this section data preparation, data integration using Dempster–Shafer theory and construction of granular tree are discussed.

4.1 Data Preparation

In this research, a dataset consisting 25 urban areas selected by an unaligned stratified random sampling approach along with the six effective factors in the assessment of seismic vulnerability is considered.

One of the most important reasons for severe losses in earthquake risks is the high level of physical vulnerability of buildings. Because of the existence of some weakly constructed high buildings in Tehran especially in the north of Tehran, it is assumed that six parameters have physical seismic vulnerability effects. Assuming that the north fault of Tehran is activated, these criteria include earthquake intensity in terms of Modified Mercalli Intensity scale (MMI) unit which describes the seismic potential of the fault for each area, slope, percentages of weak buildings less than 4 floors, percentage of equal and more than 4 floor buildings, percentage of buildings built before 1966, and percentage of buildings built between 1966 and 1988. These criteria identify a measure for the vulnerability of the area [216–19].

Table 1 illustrates the information table of this research. For simplification, the titles of criteria are summarized as follows [2; 18].

MMI	Earthquake intensity
Slop	Slope
build_less4	Percentage of Weak buildings less than 4 floors
build_more4	Percentage of equal and more than 4 floor buildings

Bef-66	Percentage of buildings built before 1966
Bet-66-88	Percentage of buildings built between 1966 and 1988

All spatial and non-spatial data on the urban area were converted to ArcGIS database format.

4.2 Combination of Experts' Point of Views

At the first stage, the seismic vulnerability of 25 urban areas was asked from five experts in the fields of civil engineering, seismology and geology in the form of a questioner. Five classes including very high vulnerable, high vulnerable, moderate vulnerable, low vulnerable and very low vulnerable are concerned for discerning levels of seismic vulnerability between the groups of urban blocks.

Table 1 Seismic vulnerability information table

Object	Slope	MMI	build_less4	Bef-66	Bet-66-88	build_more4	Class
u1	High	Medium	Very high	High	Low	High	5
u2	High	Medium	Very high	Low	High	High	5
u3	High	Medium	Very high	Low	High	High	5
u4	High	Medium	Low	Low	Very high	High	3
u5	High	High	Low	High	Medium	High	2
u6	Medium	High	Medium	Low	High	High	2
u7	High	Medium	Low	Low	High	High	1
u8	High	Medium	Very high	Low	High	High	5
u9	High	Low	Very high	Medium	Medium	High	4
u10	High	Low	Very high	High	Low	High	4
u11	High	Medium	Medium	Low	High	High	3
u12	High	Medium	Low	High	Low	High	2
u13	High	Medium	High	High	Low	Medium	3
u14	High	Medium	Medium	Low	Very high	High	3
u15	High	Medium	Very high	High	Low	High	5
u16	High	Low	High	Medium	High	High	4
u17	Very high	Very high	Very high	High	Low	High	3
u18	Medium	Very high	Medium	Medium	Medium	High	3
u19	Medium	Very high	Medium	Medium	Medium	High	3
u20	Very high	Very high	Medium	Medium	Medium	High	4
u21	High	High	High	Low	Very high	High	4
u22	Very high	High	Low	Low	High	High	3
u23	High	High	High	Low	High	High	4
u24	High	Low	Low	Low	Very high	Medium	3
u25	High	Low	Very high	Low	High	High	5

The Beynon formula has been employed to normalize experts' point of views. Equation 10 is used to calculate the Basic Probability Numbers (BPN) for each group of sample region based on experts' point of views [3]:

$$m(si) = \frac{ai\,p}{\sum_{j=1}^{d} aj\,p + \sqrt{d}} \quad i = 1, 2, \ldots d \tag{10}$$

$$m(\phi) = \frac{\sqrt{d}}{\sum_{j=1}^{d} aj\,p + \sqrt{d}} \quad (\phi - \text{ambiguity})$$

where, p is the weight of experts' point of view, d is the number of group blocks based on experts' point of view, a_i is the relative vulnerability value of each group and s_i defines group of blocks.

Each expert is considered as an independent source (evidence) and by using the Dempster–Shafer rule of combination shown in Eq. 4, experts' point of views were combined to reach a unique degree of vulnerability.

To clarify stages of this work, an example which shows combining of three experts' point of views for four sample areas is presented in Table 2:

Regarding each expert's idea and assuming the experts' point of views have the same weights, BPN is computed as shown in Table 3.

In the same way, for the other two experts, BPN were calculated as shown in Table 4:

Combination of different experts' point of views are started by combining the first and second experts' ideas. Then, the result of combination is combined by the third experts' point of view. The result of combination of expert 1 with expert 2′s point of views is as follow:

Table 2 Experts' point of views

	u_1	u_2	u_3	u_4
Expert 1	4	2	3	4
Expert 2	4	3	2	3
Expert 3	5	3	3	3

Table 3 BPN of Expert 1

Subsets	Calculation	BPN
$\{u_1, u_4\}$	$\frac{(4)(1)}{\sum (4)(1) + (2)(1) + (3)(1) + \sqrt{3}}$	0.3727
$\{u_2\}$	$\frac{(2)(1)}{\sum (4)(1) + (2)(1) + (3)(1) + \sqrt{3}}$	0.1864
$\{u_3\}$	$\frac{(3)(1)}{\sum (4)(1) + (2)(1) + (3)(1) + \sqrt{3}}$	0.2795
$\{\phi\}$	$\frac{\sqrt{3}}{\sum (4)(1) + (2)(1) + (3)(1) + \sqrt{3}}$	0.1614

Table 4 BPN of Expert 2 and Expert 3

Expert 2		Expert 3	
Subsets	BPN	Subsets	BPN
$\{u_1\}$	0.3727	$\{u_1\}$	0.5311
$\{u_2, u_4\}$	0.2795	$\{u_2, u_3, u_4\}$	0.3187
$\{u_3\}$	0.1864	$\{\phi\}$	0.1502
$\{\phi\}$	0.1614		

$$\{u_1\} = 0.2130,\ \{u_4\} = 0.1845,\ \{u_2\} = 0.1417,\ \{u_3\} = 0.1417,\ \{\phi\} = 0.2652$$

The methodology of calculating BPN corresponding to u_1 is as follows:

$$m12(\{u_1\}) = \frac{\sum_{A\cap B=u1} m1(A)m2(B)}{1-\sum_{A\cap B=\phi} m1(A)m2(B)} =$$

$$\frac{(.3727)(.3727)+(.3727)(.1614)+(.3727)(.1614)}{1-[(.3727)(.1864)+(.1864)(.3727)+(.1864)(.1864)+(.2795)(.3727)+(.2795)(.2795)]}$$

In the same way, the result of the combination of the first and second experts' point of views is combined by that of the third one. The result of this combination is presented in the last column of Table 1.

4.3 Induction of classification rules by granular tree

Information table of this research (Table 1) is constructed from 25 statistical units and 6 attributes that are introducing objects and unique attribute class which is the grade of seismic vulnerability filled by the result of Dempster–Shafer integration. The objective of this paper is to determine the degree of seismic vulnerability using granular computing model. Vulnerability degrees are defined as Very high vulnerable = 1, high vulnerable = 2, moderate vulnerable = 3, low vulnerable = 4 and very low vulnerable = 5.

To produce a minimum uncertainty of the seismic vulnerability classification, it is required to find a subset of attribute-value with high coverage, confidence and minimum entropy. Figure 1 shows the methodology of the tree construction for extracting the association rules.

By running the algorithm on 25 training urban statistical units, 14 rules are inferred at 4 levels. Union of inactive granules at 4 levels covers the universe set. Each node of the granule decision tree is labeled by a value of attribute and each branch is labeled by a value of the parent attribute.

Decision tree of seismic vulnerability of this research is illustrated in Fig. 2.

The accuracy of the results is computed by $\frac{k}{k+n}$ [20], where k is the number of objects correctly classified by a rule. It means that the classified objects completely are within the related class, and n is the number of objects that are incorrectly

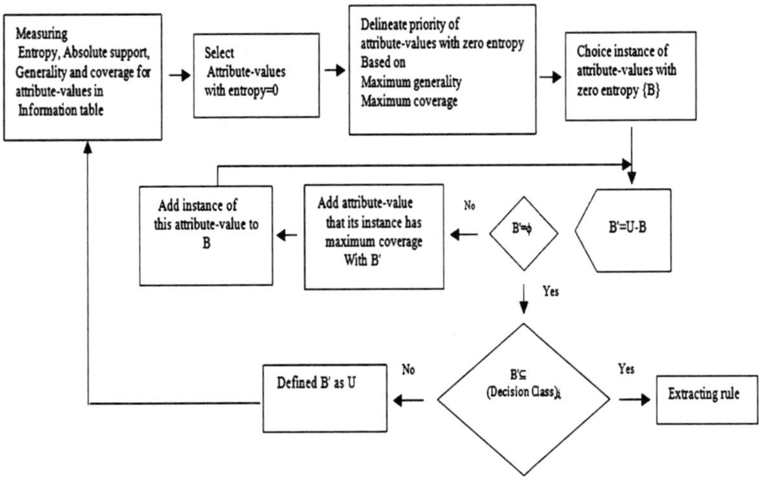

Fig. 1 Granule tree algorithm

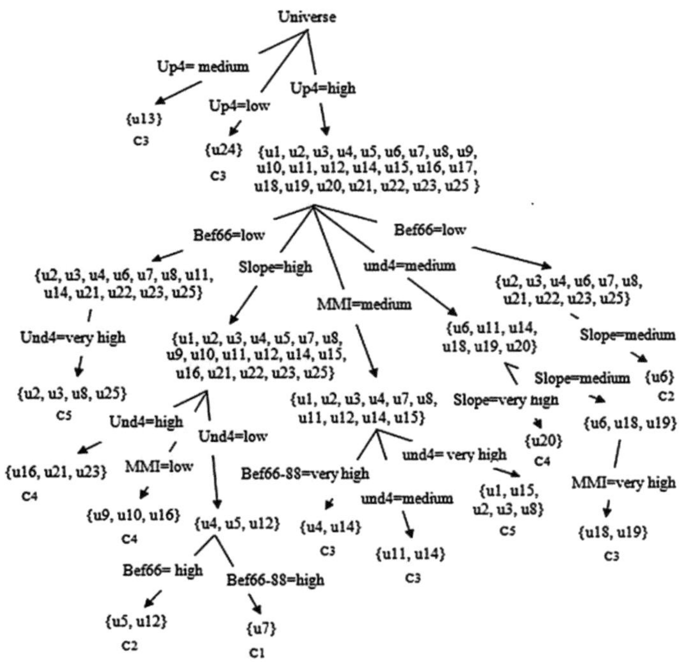

Fig. 2 The developed granule decision tree

classified. In this paper the inferred rules are employed to the 50 urban areas of north of Tehran. The average accuracy of the seismic vulnerability classification is equal to 60 % which is within the considered threshold.

5 Conclusion

This paper has proposed a new approach to classify physical seismic vulnerability assuming the northern fault of Tehran is activated. Some information about the effective parameters in the seismic vulnerability in each statistical units in Tehran including earthquake intensity in terms of MMI unit, slope, weak buildings less than 4 floors, percentage of buildings with equal and more than 4 floors, percentage of buildings built before 1966, and percentage of buildings built between 1966 and 1988 were considered for mining seismic vulnerability rules with minimum uncertainty.

Also, this paper demonstrates the prominent advantage of the integration of granular computing with Dempster–Shafer for induction of seismic vulnerability classification rules. This method infers the classification rules with maximum consistency compared to the previous research [2] which achieved 48 % accuracy of the granular tree considering only one expert's weighting and a basic granular computing for seismic vulnerability assessment.

References

1. JICA, *The study on seismic microzoning of the greater Tehran area in the Islamic Republic of Iran* (Final report, Japan International Cooperation Agency (JICA), 2000)
2. H. Samadi Alinia, M.R. Delavar, Y.Y. Yao, Support and confidence parameters to induct decision rules to classify Tehran's seismic vulnerability, in *Proceedings of the 6th International Symposium on Geo-Information for Disaster Management* (Gi4DM 2010), 2–4 Feb 2010, Torino, Italy(2010a), p. 5
3. M. Beynon, D. Cosker, D. Marshall, An expert system for multi-criteria decision making using Dempster Shafer theory. Expert Syst. Appl. **20**, 357–367 (2001)
4. A.P. Dempster, A generalization of Bayesian inference. J. R. Stat. Soc. **30**, 205–247 (1968)
5. G.A. Shafer, *Mathematical theory of evidence* (Princeton University Press, Princeton, 1976)
6. B. Bloch, Some aspects of Dempster–Shafer evidence theory for classification of multi-modality images taking partial volume effect into account. Pattern Recogn. Lett. **17**, 905–919 (1996)
7. L.A. Zadeh, Towards a theory of fuzzy information granulation and its centrality in human reasoning and fuzzy logic. Fuzzy Sets Syst. **90**, 111–127 (1997)
8. L.A Zadeh, Some reflections on soft computing, granular computing and their roles in the conception, design and utilization of information/intelligent systems. Soft Comput. **2**, 23–25 (1998)
9. Y.Y Yao, N. Zhong, Potential applications of granular computing in knowledge discovery and data mining, in *Proceedings of World Multi Conference on Systemics, Cybernetics and Informatics*, vol. 5, Orlando, pp. 573–580 (1999)

10. Z. Pawlak, *Rough Sets: Theoretical Aspects of Reasoning about Data* (Kluwer Academic Publishers, Dordrecht, 1991)
11. W. Zhong, J. He, R. Harrison, P.C Tai, Y. Pan, Clustering support vector machines for protein local structure prediction. Expert Syst. Appl. **32**, 518–526 (2007)
12. S. Tsumoto, Modeling medical diagnostic rules based on rough sets. Rough sets and current trends in computing. Lecture notes in artificial intelligence, vol. 1424 (Springer-Verlag, Berlin 1998), pp. 475–482
13. Y.Y. Yao, On Modeling data mining with granular computing, in *Proceedings of the 25th Annual International Computer Software and Applications Conference* (COMPSAC 2001), Chicago, 8–12 Oct 2001, pp. 638–643
14. Y.Y. Yao, J.T Yao (2002) Granular computing as a basis for consistent classification problems, in *Proceedings* of *PAKDD'02 Workshop* on *Foundations* of *Data Mining*. Taiwan (2002), pp. 101–106
15. J.R. Quinlan, Learning efficient classification procedures and their applications to chess end-games, in *Machine Learning: An Artificial Intelligence Approach*, vol. 1 (Tioga Press, Palo Alto, 1983), pp. 463–482
16. A.R. Amiri, M.R. Delavar, S.M. Zahrai, M.R. Malek, Tehran seismic vulnerability assessment using Dempster–Shafer theory of evidence, in *Proceedings of the Conference Map Asia* , Kuala Lumpur, Malaysia, 14–16 Aug 2007, p. 9
17. T. Silavi, M.R. Delavar, M.R. Malek, N. Kamalian, K. Karimizand, An integrated strategy for GIS-based fuzzy improved earthquake vulnerability assessment. The Second International Symposium in Geo-Information for Disaster Management, ISPRS, Goa, 25–26 Sep 2006, p. 6
18. H. Samadi Alinia, M.R. Delavar, Granular computing model for solving uncertain classification problems of seismic vulnerability, in *Spatial Data Quality from Process to Decision*, ed. by R. Devillers, H. Goodchild., CRC Press, Taylor & Francis, Boca Raton (2010b), pp. 132–134
19. H. Samadi Alinia, M.R. Delavar, Tehran's seismic vulnerability classification using granular computing approach. Int. J. Appl. Geomatics **3**(4), 229–240 (2011)
20. Y.Y. Yao, Y. Zhao, in *Explanation-oriented Data Mining*, ed by J. Wang. Encyclopedia of Data Warehousing and Mining, Idea Group Inc., pp. 492–297 (2005)

Managing Satellite Precipitation Data (PERSIANN) Through Web GeoServices: A Case Study in North Vietnam

Maria A. Brovelli, Truong Xuan Quang and Gérald Fenoy

Abstract Rainfall is one of the most important factors affecting various types of hazards such as: landslides, floods, sea level rise and so on. With the availability of satellite rainfall estimates at fine time and space resolution, it has also become possible to mitigate such problems over the world. But satellite rainfall needs to be monitored before use, because the satellite data does not reflect the strong influences on precipitation of topography in some cases. Relief of study area is very complex including mountain and plain areas. In this paper we present a Decision System and an intelligent geoportal for North Vietnam based on Web Service allowing users to investigate satellite rainfall by means of a direct comparison and of the Revised Universal Soil Loss Equation (RUSLE) model. The comparison method uses data from Precipitation Estimation from Remote Sensing Information using Artificial Neural Network (PERSIANN) and rain gauges (RG) to investigate the interpolation of RG data. Furthermore, we also estimate a correlation and examined a percentage of simultaneous rain or no-rain between them. We realize that correlation between PERSIANN and gauge data meets expectation value when we investigate monthly data. The RUSLE model for computing the soil loss, which requires a huge amount of information and data, was handled for both PERSIANN product and rain gauges data to estimate the difference due to the usage of the two data sources.

M. A. Brovelli (✉) · T. X. Quang
DIIAR, Politecnico di Milano—Como Campus, Como, Italy
e-mail: maria.brovelli@polimi.it

T. X. Quang
e-mail: xuan.truong@mail.polimi.it

G. Fenoy
GeoLabs SARL, Futur Building 1, 1280, Avenue des Platanes 34970 Lattes, France
e-mail: gerald.fenoy@geolabs.fr

S. Zlatanova et al. (eds.), *Intelligent Systems for Crisis Management*, Lecture Notes in Geoinformation and Cartography, DOI: 10.1007/978-3-642-33218-0_12,

1 Introduction

Intense rainfall is the main reason for many types of hazards and major hazard frequently occurs in developing countries. Networks of ground-based hydro-meteorological observation are sparse in developing countries and the situation is not improving. Vietnam between 1980 and 2010 has had about 16000 people who lost their lives as a result of storms, floods and landslides [1]. Recent study of Oxfam estimated that 70 % of the country's population live in areas subject to water related natural disasters [2]. The study area on which we concentrated lies in a tropical zone; in addition, it includes the two biggest hydroelectric dams in the South East Asia, and also contains a number of faults, the most active of which with a maximum shaking intensity of 8–9 (MSK scale). To mitigate natural disaster problems in the Northern part of Vietnam, one possibility compliant with the vision of 'Digital Earth' [3], is to build up a Spatial Data Infrastructure (SDI) for monitoring the hydrological status. The area is affected by natural calamities like floods, landslides, sea level rise, etc. The goal of the SDI is to make immediately and in real time available the data for all users and particularly for scientists. A geoportal typically provides access to geospatial data. On the opposite an intelligent geoportal provide access to resources including datasets (raster and vector data, imagery) and all the relevant Web GeoServices, which facilitate the discovery, display, editing and analysis of data. Spatial Information Infrastructure (SII) provides access to information, i.e.: data from SII has been processed, organized, and then presented through a specific user interface [4]. An SII geoportal therefore requires intelligence to automatically coordinate the web services that prepare, discover and present information, instead of data, to the user. Nowadays, current advantages of computation speed, storage capacity, and specific software provide great opportunities to develop decision support system (DSS) which can be run through intelligent geoportals. Recently, many DSS have been developed in various study areas over the world for watershed management [5], for earthquake disaster reduction [6], for flood prediction and monitoring, integrating hydrological modeling and GIS [7]. The advantage of the proposed DSS consists in the fact that it is based on standard web geoservices and a relevant intelligent geoportal suitable for querying and analyzing the data. Due to the interoperable components of the system, it can be easily extended in order to integrate it with other data or other models for processing the data. In its implementation the standard of Open Geospatial Consortium (OGC) are taken into account. OGC is a non-profit, international industry consortium composed by 466 companies, government agencies and university. It proposes standards for Web Services (OWS) such as the well known Web Map Service (WMS), Web Feature Services (WFS) and Web Coverage Service (WCS) which allow the generation of maps made available respectively as images, geographic entities and coverages. These services are currently widely diffused and successfully applied in many projects. Beside these "classic" services, the almost new OGC Sensor Web Enablement [8] defines opportunities for connecting in real-time heterogeneous sensors over the internet. This allows a better exploitation in a synergic way of the existing networks of remote, in situ, and proximal sensors to monitor the status of the Earth. The increasing rate and severity of damages due to floods, storms, forest fires, and other natural hazards attributed to climate change has clearly shown the shortcomings of existing environmental monitoring and information systems. The observed inefficiency is mainly a consequence of historical and organizational factors. A great amount of work on data and service standardization would be required to build more efficient information systems using state-of-the-art technology. The Earth wears an electronic skin consisting of millions of embedded electronic measuring devices. We can use the Internet as a platform to support and transmit its sensations [9]. The goal of the Sensor SWE is to enable all types of Web and/or Internet compatible sensors, instruments, and imaging devices to be accessible and, where possible, controllable via the Web [10]. Furthermore, one of the most recent interoperability standards is OGC Web Processing Service [11]. It provides rules for standardizing input and output data, presents how a client can request the execution of a process and an output can be handled for a same process.

In this paper we not only present our intelligent geoportal for the North of Vietnam but also our DSS for monitoring satellite precipitation (PERSIANN) based on Web Services. DSS was developed as a tool that can compare directly or indirectly RG and PERSIANN products at daily, monthly and yearly intervals. Users may compare them at daily interval by interpolating RG data and associate those data with visualization of PERSIANN. Otherwise one can also request statistical tool for checking correlation and percentage of simultaneous rain or no-rain. At last, statistical tools and RUSLE are able to estimate accuracy of PERSIANN comparing with RG data.

2 Data

Study area involves the Red River Basin, which is the second largest basin in Vietnam. The data used can be subdivided into two groups: the former (group 1) consists in remote sensed data (RS) including ASTER images (ground sampling interval of 30 m), used for evaluating terrain morphology (slope, aspect, etc.); Landsat TM (30 m) to create land use map and Precipitation Estimation from Remotely Sensed Information using Artificial Neural Networks (PERSIANN). The PERSIANN system for rainfall estimation is under development at the Center for Hydrometeorology and Remote Sensing at The University of California, Irvine [12]. The current operational PERSIANN system uses neural network function classification/approximation procedures to compute an estimate of rainfall rate at each $0.25° \times 0.25°$ pixel. Moreover it includes GIS (Geographic Information System) data, i.e. thematic maps such as: drainage system, vegetation covers, etc.

The latter group (group 2, see Table 1) consists of spread observations; data were provided by the Institute of Water Resource Planning of Vietnam (WRP). The original measured data consist in daily averages ranging from 1956 to 2008.

Table 1 Data in group 2

Type	Name	Sensors/gauges	Freq
Gauge or sensor	Water level	6	Daily
	Discharge	9	Daily
	Rainfall	19	Daily
	Evaporation	15	Daily
DAM	Inflow	1	Daily

3 Software Tools

The SOS enabling tool used for creating our system is the OGC compliant istSOS [13]. Our system implements not only the "read-only" mode interfaces Get Capabilities, Describe Sensor, Get Observation but also the optional transactional profile, because in our case we need also to register new sensors and their observations; transactional profile provides an access point for data uploading. The RegisterSensor operation allows adding into the SOS system a new sensor by supplying the sensor description. The InsertObservation facilitate the sensor in registering new observations for one or more sensors already registered in the SOS system and the Observations & Measurements O&M [14] encode elements containing and presenting the measured values. To get a result from a specific sensor in a period of time such as day, month, year, Get Observation request allows us to obtain the requested observation value.

The WPS software used in this study is the ZOO Project [15]. ZOO-Project includes ZOO Kernel and ZOO Services Provider. ZOO Kernel is the main core of the ZOO Project. It makes possible to create, manage and chain WPS 1.0.0 compliant Web Services by loading libraries dynamically and handling them on-demand. Especially, ZOO-Kernel is able to load GDAL/OGR library to perform basic raster and vector operations. ZOO Service Provider is a couple of Service Shared Objects (SSO) and one metadata ZOO configuration file (.zcfg) for each provided services. ZOO Configuration file contains all the metadata information and the name corresponding to the service's identifier. The SSO is the file containing the function corresponding to the related service. ZOO-Kernel was written in C, but Web Processing Services can be implemented in various programming languages such as C, FORTRAN, Python, Java, Perl, PHP and JavaScript.

4 The Intelligent Geoportal

All imagery and thematic maps can be easily published over internet by using a WMS. To generate base maps and increase speed of map rendering we use image pyramid managed by geo-web cache. For the sake of completeness, in the WMS the Get Feature Info request has already defined to query feature information on a map. Moreover the whole ESRI shapefiles available are provided through WFS.

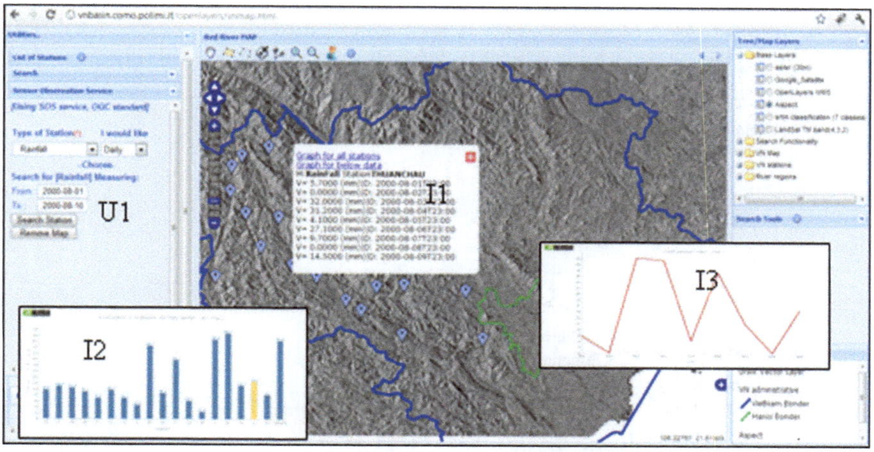

Fig. 1 The Intelligent Geoportal, I1 is a popup showing the values in the interval selected by the user through the interface (U1). I2 is a graph for all sensors of the same type. I3 is a graph showing the time sequence of observations corresponding to the station in which the user clicks

Historical series of rainfall, discharge, evaporation, water level, inflow data from 1956 to 2004 at daily, monthly and yearly measured by gauges/sensors at hydrological or hydro-meteorological stations are registered into Postgre SQL by using Register Sensor operation. The Insert Observation gives the capability of the sensor to register new observations for one or more sensors already registered in the SOS system and then the O&M encodes the containing elements and presents the measured values. Users can request any type of data at any time periods; their requests are answered by means of maps, text and statistics charts (Fig. 1). Principal statistics functionalities and charts are used for giving the users the capability of extracting information out of the hydrological and environmental data available into the PostgreSQL database. To build specific statistical functionalities we implemented new WPS services. To run these services, users only need to provide query parameters to the server. Services directly access the data in the database and then calculate all statistics indexes such as: max, min, average, mean, standard deviation without calling any SOS web service. By consequence, we avoid any network overload. The system architecture is schematically outlined in Fig. 2.

5 Web Processing of Precipitation Data

5.1 Gauge Data Interpolation and PERSIANN Visualization

WPS can access data by connecting to the PostgreSQL database. From 19 rainfall gauges data set, it is possible to estimate rainfall spatial distribution within an area. Instead of implementing an interpolation algorithm from scratch, we use the

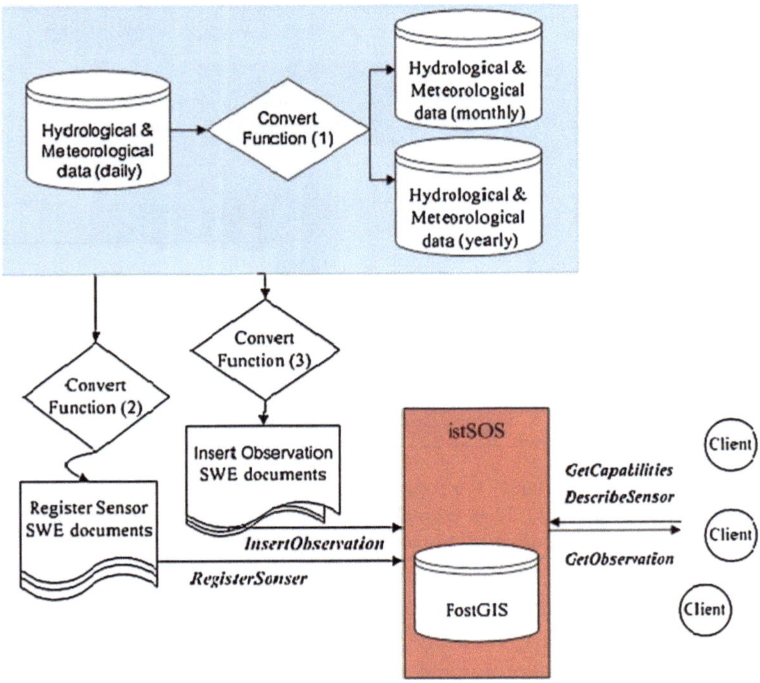

Fig. 2 The system architecture. Convert function (1), (2) and (3) are small code written in Python in order to convert data to specific format

GRASS package, a well known and mature GIS. ZOO Service can be implemented using the Python language and Python code invokes GRASS functionalities, so we are able to access those functions also from the ZOO service's code. We decided to use the inverse distance interpolation but in principle we can leave the users the freedom of choosing the algorithm they prefer among those available in GRASS.

Daily precipitation data for whole Vietnam can be downloaded though internet at the web site of PERSIANN product (see at http://chrs.web.uci.edu/persiann/data.html). To get precipitation we used "Country tool"; users can receive their data through their personal email address by using this tool. The data is received in the ESRI ASCII raster format, $0.25^\circ \times 0.25^\circ$ resolution, data available from 2000 to 2007. The name of each file representing each day's precipitation was defined as YYDOY.asc, where YY stands for year and DOY for the day of the year. To be able to provide visualization of the PERSIANN data, we used the new enhancement of ZOO Project: its support for Standard Web Services Outputs (SWSO see at http://www.zoo-project.org/trac/ticket/34). The idea of SWSO is based on the fact that the WPS main goal is to deal with GIS data; by consequence, execution of many services will result in the production of new GIS data. Or, with WPS 1.0.0, clients may ask the server, by adding a as Reference property, to store the result and to provide an URL which can be used to access it later. So, for each services which is producing GIS data as output, ZOO-Kernel handle the WMS, WFS and WCS (when applicable) publication automatically by using GDAL/OGR library for accessing raster and vector properties and the MapServer library for publishing Standard Web Services. Note, the user of the service do not have to modify its code but simply specify in the corresponding.zcfg file that it outputs GIS data and it require that it be accessible as OGC Standard Web Services. By settings, user can ask for automatic data classification of both vector and raster data. The Web Service selection will be made automatically by the ZOO-Kernel depending on the kind of output resulting of the service execution, parameters set in the.zcfg file and those provided by the client when requesting the server. The client can then access the output data by requesting MapServer as a WMS, a WFS or a WCS server. Hence, to publish and classify PRESIANN data using WMS, we created a service which returns the input data as provided by the client and defined a.zcfg file to get the WMS server available. Then we requested the WPS server to execute our service by providing URL for the data to be published and adding the as Reference attribute to the Response Document value requested. The execution for each files results in a set of Map files which are used by Map Server to access locations and properties of the data to be displayed that we can use to access our classified raster data through WMS.

5.2 *Comparisons Between Rain Gauges and PERSIANN Data*

At this point, in addition to rain gauges data, the data at the locations of the 19 stations can be extracted from the PERSIANN product. Then, they are registered by mean of the istSOS Register Sensor and inserted by using the Insert Observation request. This implementation is processed analogously to the details provided in Sect. 5.1. In our system two database sources exists: one for rain gauge RG data and the other one for the PERSIANN one. By accessing these two data sets through database system, a WPS service was created to analyze the correlation and percentage of simultaneous rain or no-rain between them.

5.3 *Soil Loss Computation*

The Revised Universal Soil Loss Equation (RUSLE) [16] is an empirical equation designed for the computation of average soil loss in total area. Many researches have implemented RUSLE model, but their implementation is based on off-line Geographic Information System (GIS) such as [17, 18]. The RUSLE is written as:

$$A = R \times K \times L \times S \times C \times P \tag{1}$$

where *A* is the estimation of average soil loss in t/ha over a period selected for *R* (usually a yearly basis); *R* is the rainfall-runoff erosivity factor. *K* is the soil erodibility factor which is a measure of the susceptibility of soil to be eroded under standard conditions and is adjusted to accommodate variation in soil moisture content [19]; it is determined by the cohesive force between the soil particles and may vary depending on the presence or absence of plant cover. *L* is the slope length factor; *S* is slope steepness factor; *C* is cover and management factor which estimates the soil loss ratio (it may be understood as a ratio comparing the soil loss from land under a specific crop and management system to the corresponding loss from continuously fallow and tilled land). *P* is the ratio of soil loss with a support practice like contouring, strip-cropping, or terracing to that with straight-row farming up and down the slope [20]. Given *LS* as the combination of the slope steepness (*S*) and slope length (*L*) measurements, it is determined using the equation presented in [21]:

$$LS = (m + 1)[A(r)/22.1]^{m}[\sin b(r)/b_0]^{n} \quad (2)$$

where A(r): is upslope contributing area per unit contour width, b is the slope, m is 0.4–0.6 and n = 1–1.4.

The land cover factor was determined by using the Vietnam Atlas in 2004. The C factor includes: water body, rivers or deep zones, forest, dense vegetation, city, etc.; their values vary from 0 to 0.85. The soil erodibility factor (K) was obtained from Vietnam Atlas in 2004. Soil profile in study area includes: fine sand, very fine sand, loamy, clay, etc.; values range from 0.10 to 0.42. The P factor was calculated based on the land use and slope map in percent. Slope classes vary from 0–5 to 50–100 % and the corresponding values are from 0.10 to 0.33 in case of land use is agriculture [22]. For the determination of the R factor, the [16] equation was used (Eq. 3). Average of rainfall annually over 40 years ranges from ~ 1400 to 3000 mm.

$$R = 587.8 - 1279P + 0.004105P^2 \quad (3)$$

In our study, LS, C, K, P factors were derived from Atlas-Vietnam maps in the years 2000/2004 and ASTER DEM. Finally, according to (2), we build a dedicated Web Service able to query the two database sources (PERSIANN and RG) and invoke GRASS functions in order to use *r.mapcalc* to convert rainfall annual to R factors by using (3). At last, we added the new R factor to (1) to calculate soil erosion hazard map (see Fig. 4). To report the results of calculation, GRASS's statistical report functionality was used for each area affected by soil erosion problems over the whole area. All processes were made available as GRASS-based WPS (Figs. 3 and 4).

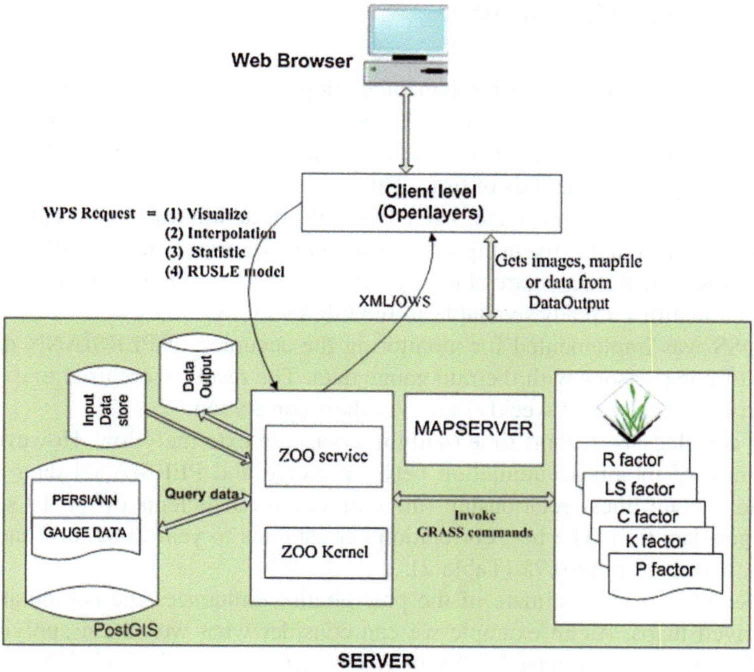

Fig. 3 Showing architecture of main component of DSS

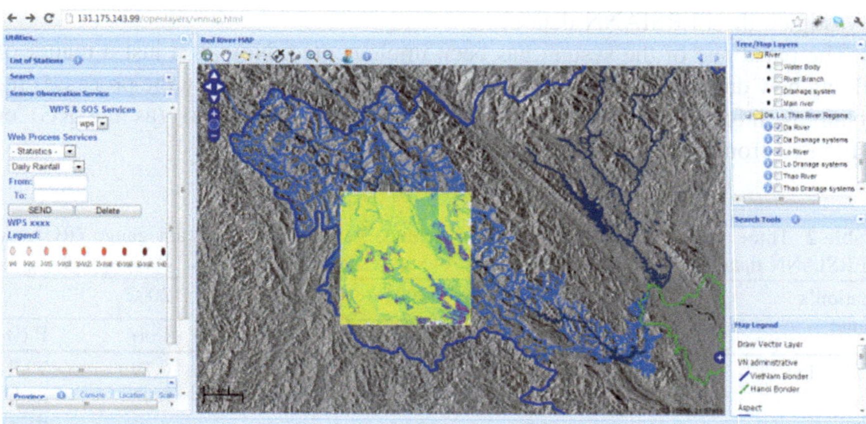

Fig. 4 Showing erosion hazard map of sub-area (A4) based on PERSIANN in year 2000

6 Results and Conclusion

The Intelligent geoportal we implemented supports, besides the traditional functionalities available for maps within WebGIS (zoom, pan, search, …), the visualization under the form of maps, tables and charts of sensor data series corresponding to long periods of time. This was realized connecting the client with a OGC compliant SOS package, named istSOS. In addition to the simple visualization of sensor data, by using a second kind of service, named WPS, and a relevant tool, the ZOO, we are able to process our data through the web exploiting the functionalities already available in the GRASS GIS.

A DSS was implemented for monitoring the accuracy of PERSIANN data by means of comparisons with the rain gauge data. The results were used to examine in short-term (daily and weekly) interval their correlation.

The correlations in short time (daily interval) are extremely low. However, the correlation of month accumulation between gauge and PERSIANN data almost indicates strong linear relationship (the correlation coefficients of the 19 stations are larger than 0.7). The best correlation corresponds to year 2002, at Tuan Giao station, and it is about 0.95 (Table 2).

Differences in the estimate of the precipitation influences the computation of the derived maps. As an example we can consider what we obtain applying the *RUSLE* model to the interpolated rain gauge map or to the PERSIANN one.

Applying *RUSLE* in the study area with the two datasets, the difference in percentage varies from 0.02 to 0.04 % bringing to a soil loss rate for the area smaller than 100 $t.ha^{-1}$ $year^{-1}$, see (Table 3).

In our future work we would like to select and apply some hydraulic models to enrich the functionalities of the DSS and, at the same time, to continue to investigate the PERSIANN dataset.

The backbone of the system has been implemented. Now it is just a matter of adding new data and functionalities for enriching it, orchestrating the processes for obtaining information from the raw data collected by sensors (as the RG) or estimated from models (PERSIANN).

Table 2 Three years data (2000–2002) of difference between monthly rain gauge (RG) and PERSIANN data (PN) for six stations in the A4 study sub-area

Station's name	2000		2001		2002	
	Corr	P (%)	Corr	P (%)	Corr	P (%)
Nam Muc	0.73	72.3	0.87	75.6	0.79	75.8
Than Uyen	0.91	73.6	0.90	78.75	0.93	69.9
Tuan Giao	0.67	67.2	0.75	76.2	0.95	71.4
Hoa Binh	0.59	66.1	0.81	64.5	0.71	71.4
Quynh Nhai	0.81	68.5	0.70	75.1	0.91	72.58
Son La	0.81	70.6	0.75	77.25	0.82	75.4

Corr: correlation index between RG and PN; P: concordance level (%) in the case of both rain and both no-rain

Table 3 RUSLE model for A4 sub-area

Years	2000		2001		2002	
Name	PERSIANN	RG	PERSIANN	RG	PERSIANN	RG
hectares	886240	849013	89332	837263	858301	840011
Area (ha)	1,147,943					
P (%)	0.76	0.74	0.78	0.73	0.75	0.73
Def (ha)	37227		56059		18290	
Def (%)	0.02		0.05		0.02	

P (%) is the percentage of the total effective area and the total land. Def (ha) and Def (%) are difference between PERSIANN and RG in hectare and in percentage

References

1. Prevention Web [PW], International disaster database (1978), http://www.preventionweb.net/english/countries/statistics/?cid=190. Accessed 06 June 2012
2. Oxfam, Viet nam climate change, adaptation and poor people. Oxfam International, ISBN 978-1-84814-055-4 (2008), www.oxfam.org.uk/publications
3. M. Craglia, K. be Bie, D. Jackson, M. Pesaresi, G. Remeyer, C. Wang, A. Annoni, L. Bian, F. Campbell, M. Ehlers, J.van Genderen, M. Goodchildi, H. Guo, A. Lewis, R. Simpson, A. Skidmore, A. Skidmore, P. Woodgate, Digital earth 2020: towards the vision for the next decade. Int. J. Digit. Earth **5**(1), 4–21 (2012)
4. A. Iwaniak, I. Kaczmarek, T. Kubik, J. Lukowicz, W. Paluszyński, D. Kourie, A. Cooper, S. Coetzee, An intelligent Geoportal for spatial planning. 25th International Cartographic Conference, Paris, 4–8 July 2011
5. J.-Y. Choi, B.A. Engel, R.L. Farnsworth, Web-based GIS and spatial decision support system for watershed management. J. Hydroinform. **7**(3), (2005)
6. J. Bo, T. Xiaxin, L. Ping, W. Yanru, WebGIS based information and decision-marking support system for earth disaster reduction. FSKD'09 Proceedings of the 6th International Conference on Fuzzy Systems and Knowledge Discovery, vol. 7 (IEEE Press, Piscataway, 2009), pp. 397–401
7. D. Mioc, B. Nickerson, E. MacGillivray, A. Morton, F. Anton, D. Fraser, P. Tang, G. Liang, Early warning and mapping for flood disasters. WebMGS: 1st International Workshop on Pervasive Web Mapping, Geoprocessing and Services, Como, Italy, 26–27 Aug 2010
8. SWE Common Data Model Encoding Standard 1.0, OGC, http://www.opengeospatial.org/standards/swecommon. Accessed 20 Sept 2012
9. N. Gross, The Earth will don an electronic skin: Interview with Cherry Murray. Business week online (1999), http://www.businessweek.com/1999/99_35/b3644024.htm. Accessed Oct 2012
10. M. Botts, A. Robin, Bringing the sensor web together. WWW document (2007a), http://www.brgm.fr/dcenewsFile?ID=473. Accessed Oct 2012
11. WPS. OGC web processing service 1.0.0, OGC, http://www.opengeospatial.org/standards/wps. 2009
12. CHRS, Center for hydrometeorology & remote sensing, University of California, Irvine, http://chrs.web.uci.edu. Last Accessed Jun-2012
13. M. Cannata, M. Antonovic, ISTSOS: investigation of the sensor observation service. WebMGS 1st international workshop on pervasive web mapping, geoprocessing and services, Como, Italy, 26–27 Aug 2010
14. M. Botts, A. Robin, OpenGIS® sensor model language (SensorML) implementation specification, version 1.0, OGC document 07-000 (2007b), http://www.opengeospatial.org/standards/sensorml

15. G. Fenoy, N. Bozon, V. Raghavan, ZOO-Project: the open WPS platform. Appl. Geomat. Jan 2012
16. G.K. Renard, G.R. Foster, G.A. Weesies, J.P. Porter, RUSLE-revised universal soil loss equation. J. Soil Water Conserv. **46,** 30–33 (1991)
17. A.H. Sheikh, S. Palria, A. Alam, Integration of GIS and universal soil loss equation (USLE) for soil loss estimation in a himalayan watershed. Recent Res. Sci. Technol. **3**(3), 51–57 (2011)
18. L. Yaolin, L. Zhijun, A study on estimation of the amount of soil erosion in small watershed based on GIS: A case study in the three gorge area of China, in *Proceedings of Geospatial Information, Data Mining, and Applications* (2005)
19. A.A. Millward, J.E. Mersey, Adapting the RUSLE to model soil erosion potential in a mountainous tropical watershed. Catena **38**(2), 109–129 (1999)
20. Agriculture Handbooks [AH], Predicting rainfall erosion losses: a guide to conservation planning, (USA), No. 537 (1978)
21. H. Mitasova et al, Geographic modeling systems lab website, http://skagit.meas.ncsu.edu/~helena/gmslab/reports/CerlErosionTutorial/denix/denixstart.html. Last Accessed Aug 2012
22. W.H. Wischmeier, D.D. Smith, Predicting rainfall erosion losses, a guide to conservation planning, Agriculture, (Washington D.C., 1978), pp. 55–57

Applying GIS in Seismic Hazard Assessment and Data Integration for Disaster Management

Rumiana Vatseva, Dimcho Solakov, Emilia Tcherkezova, Stela Simeonova and Petya Trifonova

Abstract Among the many kinds of natural and man-made disasters, earthquakes dominate with regard to their socially and economically impact on the urban environment. Seismic hazard assessment for industrial objects is of a substantial importance, because it provides valuable information for seismic safety and disaster mitigation. The main objective of the study is to integrate basic geo-datasets in thematic mapping products and to assess the seismic hazard using GIS techniques to provide a basis for disaster management of the case study of Ada Tepe in Bulgaria. GIS is applied as a valuable tool to support an effective decision-making by managing, structuring and utilizing comprehensive data for disaster prevention. Web GIS is of interest to us in the future work, because it is provided the ease for real-time access and simultaneous informed decision-making process for all stakeholders involved.

Keywords Geographic information system (GIS) · Data integration · Seismic hazard assessment · Disaster management · Bulgaria

R. Vatseva (✉) · D. Solakov · E. Tcherkezova · S. Simeonova · P. Trifonova
National Institute of Geophysics, Geodesy and Geography, Bulgarian Academy of Sciences, Acad. G. Bonchev Str., Bl. 3, Sofia, Bulgaria
e-mail: rvatseva@gmail.com

D. Solakov
e-mail: dimos@geophys.bas.bg

E. Tcherkezova
e-mail: et@geophys.bas.bg

S. Simeonova
e-mail: stelas@geophys.bas.bg

P. Trifonova
e-mail: petyageo@mail.bg

S. Zlatanova et al. (eds.), *Intelligent Systems for Crisis Management*, Lecture Notes in Geoinformation and Cartography, DOI: 10.1007/978-3-642-33218-0_13,

1 Introduction

Earthquakes are the most deadly of the natural disasters affecting the human environment, indeed catastrophic earthquakes have marked the whole human history. Global seismic risk and particularly vulnerability are increasing steadily as urbanization and development occupy more areas that are prone to effects of strong earthquakes. Additionally, the uncontrolled urban growth in highly seismic areas is often associated with the construction of seismically unsafe buildings and infrastructures, and undertaken with an insufficient knowledge of the regional seismicity peculiarities and seismic hazard. The assessment of seismic hazard is the first link in the prevention chain and the first step in the evaluation of the seismic risk. The implementation of the earthquake risk estimates into the policies for seismic risk reduction will allow focusing on the prevention of earthquake effects rather than on overcoming/following disasters.

Effective implementation of GIS in disaster management requires research and development in many areas, including data collection as a first step. Seismic hazard assessment includes detailed quantitative and qualitative data analysis and evaluation of physical, environmental, social, and economic factors and consequences. Collecting, structuring and utilizing comprehensive spatial data is essential for disaster prevention. The development of GIS applications and digital maps is needed as the bases for preparedness and evaluation of the region's vulnerability regarding earthquake hazard.

The main objective of the present study is to integrate basic spatial geo-datasets in thematic mapping products and to assess the seismic hazard using GIS techniques to provide a basis for disaster management of the Ada Tepe region in Bulgaria. The GIS approach applied for hazard assessment is based on analysis and processing of the source-geometry, earthquake occurrence model, characteristics and statistics of seismicity in each source, and the appropriate attenuation relations. Thus the seismic hazard at the Ada Tepe site is obtained by integrating the effects of ground motion from earthquakes of different size occurring at different locations within different seismic sources, and with different frequencies of occurrence.

2 Case Study

The Ada Tepe region was chosen as a suitable case study for applying GIS in geodata integration and seismic hazard assessment for disaster management. It is located in South Bulgaria, the Eastern Rhodope Mountain, Krumovgrad municipality (Fig. 1).

The investigated geographical area vary according to requirements of the International Atomic Energy Agency (IAEA) [1] for scales of investigations needed to evaluate and resolve all hazard associated with earthquakes. In this study

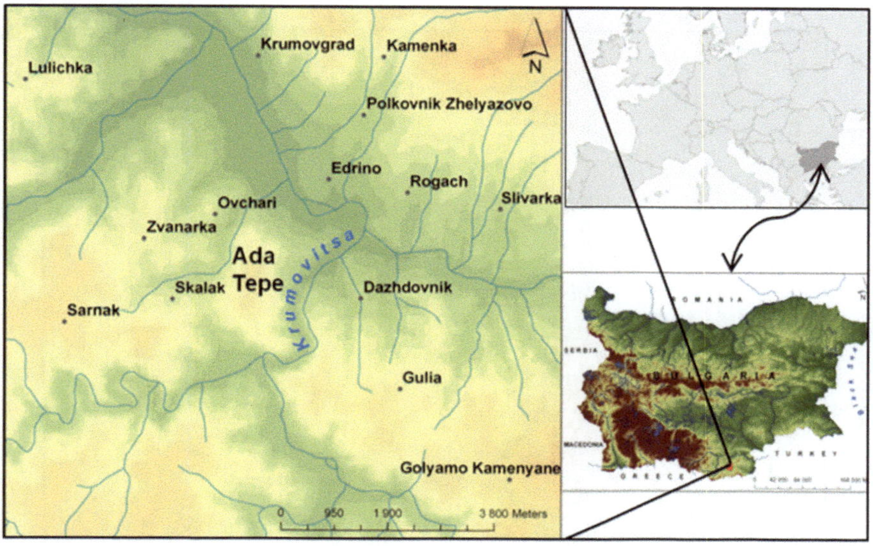

Fig. 1 The study area of Ada Tepe, Bulgaria

the geological, geophysical, and seismological data were developed at regional scale (within a radius of 200 km), and the geographical data were created at site vicinity scale (within a radius of 5 km). The case study of Ada Tepe at site vicinity scale covers an area of 100 sq. km in a square of 10 × 10 km, and is approximately bounded by the coordinates: 41°23′32″–41°28′59″N and 25°35′50″–25°43′12″E (Fig. 1). It has an elevation from 220 to 580 m above sea level.

Ada Tepe (Khan Krum deposit) is a typical low-sulfidation epithermal gold deposit and is hosted in Maastrichtian–Paleocene sedimentary rocks above a detachment fault contact with underlying Paleozoic metamorphic rocks [2]. According to the current archaeological investigations Ada Tepe is the oldest gold mine in Europe with Late Bronze and Early Iron age [3]. The seismic hazard assessment and geo-datasets for disaster management are needed for the design and construction of tailings dam in the gold mine Ada Tepe.

3 Materials and Methods

Spatial pattern of seismogenic structures in the Ada Tepe region (within a radius of 200 km) was identified by integration of geological and geophysical data with the seismological information (historical and instrumental seismicity). The correlation (or lack of correlation) of historical and instrumental earthquakes with geological and geophysical peculiarities of the structures can be particularly important in identifying seismogenic structures. From plate-tectonic point of view the territory of Bulgaria and the Balkans belongs to the southern edge of the Eurasian plate.

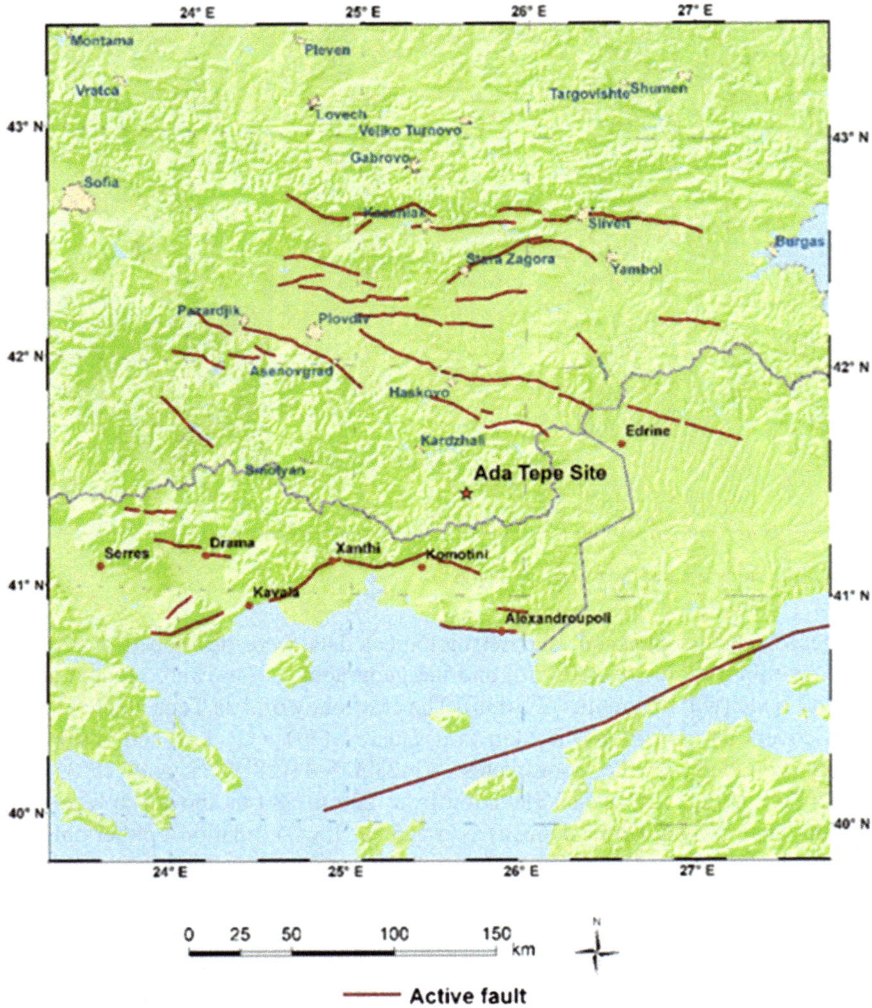

Fig. 2 Active faults in the Ada Tepe region (modified from [4])

The geodynamics of the region is determined mainly by the subduction of the African plate in the Aegean subduction zone and the collision of the Arabe plate in Eastern Anatolia. The contemporary active faults in the considered region [4] that are related to large earthquakes occurrence are presented in Fig. 2.

Geophysical data consist of grids with a cell of 1 km for gravity and 3 km for magnetics. Fifteen profiles having NE–SW orientation were also used to constrain gravity and magnetic interpretation results. First, gravity data (Bouguer reduction) were transformed to the magnitude (modulus) of the Total Horizontal Gradient [5]. The map (Fig. 3) indicates the axes of steep gravity transitions by the lines of maximum gradient values, given in dark colors. The most intensive among them

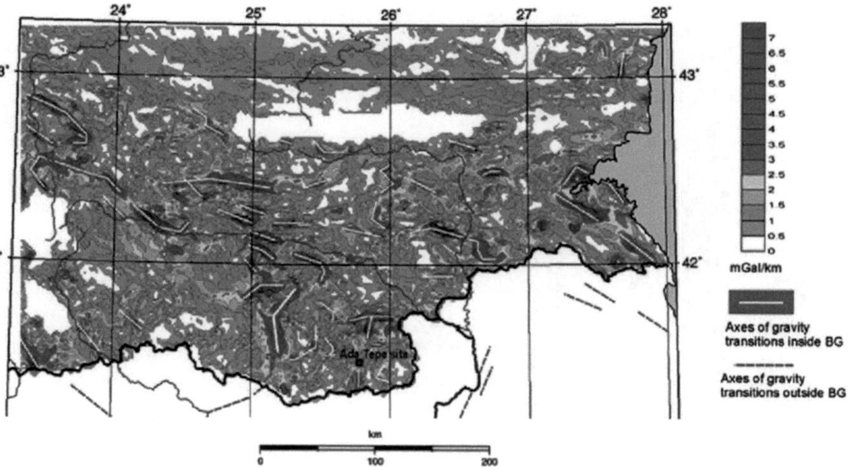

Fig. 3 Total horizontal gravity gradient

are marked by white lines as potential axes of faults, flexures and other structures of dislocation.

Second, Euler deconvolution [6] and Werner deconvolution [7] were applied to grid and profile data. These inverse methods yield valuable information for investigation concerning multiple source anomalies interpretation and need of averaged depth/position results. Observed magnetic anomalies in Ada Tepe region are related mainly to magnetic products from volcanic activity with different age and less commonly to varieties of metamorphic rocks. Some of them are connected to the post collisional faulting [8].

Regional seismicity pattern is illustrated in Fig. 4. The map suggests that seismicity in the region (within a radius of 200 km) is not uniformly distributed in space. Therefore, the seismicity is described in distributed geographical zones. Its own specific tectonic, seismic, and geological features characterize each zone. The proposed seismic zonation corresponds to the seismotectonic model of Bulgaria and Northern Greece as one seismotectonic unit and can be considered to influence the seismic hazard for Ada Tepe site.

The local geo-datasets were collected and integrated in basic thematic maps commonly used for the development of GIS for disaster management. Available digital topographic maps (1:50.000) [9], colour digital orthophotos (year 2006) [10], and CORINE Land Cover (CLC) 2006 database [11] were used for producing basic local geographical datasets for transportation, settlements, hydrography, elevation, land cover and land use. On the basis on SRTM digital elevation data [12] a digital elevation model (DEM) with resolution 30×30 m was generated.

A geodatabase was developed for seismic hazard assessment. The geodatabase design includes an identification of data themes, specification of the contents and representation of each thematic layer, additional spatial and database elements, such as spatial and attribute relationships. In this study, the design process of

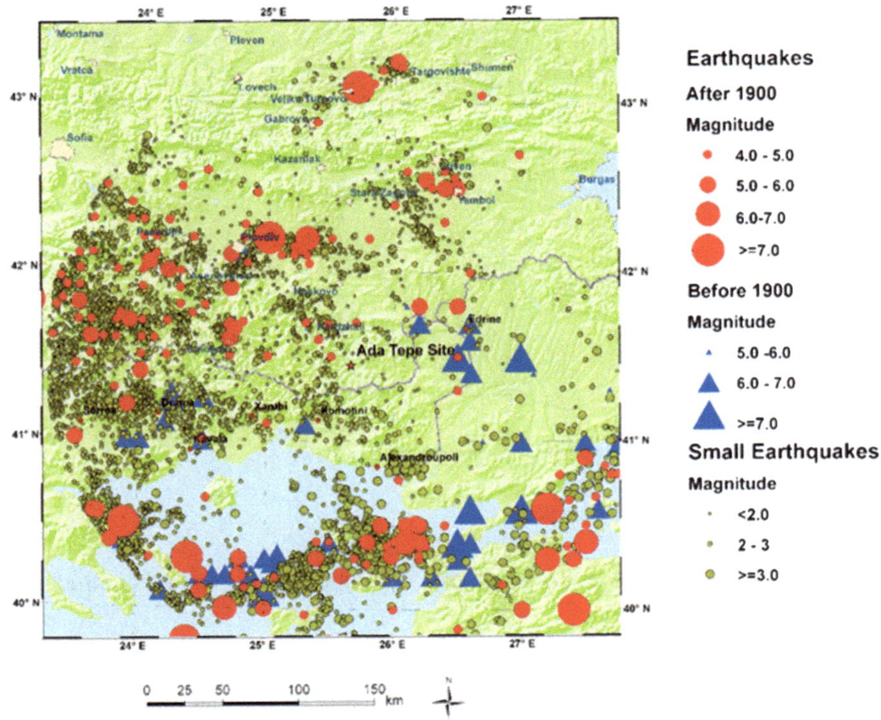

Fig. 4 Regional seismicity pattern (historical and instrumental earthquakes)

geodatabase data model for seismic hazard assessment involves a development of data structure, defining relation classes and domains, building up spatial relations, connectivity, relation rules, and elaboration of methodological reference document. The geo-datasets were stored in an ArcGIS file geodatabase.

4 Results and Discussion

Seismic hazard is the probability that various levels of strong ground motion will be exceeded during a specified time period at a site. The ground motion levels (quantity measure used to characterize the ground motion) may be expressed in terms of peak ground acceleration (velocity, displacement) and/or peak response spectral amplitudes for a range of frequencies. Probabilistic estimates of design ground motion levels are derived directly from the hazard analysis. In this study a key component of seismic hazard assessment was the development of seismic source model, which presents seismotectonic information as a spatial approximation of earthquake location and recurrence (Fig. 5). The created model consists

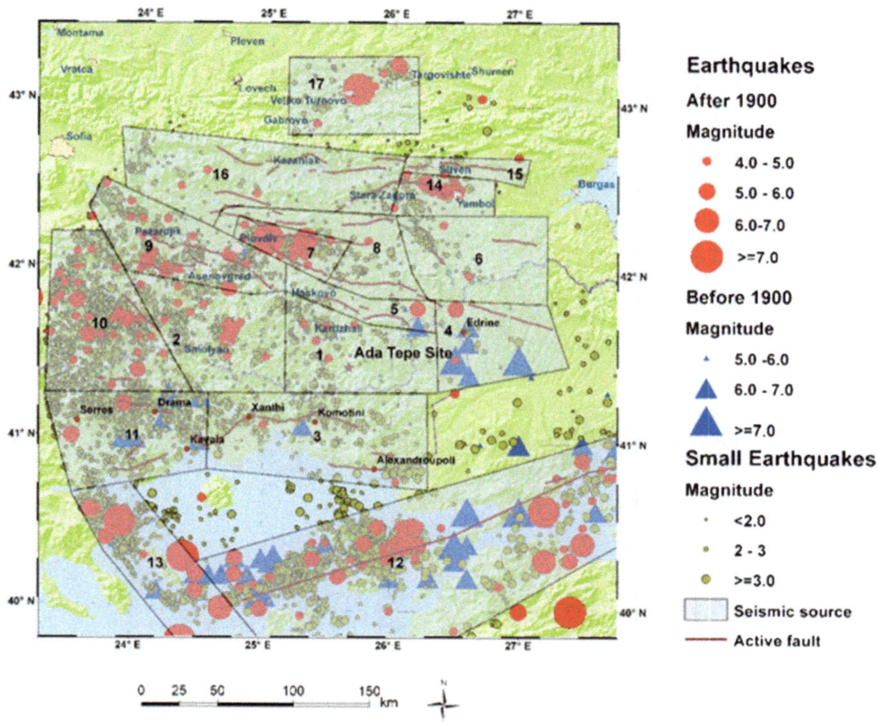

Fig. 5 Seismic source model

of 17 seismic sources: 4 area sources and 13 mixed type sources (including area and fault sources with different minimum and maximum magnitudes).

For each seismic source are defined: geometry, earthquake distribution in the source, earthquake recurrence frequency and the maximum potential earthquake magnitude. The area sources define regions that are assumed to have similar seismicity characteristics that are distinct from neighboring zones, and are exclusive of identified active faults [13]. This source type displays low to moderate disperse seismicity. A mix-source type is a seismic source in which one or more active faults are defined within a boundary of area source [13].

For seismic hazard assessment were used modeled seismic sources and their seismic parameters: *b*—Gutenberg-Rihter *b* value; *a*—annual number of events with $M \geq Mmin$; and *Mmax*—maximum expected magnitude (Table 1).

A homogenized catalogue for the region with surface-wave magnitude (M_S) was used for evaluation of *a* and *b* values. The period of completeness of the catalogue for different magnitude intervals was calculated using Stepp's method [14]. The conversion M_S-Mw (moment magnitude) was achieved using the relation presented in Scordilis [15]. The annual number of events in each source was assessed based on the number of events in the catalogue for two threshold magnitudes (4.5 and 6.0).

Table 1 Seismic sources and their seismic parameters

N	Source	*a* normalized to N (Ms >= 4.5), for a period of 110 years				*a* normalized to N (Ms >= 6.0), for a period of 300 years				M_{max}
		$b = 0.7$ (Ms)	$b = 0.8$ (Ms)	$b = 0.85$ (Mw)	$b = 1.0$ Mw)	$b = 0.7$ (Ms)	$b = 0.8$ (Ms)	$b = 0.85$ (Mw)	$b = 1.0$ (Mw)	
1	Eastern Rhodope	1.71	2.15	2.93	3.66	–	–	–	–	6
2	Central Rhodope	2.19	2.62	3.4	4.14	–	–	–	–	6
3	Ksanti	1.09	1.53	2.28	3.02	1.76	2.35	2.65	3.54	7.5
4	Edrine	1.38	1.82	2.91	3.33	2.24	2.83	3.25	4	7.5
5	Haskovo	1.38	1.82	2.57	3.31	1.78	2.35	2.64	3.51	7
6	Elhovo	1.09	1.53	2.29	3.04	–	–	–	–	6.8
7	Marica_75	2.12	2.56	3	3.7	2.03	2.61	2.9	3.79	7.5
8	Marica	1.09	1.53	2.28	3.02	–	–	–	–	7.5
9	Plovdiv	2.12	2.56	3.3	4.05	2.37	2.95	2.64	3.51	7
10	Western Rhodope	2.32	2.76	3.52	4.27	–	–	–	–	6.5
11	Drama	1.56	2.03	2.91	3.55	2.37	2.95	3.25	4.14	7
12	Aegean sea	2.72	3.16	3.91	4.67	2.99	3.59	3.89	4.78	8
13	Thessaloniki	2.09	2.48	3.23	3.99	2.61	3.19	3.49	4.38	7.5
14	Jambol	1.56	2.01	2.76	3.52	2.03	2.61	2.9	3.79	7
15	Sliven	1.09	1.53	2.28	3.02	–	–	–	–	7.5
16	Balkans	1.21	1.61	2.44	3.15	–	–	–	–	7.5
17	Gorna Orjahovitca	2.08	2.52	3.28	4.03	–	–	–	–	7

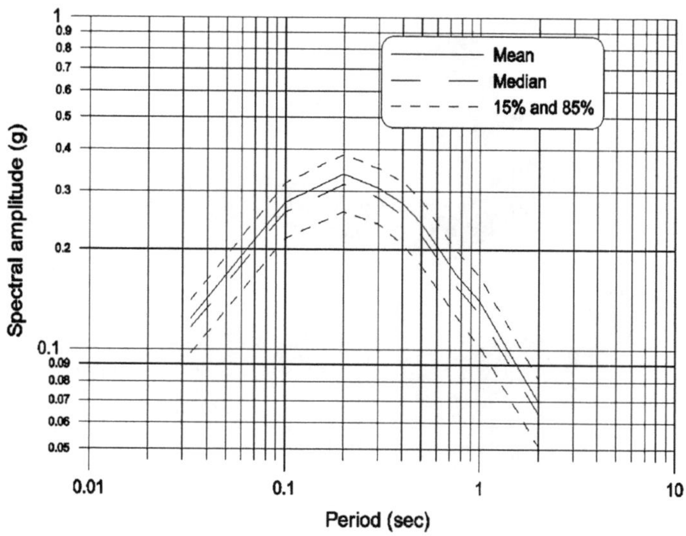

Fig. 6 Uniform hazard spectra for probability of exceedance 2.11×10^{-3}

Two attenuation relationships were used to assess the seismic hazard: AMB96 model [16] for M_S and CB-NGA model [17] for Mw. The uniform hazard response spectrum was developed based on the seismic hazard assessment. The attenuation of the response spectrum ordinates corresponding to different vibration periods were used for evaluation. Response spectra with different probability of exceedance for all periods were considered.

Seismic hazard analysis for the Ada Tepe site was performed using a combination of a logic tree [18, 19] and Monte Carlo [20] approaches. As a result a set of 192000 hazard curves was obtained. On the base of the calculated 192000 hazard curves were estimated mean, median 15 and 85 % values for PGA (peak ground acceleration), and different spectral ordinates under assumption for Log normal distribution of the probability of exceedance. More detailed results are presented in [21]. In Fig. 6 are presented the mean, median, 15 and 85 % uniform hazard spectra for probabilities of exceedance 2.11×10^{-3} (return period of 475 years).

The basic geo-datasets needed for disaster management at local level were collected, updated and integrated using GIS techniques for matching multi-scale input data, combining vector and raster data, and connecting spatial databases. The procedures for the integration of the datasets were based on data geometry, topology, semantics, or a combination of these themes.

The basic datasets consist of commonly used geographic data that are relatively "static" and can maintain before the occurrence of a disaster. For Ada Tepe region was collected multi-source and multi-scale local spatial data, which was processed and updated to produce the following basic geo-datasets: geology, geophysics,

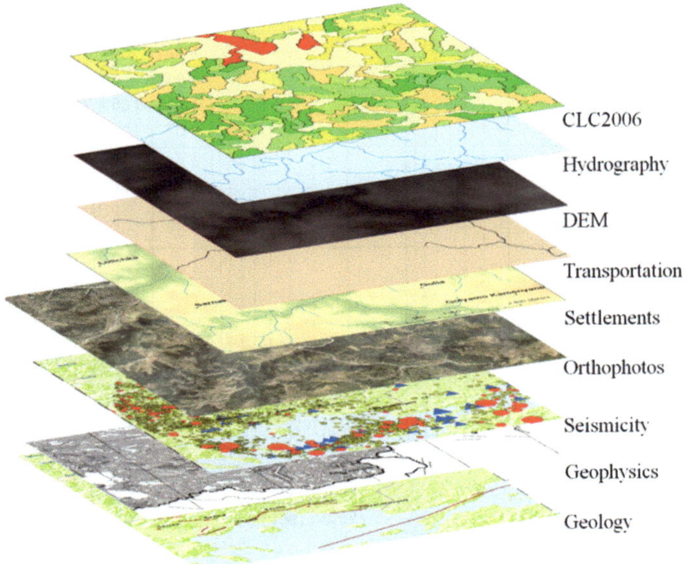

Fig. 7 Geo-datasets for Ada Tepe region

seismicity, elevation, hydrography, settlements, transportation, digital orthophotos, land cover and land use (CLC), and administrative boundaries (Fig. 7).

Geo-datasets were integrated for developing basic thematic maps as part of a geographic information services for disaster prevention. Data integration includes several GIS procedures for providing data consistency, format compatibility, geometrical integration, geodatabase design and implementation, etc. A geodatabase schema defines the structure and organization of geo-datasets (Fig. 8).

The geodatabase data model was developed to ensure an integral and comprehensive data structure and to integrate GIS solutions in seismic hazard assessment. It includes the following components: geo-datasets, relationships, and seismic hazard assessment models. A data dictionary was constructed in order to create an efficient repository of information that supports different stakeholders for disaster management. The data dictionary contains description of all the features and attributes currently used. It describes the collection of datasets and includes information about object type, feature definition, attribute type, attribute case, data accessibility, valid values and any rules or comments regarding the feature. It is important to note that the data layers listed in the data dictionary are captured in geodatabase format. The geodatabase data model was implemented in a physical ArcGIS file geodatabase "SHA—Seismic Hazard Assessment" (Fig. 8).

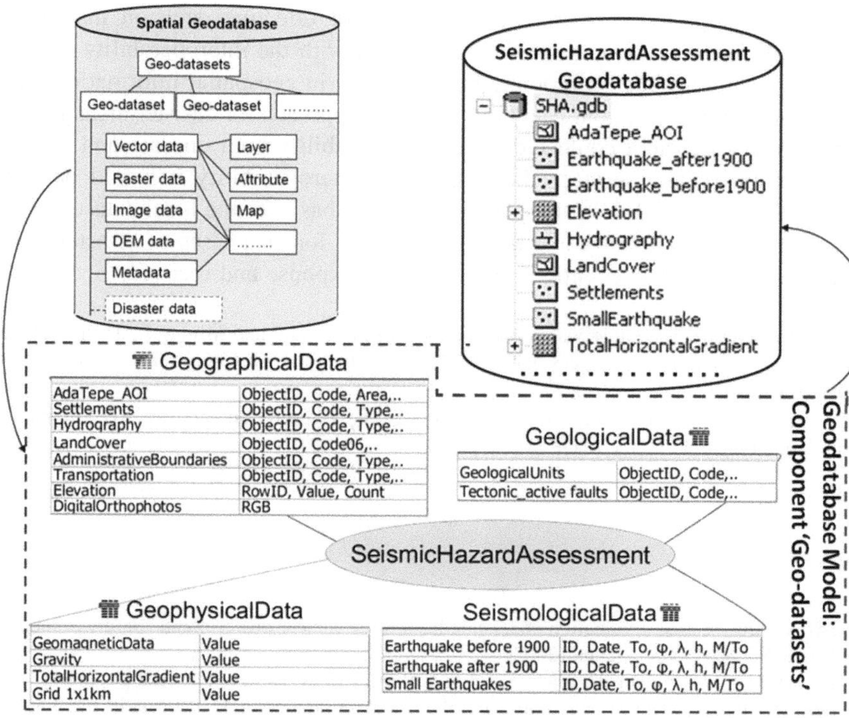

Fig. 8 Geodatabase schema

5 Conclusion

A set of 192000 hazard curves is the result from the seismic hazard analysis performed for Ada Tepe site. The family of hazard curves and their associated weights contain the information about the seismic hazard at the site and its uncertainties. Computation of uniform confidence response spectra is performed simultaneously with PGA evaluation using the same procedure. The spectral characteristics of ground motion are determined according to the relative influences of the seismic source characteristics and the attenuation characteristics of the propagation path of the seismic waves. The seismic hazard assessment results are considered in the design and construction of the tailings dam in the gold mine Ada Tepe in Bulgaria.

In GIS data integration key challenge is to ensure the interoperability of data, when combining, processing and visualizing existing heterogeneous datasets in different formats from different sources. Next challenge is to provide appropriate disaster information not only from existing data, but also from operational data. Web GIS with scalable structure is of interest to us in the future work, because it will provide the opportunity for real-time access and simultaneous informed decision-making process for all stakeholders involved. Further collective efforts are needed to produce an interoperable GIS for effective decision making in disaster management. One possible way to deal with the interoperability issue is the use of ontology for overcoming heterogeneity in geospatial information, data sharing and data integration. An ontology-based approach for disaster management in presented local area could facilitate interoperability both on data and organization of emergency services. For enhancing interoperability it would be very beneficial to use INSPIRE data specifications as a basis for the local datasets. This will foster data availability and data applicability for all phases of disaster management: mitigation, prevention, preparedness, response and recovery.

Acknowledgments The study was supported by Balkan Mineral and Mining EAD through the contract for provision of services with the National Institute of Geophysics, Geodesy and Geography—BAS.

The authors are very grateful to anonymous reviewers for useful and constructive comments, which helped to improve the manuscript.

References

1. IAEA Safety Guide, Evaluation of seismic hazard for nuclear power plants. Safety Standards Series No NS-G-3.3, Vienna (2002)
2. P. Marchev, B.S. Singer, D. Jelev, S. Hasson, R. Moritz, N. Bonev, The Ada Tepe deposit: a sediment-hosted, detachment fault controlled, low-sulfidation gold deposit in the Eastern Rhodopes, SE Bulgaria. Schweiz. Mineral. Petrogr. Mitt. **84**, 59–78 (2004)
3. H. Popov, A. Jockenhövel, Z. Tsintsov, S. Iliev, Montanarchäologische Forschungen in den Ostrhodopen, in *Interdisziplinäre Forschungen zum Kulturerbe auf der Balkanhalbinsel*, ed. by V. Nikolov, K. Bachvarov, H. Popov. Sofia (2011)
4. Report GI, Activity I: Analysis of the geological and neotectonic setting in the locality of "Ada Tepe" deposit. Fund material (2012)
5. P. Stavrev, D. Solakov, S. Simeonova, P. Trifonova, Regional set of dislocations in the Earth's crust of Bulgaria according to gravity data, in *Proceedings of 5th Congress Balkan Geophysical Society*, Bel-grade (2009)
6. A.B. Reid, J.M. Allsop, H. Granser, A.J. Millett, I.W. Somerton, Magnetic interpretation in three dimensions using Euler deconvolution. Geophysics **55**, 80–91 (1990)
7. W.M. Telford, L.P. Geldart, R.E. Sheriff, *Applied Geophysics* (Cambridge University Press, Cambridge, 1991)
8. P. Trifonova, Zh Zhelev, T. Petrova, K. Bojadgieva, Curie point depths of Bulgarian territory inferred from geomagnetic observations and its correlation with regional thermal structure and seismicity. Tectonophysics **473**, 362–374 (2009)
9. http://web.uni-plovdiv.bg/vedrin/. Accessed 21 Sept 2011
10. Ministry of Agriculture and Food, Colour digital orthophoto map of 2006 for the identification system for agricultural parcels (2011)
11. http://eea.government.bg/bg/nsmos/soil/clc-2006/index.html. Accessed 19 Oct 2010
12. A. Jarvis, H.I. Reuter, A. Nelson, E. Guevara, Hole-filled SRTM for the globe Version 4, available from the CGIAR-CSI SRTM 90 m Database (2008), http://srtm.csi.cgiar.org. Accessed 11 July 2011
13. P. Thenhaus, K. Campbell, Seismic hazard analysis, in *Earthquake Engineering Handbook*, ed. by W. Chen, C. Scawthorn (CRC Press, Boca Raton 2003)
14. J. Stepp, Dissertation. An investigation of earthquake risk in the Puget sound area by use of the type I distribution of large extremes, Pennsylvania University, 1971

15. E.M. Scordilis, Empirical global relations converting MS and mb to moment magnitude. J. Seismol. **10**, 225–236 (2006)
16. N.N. Ambraseys, K.A. Simpson, J.J. Bommer, Prediction of horizontal response spectra in Europe. Earthq. Eng. Struct. Dyn. **25**, 371–400 (1996)
17. K. Campbel, Y. Bozorgnia, NGA ground motion relations for the geometric mean horizontal component of peak and spectral ground motion parameters. Technical Report, PEER 2007/02 (2007)
18. R. Kulkarni, R. Youngs, K. Coppersmith, Assessment of confidence intervals for results of seismic hazard analysis, in *Proceedings of the 8th World Conference on Earthquake Engineering*, vol. 1 (San Francisco, 1984), p. 263
19. K. Coppersmith, R. Youngs, Capturing uncertainty in probabilistic seismic hazard assessments within intraplate environments, in *Proceedings of the 3rd World Conference on Earthquake Engineering*, vol. 1 (Charleston, 1986), p. 301
20. H. Bungum, P. Swearingen, G. Woo, Earthquake hazard assessment in the North Sea. Phys. Earth Planet. Inter. **44**, 201–210 (1986)
21. Report NIGGG, Activity IV: Seismic hazard assessment for “Ada Tepe” site. Fund material (2012)

Methodology for Landslide Susceptibility and Hazard Mapping Using GIS and SDI

Tomás Fernández, Jorge Jiménez, Jorge Delgado, Javier Cardenal, José Luis Pérez, Rachid El Hamdouni, Clemente Irigaray and José Chacón

Abstract In this work a methodology for preparing landslides susceptibility and hazard maps is presented, based in a bivariate analysis between past movements and determinant factors. The methodology for determining the susceptibility is an adaptation of the matrix method to a GIS, and it has been tested and validated in different zones and environments of Andalusia (southern Spain). The text also discusses the availability of information layers in Spanish SDI to developing these susceptibility maps. For the hazard evaluation, we propose a methodology of determining the susceptibility in different return periods from inventories of landslides that show activity in these considered periods. The activity was estimated from stereoscopic and monoscopic analysis of aerial photographs from different

T. Fernández (✉) · J. Delgado · J. Cardenal · J. L. Pérez
Department of Cartographic Geodetic and Photogrammetric Engineering Campus de las Lagunillas, University of Jaen, 23071 Jaen, Spain
e-mail: tfernan@ujaen.es

J. Delgado
e-mail: jdelgado@ujaen.es

J. Cardenal
e-mail: jcardena@ujaen.es

J. Jiménez · R. E. Hamdouni · C. Irigaray · J. Chacón
Department of Civil Engineering, Campus de Fuentenueva s/n, University of Granada, 18071 Granada, Spain
e-mail: jorgejp@ugr.es

R. E. Hamdouni
e-mail: rachidej@ugr.es

C. Irigaray
e-mail: clemente@ugr.es

J. Chacón
e-mail: jchacon@ugr.es

S. Zlatanova et al. (eds.), *Intelligent Systems for Crisis Management*, Lecture Notes in Geoinformation and Cartography, DOI: 10.1007/978-3-642-33218-0_14,

dates, using geological and geomorphic criteria, and the study of rainfall time series. Since all, four periods were considered in a logarithmic scale of 10 years (approximate return period of rainfall generating instability in the area), 100, 1000 and 10000 years. After determining the susceptibility, it was transformed into annual hazard dividing by the number of years of the return period. Finally, a total hazard map was obtained by determining at each point the maximum value of hazard of the different periods and it is expressed in several intervals.

Keywords Landslides · Susceptibility · Hazard · Mapping · GIS · SDI

1 Introduction

Natural hazards are those processes that occur naturally and in its development cause damages or affect people, properties or the environment [1, 2]. Broadly, it is estimated that in the 20th century there were more than 4.5 million of victims and 200 million of people affected by natural hazards.

In Spain, according to a study by the Geological Survey [1], estimated losses for the period 1986–2016 in a maximum risk scenario are about 48,000 million €, being floods and earthquakes the more costly risks. In a medium risk scenario, losses arising around 30,000 million €, remaining flood risks as more expensive followed by erosion and landslides. The latter is one of the more costly risk process in medium risk hypotheses because although large catastrophic events are rare, small but very frequent landslides cause in the long-term a lot of economic losses affecting populations, infrastructures and other goods. Estimated losses by landslides reach about 4,600 million € and 70 human lives for the period 1986–2016.

In this way, it is necessary an impulse to develop measures to mitigate risks because of the large impact that they represent in the current and future society. To achieve this, several structural and nonstructural remedies both in the pre-event and in the post-event can be applied. Among the preventive and non-structural measures highlights risk analysis, as it allows to know the causes of a natural hazard and to assess the consequences of its occurrence. Quantitative analysis or risk assessment must take into account social, economic and environmental risk and it is computed from its components, through the general equation of risk [3], adopted by the United Nations Disaster Relief Organization (UNDRO):

$$R = \sum (Pi * Ei * Vi)$$

R is the risk, and H (hazard), E (exposition), V (vulnerability) its components.

However, since natural hazards are spatial phenomena, the analysis must also be spatial or cartographic. In this sense, risk mapping has been one of the most used tools for risk mitigation, because it shows very clearly the localized areas where certain elements at risk are threatened by some danger, becoming the basis for other measures, such as planning, management of post-disaster or even the placement of structural measures.

In risk maps there are different levels, depending on the considered component [3, 4]. The basic level is that of the inventory maps or databases, in which together with the space–time coordinates some information about the type or intensity of phenomena can be found. The next level are the susceptibility and hazard maps as expression of the probability of occurrence of a certain event in a location and period of time. The higher level corresponds to maps that take into account not only the phenomena themselves and their probability but how they affect people and goods (exposure) and its economic, psychological and social valuation, or in terms of the degree in which they are affected (vulnerability); finally, the product of the hazard, elements at risk valuation and vulnerability results in risk assessment.

The analysis is based today in overlapping information layers of different source in Geographical Information Systems, resulting in models of different levels, depending on the used information [4, 5]. The widespread use of GIS from the 1980 to 1990s, was a major impulse to the environmental and thematic mapping, as it provided tools for the analysis and modeling of the data that previously were quite limited. However, the emergence of Spatial Data Infrastructures (SDI) is which emphasizes conclusively the importance of environmental data and maps and relates them to basic cartographies. Initiatives such as the INSPIRE at European level and the national, regional or local SDI highlight the importance of geospatial information in today's society and establish policies, standards and technologies that integrate in a coherent geographic way information of different thematic.

SDI allow the availability of the updated data in the web, promote the production of new sets of information and ensure the presence of metadata to report back on the quality of the input data and the subsequent analyses. Nevertheless, they are not limited to storage, display or data exchange, but should provide services that represent the greatest novelty respect to traditional cartography. In this way, data can be used in different ways, depending on the services applied. The more interesting and consistent with the spirit of the SDI is the use of Web File Services (WFS) and Web Coverage Services (WCS). In this case, the connection from a GIS with these services allows work with these information layers in the same way as in a conventional GIS. The next step is to develop processes services (WPS), which would be of large use and application in the risk studies.

This paper discusses the possibility of risk analysis in the field of landslides and examines the different databases available on the web.

2 Landslides Susceptibility and Hazard

The first factor of the general equation of risk, hazard, is defined as the probability of occurrence of a potentially risk phenomenon with a given severity, and in a certain period of time. To establish the likelihood in an accurate way it is necessary to handle some concepts such as the severity [2] or intensity of the phenomenon, the return period (number of years considered for evaluation) or the annual probability of exceeding a given intensity, so that it is necessary not only of spatial location of phenomena but its dating.

In some processes of a great severity and affecting large areas, as floods or earthquakes, it is easy to have a reliable historical record; on the contrary, in other phenomena such as landslides, it is more difficult to have this record so other methods have to be used such as radioactive isotopes , dendrochronology, etc., to date them. Furthermore, in most of cases, the continuous nature of such diachronic processes obstructs the use of methods to determine accurately the date of an event.

Thus, in landslides the concept of susceptibility or spatial probability is very used and all that can be determined from the available data, although it provides a lower level information than hazard does. According to Brabb and other authors in the USGS[6], who defines this term in the field of landslides, susceptibility is the probability or likelihood that a risk phenomenon happens in a specific area and in a not determined date, based on the correlation of the conditioning factors with the distribution of past events. However, the simple mapping of susceptibility is a very effective planning tool because it shows to planners and decision makers the location of the areas where has occurred or may occur a potentially damage process.

Several methods have been developed for the landslide susceptibility analysis [4, 14, 15], that can be classified into:

- Empirical heuristic methods, suitable for small-scale studies. They are based on the experience of the scientist responsible of the analysis.
- Quantitative statistical methods, suitable for medium-scale regional studies, ensure higher objectivity than the heuristic ones. In these methods, the combination of factors that determines the movements in the past is identified statistically and quantitative predictions can be made for areas currently free of movements. In the multivariate statistical analysis, the parameters of unstable zones are analyzed by multiple regression techniques [8]; alternatively, determinant factors maps are crossed with landslides distribution maps and the correlation is established in stable and unstable areas, using discriminant analysis [9]. In the bivariate statistical analysis, determinant factor maps also intersect with landslides distribution maps, but in this case the correlation coefficients are calculated for each factor. There are different varieties, among them is the matrix method [10, 11], used in the present work. In all cases, the key is in the inventory of landslides usually made from the interpretation of aerial photography or satellite imagery, complemented by fieldwork.
- Deterministic methods are models based on physical parameters or physical processes for detailed studies. They consist of slope stability analysis generally focused on the evaluation of the safety factor.

GIS provide important support in these studies, as they allow territorial analysis in which several layers of information are employed, both phenomena inventory as determinant factors. In recent years, several specific tools for the estimation of susceptibility have been implemented, with models based both on deterministic as on statistical methods such as the matrix method [12].

Regarding SDI, there is enough information related to the determinant factors (DTM at different resolutions, geological and soils layers, hydrography, etc) and the triggering factors (weather and seismic geo-referenced data) of landslides; besides, although most of the information from the SDI is in WMS services, in this case it is also available in WFS or WCS services, or is freely downloadable. On the contrary, there is little information about landslides inventories and limited to small scale maps, although orthophotos and satellite images of different resolutions and dates are usually available [13].

In terms of progress to estimate landslide hazard at regional level by statistical methods, there are many reviews in the literature [3, 4, 6, 14, 15], although there is no universally accepted standard methodology or procedure. The hazard can be evaluated by calculating the probability of slope rupture or, more commonly, through the analysis of landslides frequency (or their reactivation) in the past [6]. Moreover, the magnitude or intensity (a concept appearing in hazard definition) of a slope movement is not a direct parameter as it is the seismic magnitude, although the mobilized area used in the matrix method, can be a reasonable approximation especially in certain types of movement [15].

As noted above, the information about historical record of landslides over the last 100 years, differently to other natural processes, is very limited and generally reduced to broad and imprecise general news in the press. For a more detailed historical record the photographic resources have to be used [16]. In this sense photogrammetric techniques and image analysis both monoscopic as stereoscopic are the only ones that can address a regional study. Complementary techniques may be used that enable estimating individual slope movements activity or date, such as conventional topography, GPS or laser scanner, wireless sensor networks (WSN) dendrochronology and cosmogenic radioisotope dating in landslide scarps and organic matter dating in buried soils, sampled by boreholes. Finally, geological and geomorphic observations allow an estimation of the activity in qualitative terms [17], both for individual movements and regional analysis.

One difficulty with landslides dating is the repetitive and often almost continuous nature of the events along the time. In this regard it is interesting to introduce the concept of diachroneity degree [18]. So, it is proposed to distinguish monochronic movements as those reaching its final state in few seconds or minutes, and diachronic movements whose temporal evolution extends for several minutes, hours, tens or centuries, and is expressed in various stages of development. Diachroneity appears as a concept broader than activity, embracing the status, distribution and style of activity. A scale from I to XII is proposed, assigning a degree of diachroneity according to the displacement rate of movement and also a classification of landslide age scale in four groups: contemporary (present-1900); historical (1899-0); pre-historical (0-5000 BP) and hyper-historical (>5000 years BP) [18].

Another approach adopted in many of the published hazard maps is to determine the frequency or return period of the triggering factors and indirectly of the slope movements. Thus, different types of maps can be differentiated based on the relationships between landslides and triggering rainfall time series or triggering earthquakes of known magnitude and return period. This paper adopts a mixed methodology based on the frequency of high intensity rainfalls as the most widespread triggering mechanism, along with the analysis of historical aerial photographs and geomorphic evidences.

Just as it happens with susceptibility, hazard mapping is obtained from GIS environments, where the information layers are fully integrated to enable analysis and modeling. Meanwhile, in the SDI there are available information layers to develop hazard maps, which are essentially the same as in the susceptibility analysis. In this case, it highlights the presence of orthophotos and images from different periods and meteorological time series, since the 1940s [13].

3 Methodology for Landslides Susceptibility Mapping

Susceptibility or spatial probability mapping is here estimated using bivariate correlation analysis between a given inventory of landslides mapped in an area and a selection of factors that determine the instability. Thus, the methodology for the preparation of the maps comprises the following operations [4, 11, 19, 20]:

1. Landslides inventory from aerial photography and field work.
2. Analysis of determinant factors, using cross-correlation between slope movements and factors derived from DTM (altitude, slope, aspect, curvature, etc) and other thematic layers (lithology, structure, soil units, hydrography, etc).
3. Modeling the susceptibility for which a matrix approach is used [10, 19].

The necessary information was obtained from the Spanish SDI of the National Geographic Institute, as well as some other services such as the Geological Survey, the Meteorological Agency, the National Institute of Statistics and the General State Administration (agriculture, environment, estate, etc.). Besides, regional services such as the Andalusian SDI and the Environmental Information Network (REDIAM) were also considered. As noted above, the information related to landslides, both the basic cartography, the DTM, the images and orthophotos, the geological and meteorological data is all available in WFS or WCS services or is freely downloadable.

In this work the inventory has been own elaborated; although in the Spanish SDI there is also some information about landslides, it is at a very low scale and resolution. Regarding to determinant factors, the selection of factors was made according to previous works in the area and its environment [11, 13, 18–22]. So, different DTM derivates and thematic maps (lithology and rainfalls) were used, and others were dismissed because of their low significance at this scale (soils, fractures density or orientation ...) or their lack (vegetation in a semiarid zone).

DTM of 25 m resolution was obtained from the National SDI. Although it is possible to work on-line, by means WCS, we decided download data in ASCII format. These files were converted to raster format and then models of slope, aspect and curvature were obtained.

Geological, geomorphological, geotechnical and soils information is also available at a several scales as WMS in the Geological and Mining Institute (IGME) or the Environmental Information Network of Andalusia (REDIAM), but again downloading it was preferred. In this way the geological map at 1:400000 scale was obtained from the last mentioned service. Meteorological information is also available in REDIAM.

Factor analysis was made by cross correlation between factors and landslide inventory through the GIS, in which correlation coefficients was calculated. In Table 1 an estimation of lineal correlation coefficient derived from contingence coefficient [11, 20] is shown. Rock falls appear conditioned by slope, curvature, lithology and rainfalls; rock and earth slides are conditioned by height, slope, curvature and lithology; earth flows are conditioned by height, curvature and lithology; and finally, debris flows are conditioned by slopes, curvature and lithology [13].

The matrix method is based on the determination of all possible combinations among the classes of the considered factors, and synthetically can be summarized in the following steps:

- The total surface of the study area matrix (TSM) is obtained by calculating the area occupied by each of the class combinations of the selected factors.
- The landslide matrix (LM) is made from the inventory, calculating the total area of each combination of factors that is affected by landslides.
- The susceptibility matrix is obtained by dividing for each combination the surface occupied by landslides (LM) and the total surface (TSM).

The result is expressed in percent as the probability of observation of a slope movement at each location of the territory. It is convenient to perform the mapping by landslides types (rock falls, planar and rotational slides, flows) since the conditions in which they occur, and therefore the factor combinations in which there is a greater proportion of slope movements, are different for each type [20]. The map was classified into 5 levels of susceptibility: very low (0–1 %), low (1–5 %), moderate (5–10 %), high (10–25 %) and very high (>25 %).

Once the susceptibility map has been made, it is appropriate to make a validation assessment for which there are several possibilities: using different samples of the landslide inventory to make both the mapping and the validation; application of the resultant analysis obtained in a given area to an adjacent one [16]; or a temporary

Table 1 Estimated correlation coefficients resultant of determinant factors analysis

Landslides	Height	Slope	Aspect	Curvature	Lithology	Rainfalls
Rock falls	0.451	0.795	0.253	0.534	0.615	0.482
Slides	0.515	0.481	0.110	0.554	0.597	0.392
Debris flows	0.190	0.465	0.066	0.568	0.545	0.183
Earth flows	0.813	0.382	0.274	0.493	0.992	0.606

Fig. 1 Susceptibility map of the study zone

validation using a inventory of landslides triggered after the preparation of the susceptibility map [11, 19, 20], which has been the technique used in this work.

The analysis is executed in a GIS environment, and the methodology has been fully developed and tested [11, 19, 20]. Even for the susceptibility modeling, an application has been implemented using the ArcGIS ModelBuilder^{TM} in two models, one for the development of susceptibility maps and other for validation [12]. This application is a patent available for free and direct download, in the following link: http://www.ugr.es/local/ren03366/susc_model.rar. For susceptibility maps, the application simply requires enter the inventory of landslides, the digital terrain model and the geological map, and for the validation only requires the susceptibility map and the new inventory.

Figure 1 shows the susceptibility map for rock slides in an area located in the south of the Granada province, between Sierra Nevada and the Mediterranean Sea.

4 Methodology for Landslides Hazard Maps

The hazard estimation requires determination of the space–time probability, i.e., the probability that a phenomenon occurs in an area at a given return period of time. With this premise, the method consists in applying the same procedure used in the susceptibility mapping but with several inventories, each corresponding to a given return period. Therefore, it is necessary to date the slope movements or at least to determine their activity within different return periods; in monochronic landslide, their age should be established or limited to a time interval; meanwhile, in diachronic landslides, their reactivation should be dated or at least should be established whether they had a certain activity in a given period.

To date landslides, here a mixed methodology based on the use of aerial photography and intense rainfall time series analysis, as the main triggering mechanism, was adopted. The accuracy of dating a landslide by aerial photography is given by the interval between pictures, although the rainfalls timing analysis can help to attain much better precision; nevertheless, in case of hazard related to given return periods, this approach is usually sufficient.

For the most recent slope movements it can be made a more accurate dating based on applying photogrammetric techniques and stereoscopic interpretation on historical photographs time series, which in the case of Spain dates back to 1956–1957 (first Spanish coverage of aerial photographs made by the USA army); since then, there are increasing series of aerial photographs, and in recent years satellite images, that allow constrain the date of generation or reactivation of movements. In this preliminary and methodological work, aerial photographs corresponding to a flight from 1992 were stereoscopically examined and orthophotos from 2001 monoscopically interpreted; they correspond to a capture before and after one of the last heavy rainfall episodes in 1997–1998 (the last one occurred most recently in 2010). Besides, orthophotos from 1956, as the earliest flight, were also used. More detailed information has been obtained using other flights and photogrammetric techniques (comparing DTM and calculating 3D displacements) but only in small areas around major landslides [23].

For older slope movements, the activity was estimated on a qualitative approach based in geomorphic features of some landslides and its elements: crown, scarps, lateral boundaries, tension cracks, toe, etc. These observations were made by stereoscopic photo-interpretation on the 1992 flight.

Regarding the utilization of rainfall time series, the relationship between these and the slope movements is well known and established in this area [21]. From precipitation data available in the REDIAM and National Meteorological Agency for the study area, on a wide time range (1940–2010), and after a preliminary analysis, about 5 rainy periods similar to those of 1997 and 2010 were found [13]. In these periods, with a duration of 2–3 years, there are accumulated precipitations exceeding 300 mm in three months between the fall and winter, which caused major slope instability in the area; besides, other rainy periods have been observed with accumulated precipitations exceeding 200 mm, so that the frequency of rainfalls that potentially trigger movements is around 15 years, considering more intense periods and about 8 years to less intense periods. Other analyses from the mean annual precipitations, which are around 620 mm, allow find out periods of 18 years for rainfalls above 950 mm and periods of 5 years for rainfalls above 750 mm, which are the thresholds in which landslides occur with greater or lower incidence [22] depending on the susceptibility zone.

If these data are compared with climate studies that analyze the source of the rainfall in the south of Spain, we find that intense precipitations in this region are

Table 2 Activity and return period of landslides considered in this work

Activity	Period	Description	Detection
Active and suspended landslides	10 years	Active movements are those that are mobilizing now. Suspended movements are defined as those who have had continuous or intermittent activity in recent years. In areas such as Spain southeast, where the movement activity is related to rainfall, this intermittence is not seasonal, but the slope movements are suspended several years. As these periods occur at a rate of about 10–15 years, landslides that have been active at some point in the last 10 years are considered in this group	Multitemporal analysis of recent aerial photographs and satellite images, monitoring techniques and some others such as dendrochronology
Dormant landslides	100 years	Dormant landslides are those without evidence of activity in recent rainfalls cycles in the area (for the years 1996/1997 and 2009/10), but active in the last 100 years. The causes of these processes still remain unknown, so that the performance of a triggering mechanism may return them to activity at least partially	Photogrammetry and digital image analysis, including historical photographs and dendrochronology
Abandoned landslides	1000 years	Abandoned movements are those without recent activity; the causes that originated them may even have vanished, although the relatively fresh land morphology still remains and an activity not remote in time can be deduced. A new change in the conditions can reactivate them. Once discarded recent landslides (active and dormant), the criterion is to differentiate, based on their morphology, abandoned movements which retain the characteristics of unstable areas, from relict movements with eroded forms	Interpretation of aerial photography. These morphological criteria can be seen supported by isotopic dating of scarps or in boreholes
Relict landslides	10000 years	Relict movements are those that still present their characteristic features, but dismantled or eroded in part, so are estimated to be older than 1000 years. The causes of these processes have disappeared, although it is possible that a new change in conditions could reactivate them. In any case, they are movements developed after the last ice age, during the Holocene	As in the above cases, the methods are the identification on aerial photography by geomorphic criteria

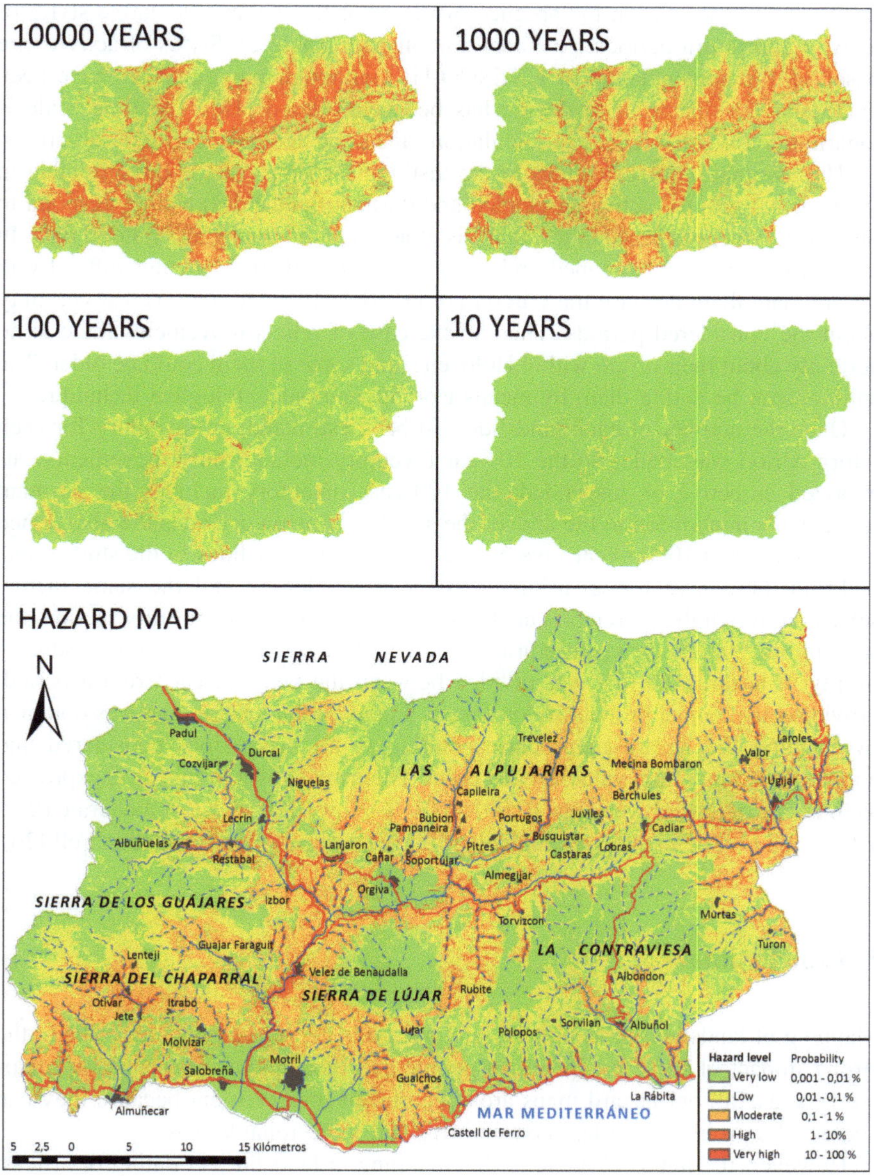

Fig. 2 Hazard map of the study zone

related to negative values of North Atlantic Oscillation index in winter (NAOi) [24]; meanwhile, positive values of this index, more frequent than negative ones, are usually related to higher precipitations in northern Europe and droughts in the south. Since the frequency of the NAO index in winter sets in about 8 years [25], there is a certain relationship with some of the rainfall cycles considered before.

On these bases and in the absence of more conclusive data, tentative styles of activity and return periods have been established (Table 2). Styles of activity are based on the nomenclature of UNESCO [17], although some changes have been made and the diachroneity degree has been considered [18]. The time scale is logarithmic for convenience in calculations and good agreement with observations.

The assignation of landslides to the distinguished intervals is quite reliable in the first two groups (active-suspended and dormant landslides) as they respond to objective evidences based in the analysis of aerial photography and rainfall data. In the last two cases (abandoned and relict landslides) there is greater difficulty to differentiate them because the criteria are less objective and there is less certainty about the considered periods. Thus, in the case of relicts movements there is no certainty about their origin within Holocene, and some of them could be older. The solution can be dating them by means isotopic and paleontological techniques.

Once the activity of each landslide had been estimated, an inventory for each return period was made. So the 10-year inventory includes only movements categorized as active or suspended; the 100-year inventory includes the dormant movements in addition to the above; the 1000-year inventory adds the abandoned movements; and 10000-year inventory comprises all landslides in the study zone.

From these inventories, using the same methodology and the same factors selected through the correlation analysis described in the susceptibility section, the susceptibility maps for each return period could be obtained. To convert susceptibility in hazard, the first was divided between the years of each return period, providing the annual probability or hazard. Finally, total hazard map was obtained overlapping the hazard maps corresponding to all the periods considered and adopting the maximum value at each point. The map, shown in Fig. 2, is expressed as intervals of hazard: very low (0.001–0.01 %), low (0.01–0.1 %), moderate (0.1–1 %), high (1–10 %) and very high (10–100 %), as it is recommended in Fell [26].

5 Conclusions

Mapping of landslides susceptibility and hazard is very necessary today due to the strong impact of these processes on people and their goods. Susceptibility maps are very useful, but hazard maps are a significant improvement because they can address the problem in a less conservative and more reliable way.

However, the major obstacle to this mapping in landslides is dating or at least determining their activity in a given period (return period). This paper proposes a methodology of determining the susceptibility in different return periods from landslide inventories with a certain activity in these periods. The activity was estimated from stereoscopic and monoscopic analysis of aerial photographs from different dates, geomorphic criteria and the study of rainfall time series.

Four periods were considered in logarithmic scale of 10, 100, 1000 and 10000 years. After determining the susceptibility, it was transformed in annual hazard dividing by the number of years of return period. Finally the overall hazard map was obtained by determining at each point the maximum value of hazard of the different periods and expressed in various intervals.

Future work will be directed to a refinement in the dating of the movements, both recent and the oldest. In the first case, we propose using photogrammetric change detection techniques from aerial photographs and satellite images of different dates. To older movements, the proposal is to support the geomorphic criteria in dating by isotopic or paleontological techniques or by comparisons with old landslides dated in other regions with similar lithologies and environmental conditions.

Acknowledgments This work is supported by the ISTEGEO project RNM-06862 funded by the Andalusian Research Plan, National Research Plan project CGL2008.04854, and Research Groups TEP-213 and RNM-221 of Andalusian Research Plan.

References

1. F.J. Ayala, E. Elizaga, L.I. González de Vallejo, Impacto económico y social de los riesgos geológicos en España. Serie Geológica Ambiental, IGME. Madrid, p. 134 (1987)
2. J. Olcina, F.J. Ayala-Carcedo, Riesgos Naturales. Ed. Ariel, Barcelona, p. 304 (2002)
3. D.J. Varnes, Landslide hazard zonation: a review of principles and practice. Commission on Landslides of the IAEG, UNESCO, Paris. Natural Hazards Series 3, p. 63 (1984)
4. J. Chacón, C. Irigaray, T. Fernández, R. El Hamdouni, Engineering geology maps: landslides and geographical information systems (GIS). Bull. Eng. Geol. Environ. **65**, 341–411 (2007)
5. J. Chacón, C. Irigaray, Previsión espacial de movimientos de ladera y riesgos asociados mediante un SIG, in *SIG en riesgos naturales y medio ambiente*, ed. by L. Laín, IGME, pp. 113–123 (1999)
6. E.E. Brabb, Innovative approaches to landslide hazard and risk mapping, *In Proceedings of the 4th International Symposium on Landslides*, vol. 1 (Toronto, Canada,1984), pp. 307–323
7. F. Guzzetti, P. Reichenbach, M. Cardinali, M. Galli, F. Ardizzone, Probabilistic landslide hazard assessment at the basin scale. Geomorphology **72**(1–4), 272–299 (2005)
8. A. Carrara, Multivariate models for landslide hazard evaluation. A black box approach. Workshop on Natural Disasters European Mediterranean Countries, Perugia, Italy, pp. 205–224 (1988)
9. C.F. Chung, A.G. Fabbri, C.J. Van Westen, Multivariate regression analysis for landslide hazard zonation, in *Geographical Information System in Assessing Natural Hazards*, ed. by A. Carrara, F. Guzzetti (Kluwer Academic Publishers, Dordrecht, 1995), pp. 107–134
10. J.V. DeGraff, H.C. Romesburg, Regional landslide-susceptibility assessment for wildland management: a matrix approach, in *Thresholds in geomorphology, 19*, ed. by D.R. Coates, J.D. Vitek (Alien & Unwin, Boston, 1980), pp. 401–414
11. C. Irigaray, T. Fernández, R. El Hamdouni, J. Chacón, Verification of landslide susceptibility mapping. A case study. Earth Surf. Proc. Land. **24**(6), 537–544 (1999)
12. J.D. Jiménez, C. Irigaray, R. El Hamdouni, J. Chacón, Building models for automatic landslide susceptibility analysis, mapping and validation in ArcGIS. Nat. Hazards **50**, 571–590 (2009)
13. T. Fernández, M.A. Ureña, J. Delgado, F.J. Cardenal, C. Irigaray, J. Chacón, Examples of natural risk analysis from SDI. International Cartographic Conference, Paris (2011)
14. C.J. Van Westen, Application of GIS to landslide hazard zonation. Ph.D Dissertation, Technical University Delft, ITC-Publication n° 15, ITC, Enschede, The Netherlands, p. 245 (1993)

15. C.J. Van Westen, The modelling of landslide hazards using GIS. Surv. Geophys. **21**(2–3), 241–255 (2000)
16. J. Remondo, A. González, J.R. de Díaz Terán, A. Cendrero, A. Fabbri, C.F. Chung, Validation of landslide susceptibility maps: examples and applications from a case study in Northern Spain. Nat. Hazards **30**(3), 437–449 (2003)
17. WP/WLI (1993). A suggested method for describing the activity of a landslide, in *International Geotechnical Societies' UNESCO Working Party on World Landslide Inventory*; Chairman Cruden DM, Bull Eng Geol Environ 47:53-57
18. J. Chacón, C. Irigaray, R. El Hamdouni, J.D. Jiménez-Perálvarez, Diachroneity of landslides, in *Geologically Active, 1: 999–1006*, ed. by A.L. Williams, G.M. Pinches, C.Y. Chin, T.J. McMorran, C.I. Massey (CRC Press/Balkema, Taylor & Francis Group, 2010)
19. C. Irigaray, T. Fernández, R. El Hamdouni, J. Chacón, Evaluation and validation of landslide-susceptibility maps obtained by a GIS matrix method: examples from the Betic Cordillera (southern Spain). Nat. Hazards **41**(1), 61–79 (2007)
20. T. Fernández, C. Irigaray , R.El Hamdouni, J. Chacón, Methodology for landslide susceptibility mapping by means of a GIS. Application to the Contraviesa area (Granada, Spain). Nat. Hazards **30**(3), 297–308 (2003)
21. C. Irigaray, F. Lamas, R. El Hamdouni, T. Fernández, J. Chacón, The importance of the precipitation and the susceptibility of the slopes for the triggering of landslides along the roads. Nat. Hazards **21**(1), 65–81 (2000)
22. J.D. Jimenez, Movimientos de ladera en la vertiente meridional de Sierra Nevada (Granada, España): identificación, análisis y cartografía de susceptibilidad y peligrosidad mediante SIG. Tesis Doctoral, University of Granada. p. 210 (2012)
23. T. Fernández, J.L. Pérez, J. Delgado, J. Cardenal, C. Irigaray, J. Chacón, *Evolution of a diachronic landslide by comparison between different DEMs obtained from digital photogrammetry techniques in las Alpujarras, GI4DM* (Antalya, Turkey, 2011)
24. D. Pozo, M.J. Esteban, F.S. Rodrigo, Y. Castro, An analysis of the variability of the North Atlantic Oscillation in the time and frequency domains. Int. J. Climatol. **20**, 1675–1692 (2000)
25. R.M. Trigo, D. Pozo, C. Timothy, J. Osborn, Y. Castro, S. Gámiz, M.J. Esteban, NAO influence on precipitation, river flow and water resources in the Iberian Peninsula. Int. J. Climatol. **24**, 925–994 (2004)
26. R. Fell, J. Corominas, C. Bonnard, L. Cascini, E. Leroi, W.Z. Savage on behalf of the JTC-1 Joint Technical Committee on Landslides and Engineered Slopes, Guidelines for landslide susceptibility, hazard and risk zoning for landuse planning. Eng. Geology **102**(3–4), 85–98 (2008)

Transport Network Vulnerability Assessment Methodology, Based on the Cost-Distance Method and GIS Integration

Dragos Toma-Danila

Abstract Considering the various effects of natural disasters, and the need for a fast intervention and recovery time, before facing the associated problems it is needed to mitigate the risks. A basic and initial step is to assess the vulnerability in high risk areas. The importance of a transport network is major, whether it is a road, railway (for access) or pipe (for resources) network. Various methods were described for analyzing their behavior to disastrous events (like earthquakes, landslides, flooding). The methodology proposed in this study integrates all related input data within a GIS software, adding by so the spatial dimension, and adapt the cost-distance method to obtain fictive costs that translate into vulnerability states for each point of a network. Also, the hot-points that can determine detour costs are taken into consideration, by means of random "What if?" scenarios that are generated by an automation model. The fact that the cost-distance method requires origins to which the costs will refer it is important, because the vulnerability values will also be related to how hard it is for an emergency intervention team to reach a certain segment of the network. Because of the various degrees of freedom in the methodology, different methods can be also added to the actual core, in order to serve the purpose, whether it is emergency route analysis, road planning or loss estimation assessment. In order to test and exemplify the methodology and the results, a road network seismic vulnerability assessment example is presented, for a Romanian County right on top of the Vrancea Seismic Area. Specific details are given about the possibilities to implement the methodology.

Keywords Network analysis · Cost-distance method · Vulnerability assessment · GIS

D. Toma-Danila (✉)
National Institute for Earth Physics, Magurele, Ilfov, Romania
e-mail: toma@infp.ro

S. Zlatanova et al. (eds.), *Intelligent Systems for Crisis Management*, Lecture Notes in Geoinformation and Cartography, DOI: 10.1007/978-3-642-33218-0_15,

1 Introduction

Each country in the world is subject to different natural disasters, which pose various risks to the population and economy. The extent of the damage (both in terms of space and effects) determines the scale of the recovery actions to follow. But one of the most basic initial steps is to have access in the affected area. Upon this depend the quick real assessment of the situation, the salvation of people requiring medical care or trapped under debris, the delivery of provisions and other important actions. The road and railway networks need to be operative as quickly as it can, making also proper links with the airports and harbors. Following a big natural disaster, the traffic flow to the affected region also greatly increases, so the support to sustain it must be assured.

As it can be seen, no matter of the disaster, the transportation network is the fundamental base for interventions. By assessing its vulnerabilities, early planning can be made in order to eliminate the threat of disrupting the important links within the network. The planning can consider numerous aspects, like the creation of alternative routes, strengthening the vital corridors, bridges or tunnels safety, reconsidering the efficiency of the emergency intervention centers and medical resources in the territory etc. The methodology proposed in this study will refer to the vulnerability assessment, with an example for the seismic risk.

In order to evaluate the vulnerability of a transport network, various methods have been previously developed [6]. According to Pinto et al. [7], three types of studies that asses the seismic performance of transportation networks can be depicted, based on the level they are referring to:

- Level I studies: the attention is focused on the functioning of the network in terms of pure connectivity—useful for rescue function right after the earthquake
- Level II studies: add the consideration of the network capacity to accommodate traffic flows
- Level III studies, that try to give a more realistic general picture, by combining direct physical damage estimates with various economic models.

The purpose of the methodology presented here might be considered as a level I study, but nevertheless the possibility of adding knowledge based on more complex procedures and on an economical approach can turn it into a more advanced study. Practically, it can adapt to the requirements of the user, keeping just the cost-distance method as a core. The cost-distance method is specific for evaluating the economic travel costs in a territory, but this cost can be considered fictive, showing how hard it is to reach a point from an origin, by accumulating vulnerability values along the way. The idea of using the Cost-Distance Method in various fields of study like ecology [2] or road planning [1] was already considered, providing proper results.

With the proposed methodology, performed with relevant and well understood input data, answers to several important questions can be obtained:

- How efficient is the transport network distributed, as support for prompt interventions in risk areas?
- What are the vulnerable areas (of the network) and what impact can have the isolation of them?
- How well are the emergency intervention centers distributed along the network? How long can it take to intervene, giving the vulnerabilities of the road?
- What are the safest and reliable routes?
- What segments are vital for the access in some isolated areas?
- How many people can be affected by the failure to provide quick reaction measures?

2 About the Methodology

The fundamental idea behind the proposed methodology was to create a completely GIS integrated complex tool, allowing different but not fixed integration of own data and procedures from different fields, in order to assess the vulnerability of a network. This network might be of roads, railways, pipes or any other network with a spatial extent. What we've tried was to involve more of the geographic component of the elements (link them in a singular system) and focus on the relations within the whole network. Studying the capabilities of the Spatial Analyst Toolbox in the ESRI ArcInfo software, especially the Cost-Distance Method, we observed that by setting the right input (create a cost vulnerability raster just for the network) you can obtain a map showing how potentially difficult is to get from an origin to any connected place, therefore if a crossed cell is vulnerable, the next cell will also be vulnerable, if there is no safer way to get to it.

As the ArcMap help on cost functions explains, "the cost values assigned to each cell are per-unit distance measures for the cell. If the cell size is expressed in meters, the cost assigned to the cell is the cost necessary to travel 1 m within the cell. If the resolution is 50 m, the total cost to travel either horizontally or vertically through the cell would be the cost assigned to the cell times the resolution (total cost = cost * 50)".

In selecting the GIS software suited for the implementation of the methodology that uses not only the cost-distance analysis but also spatial editing, raster analysis or automation of processes (as seen in Fig. 1), ESRI ArcGIS software, as a leading and worldwide used solution, with easy to use but complex features, was the proper solution and environment.

As it can be observed in Fig. 1, the first mandatory element for the assessment of the network vulnerability is the network definition. This can be done in a vector or raster format directly. Important is that after the conversion into raster, connected segments are linked, the cell size is small enough and the blank spaces near the network get the "No Data" raster value. In order to fix problems that occur

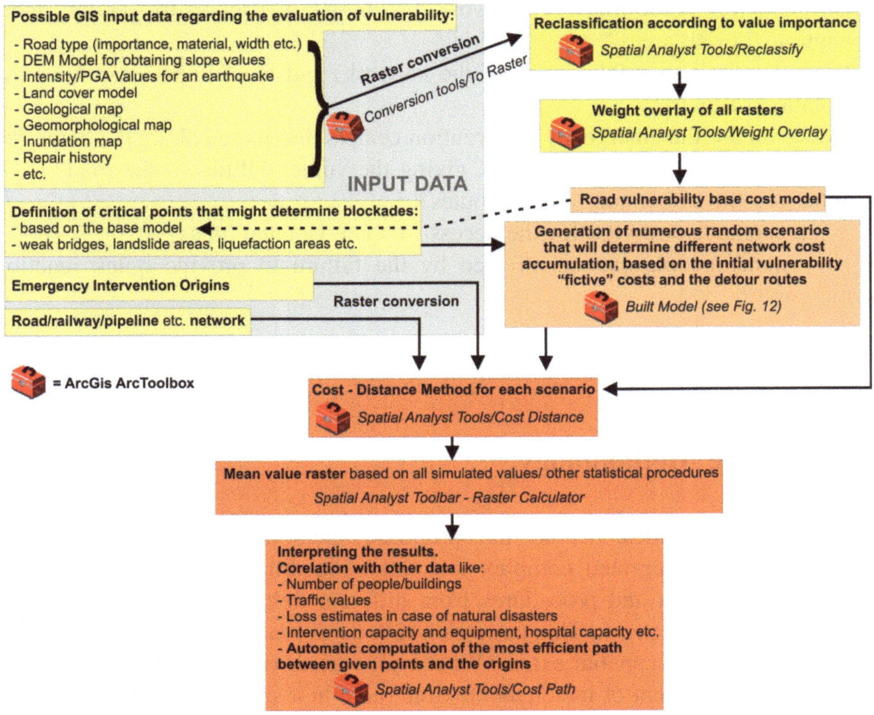

Fig. 1 The general plan of the proposed methodology

when the polyline is converted into raster, a good practice is to generate buffers around them, initially.

Another mandatory input data is the definition of Emergency Intervention Origins. In order to compute the vulnerability with the Cost-Distance Method, it is needed to know points to refer to. If a city has an Emergency Intervention Center for natural disasters, or a hospital for example, then the vulnerability of the network in this city is the smallest. There is no limit in defining these origins, but they have to be placed on top of the network cells with data. The provided data doesn't necessarily have to be real—for planning purposes, proposed points can be created and the vulnerability differences can be noticed.

One of the important steps that involve the operator's understanding is the definition of elements that can influence the network vulnerability. Each one has to decide what data to use, what the implications to the network are or what is more hazardous. In the end, the methodology requires just a cost raster, with the similar extent as the network raster. But getting to it, several steps need to be carried out. Initially, each set of data must be converted into raster. Then, depending on the importance of the numeric intervals or text description, new risk values are assigned—in general, based on a similar scale. This similar scale will reflect when all the impact rasters are compiled together, and a certain weight factor cumulating

100 % is attributed. For the case study bellow, a model of how the reclassification values are assigned is given and also proposed. In general, natural disasters influence the landscape, so data sets like the slope or the land use, which can be obtained for free from SRTM or CORINE data sets, can be added to the weighted overlay compilation.

After obtaining the cost raster, the first Cost-Distance Analysis can be performed—for a perfectly linked network situation. During a natural disaster, several points are more vulnerable than others. If these points will block the network, how will the joined segments be affected? To answer this question, the next step of the methodology involves a "What if?" approach, that tries to assess the vulnerability further, by simulating a large and random number of situations. The automated analysis created (with the aid of ArcGIS Model Builder) requires the definition of the "hot spots", which can be made based upon the cost raster created earlier, or on specific knowledge. This time, data has to be in a vector format—polygons that cut the segments perpendicular. Each simulation determines what points are considered to block the network, then changes values from the initial cost raster into "no data", and then applies the cost-distance method. After many simulations, from worst case scenario to few blockages scenario, all distance rasters can be merged as mean, into a final output of the network vulnerability, called also distance raster.

The output, together with additional data, will hopefully be able to answer realistically to the questions enumerated in the introduction. By adding a population layer and reclassifying the distance raster values, estimations about the possible number of isolated people after a natural disaster can be made. Also, the ArcGIS Spatial Analyst Toolbox offers the possibility to perform a Cost Path Method that will return the best (safest) path between any point in the network and the costly nearest origin. There are many possibilities, for different purposes. In the case study we tried to provide a good example of the methodology's capabilities.

3 Case Study: Vrancea County Road Network Seismic Vulnerability Assessment

3.1 Main Characteristics of the Analyzed Area

In order to exemplify and test the proposed methodology, a representative study area was selected—the Vrancea County. Located in Romania, this county is under a constant seismic risk generated by the Vrancea active seismic area. This area is located in the curvature of the Carpathian Mountains, at the contact between the East European plate and the Intra-Alpine and Moesic subplates, and it has been in the Twentieth century the cause of 32 intermediate depth earthquakes with Mw $\geq$ 6, including devastating events like the one 10 Nov 1940, with Mw 7.7 and h = 150 km, and the one on 3 Mar 1977, with Mw 7.4 and h = 94 km (Fig. 2).

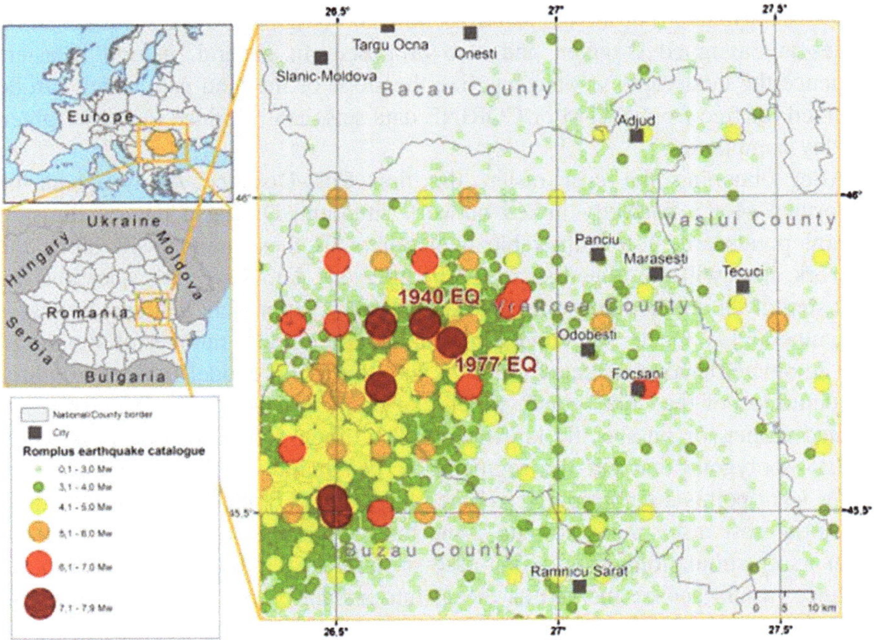

Fig. 2 The localization of the Vrancea County and the earthquake epicenters in the Romplus Catalogue [5]

These particular events highlighted major problems near but also very far from the epicenters (up to 300 km). In the Vrancea County, more than 90 % of Panciu city was destroyed during the 1940 event, a lot of buildings in Focsani City and different villages collapsed, due to the poor quality of constructions but also to landslides, liquefaction and post-earthquake phenomenon (Figs. 3 and 4). The 1940 and 1977 events also severely affected Bucharest capital city, which is $\approx$150 km away from the epicentral area; over 33 buildings and flats collapsed in

Fig. 3 Slumping and lateral spreading of the road, in Balintesti, near the Vrancea County, after the 1940 earthquake [4]

Fig. 4 Fans damaged by toe of landslide, in Slon, Prahova County, after the 1977 earthquake [4]

the 1977 earthquake. For this event, the national death toll was 1578 persons (1391 in Bucharest), and 11300 wounded persons.

In the last couple of years the concern regarding the recurrence of a new major earthquake wasn't unfortunately doubled by significant mitigation efforts. The quality of new constructions often avoided seismic design regulations, the retrofitting of old buildings proved to be a slow process, and the equipment of emergency intervention units wasn't renewed properly (especially in less economically developed counties like Vrancea, were, for example, 11 of 12 firefighter cars are more than 15 years old, and some state hospitals were closed). The purpose of studies like this one aims to highlight the importance of taking important precautionary measures, in order not to face the simulated possible deficiencies.

The main characteristics of the Vrancea County, in accordance to the followed purpose, are (as also mapped in Figs. 5, 6 and 7):

- It is situated in Romania, having a total area of 4863 km^2 and a population of 390048 in 2011 (37.7 % urban). The maximum height is 1785 m.
- The capital city of the county is Focsani, with a population of 100007 people in 2008. Other cities, much smaller, are Adjud (17554 people in 2008), Marasesti, Panciu and Odobesti. The number of villages is 331.
- The most important rivers are Siret, Putna, Milcov and Ramnicu Sarat. During spring or autumn, they can and generated severe flooding that affected many villages and the transport infrastructure.
- The road network is wide, comprising two main directions: South-North (Bucharest-Buzau-Focsani-Bacau), linked by E85, and West-East (Brasov-Focsani-Husi or Galati), linked by DN2D and E581. In the mountainous and hillside part of the county, villages are connected by local roads, most of them very narrow and not asphalted. Some bridges were newly constructed or consolidated after the floods in 2005 and 2010, especially on Putna and Milcov.

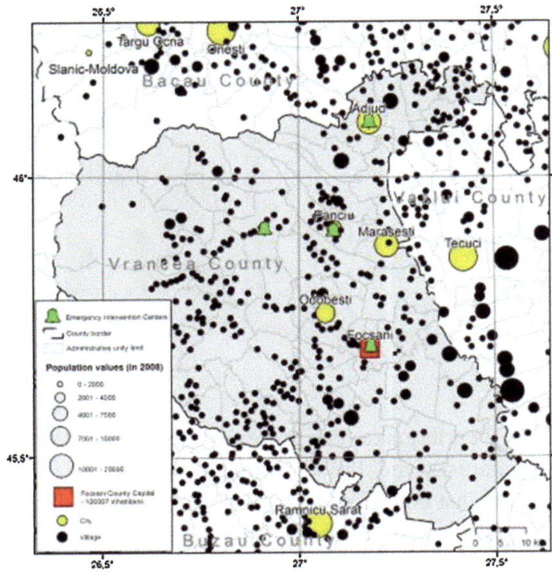

Fig. 5 Map showing the population distribution and the emergency intervention centers in case of a natural disaster

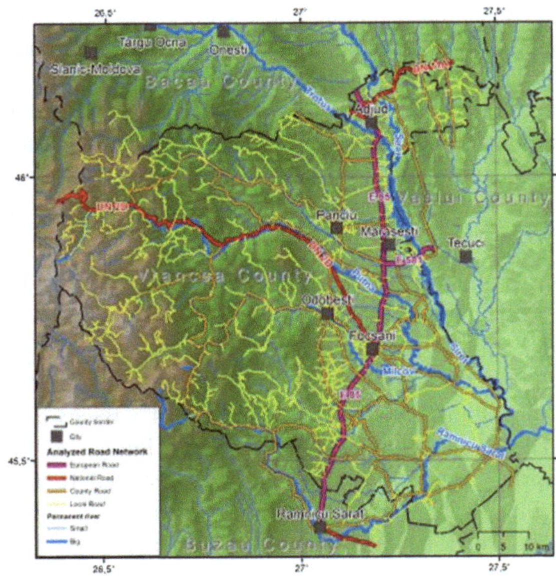

Fig. 6 General map of the county, showing the relief, rivers and road network

3.2 Performing the Cost-Distance Spatial Analysis

As the first part of the article highlighted, a few mandatory input data files are required, in order to perform a basic Cost-Distance Analysis. The selection below is meant to show how free data converted into GIS can be used, and how important is the association of proper weight factors. The implementation must be considered

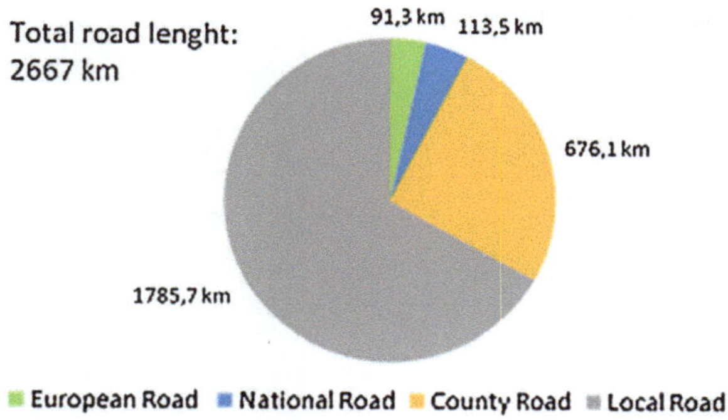

Fig. 7 Pie chart showing the length of the road network in Vrancea County

mainly as an exercise, because each region and set of data provides different challenges in assigning risk priorities.

For the definition of the road network, a vectorial GIS road database was created from a vectorial database digitized from declassified military maps from 1996, actualized in Google Earth with the aid of Google/Yahoo/Bing Maps overlays (thanks to the http://www.mgmaps.com service). Together with the geometry, the classification of the road was keeps as an attribute, being later considered in the description of the road vulnerability.

The locations of the emergency intervention centers were obtained from the official site of the General Inspectorate for Emergency Situations (www.igsu.ro). The county headquarter is in Focsani; other locations are in Adjud, Panciu and Vidra. For the basic purpose of this study, all of these centers were considered equal, as intervention potential efficiency.

For the description of road start-up vulnerability to earthquakes, several data, converted into a raster format, was used. As said before, the type of road was considered. Four classes were identified, and assigned numbers from 1 to 4 (1 for the European Roads, 2 for National Roads, 3 for County Roads and 4 for Local Roads).

The impact of slope on the road must always be considered, not just because of the actual road angle, but also because of the potential hazardous versants and roughness of the terrain. For computing the slope angle, a Digital Elevation Model (DEM) can be used. Based on the SRTM 3 Arc-Seconds free GIS data (http://srtm.csi.cgiar.org/), slope angle values were computed for the Vrancea County, varying from 0^0 to 31.5^0 (Fig. 8). The reclassification was based on equal intervals, as it can be seen in Fig. 9. The highest the angle, the more vulnerable is considered the road cell.

In case of an earthquake, the intensity defines how the earthquake was felt at the surface. Each intensity scale links observed effects to a value, most common between 1 and 12 (Modified Mercalli, Medveded–Sponheuer–Karnik scale = MSK and EMS-98). In the case of the road network, high intensities can determine the disrupting of road segments, by cracking, structure failures (bridges, tunnels etc.), soil liquefaction, landslides etc. Giving that Vrancea County is so close to the hypocenter, expected high intensities are not to be neglected. That is

Fig. 8 Map of the slope angle values

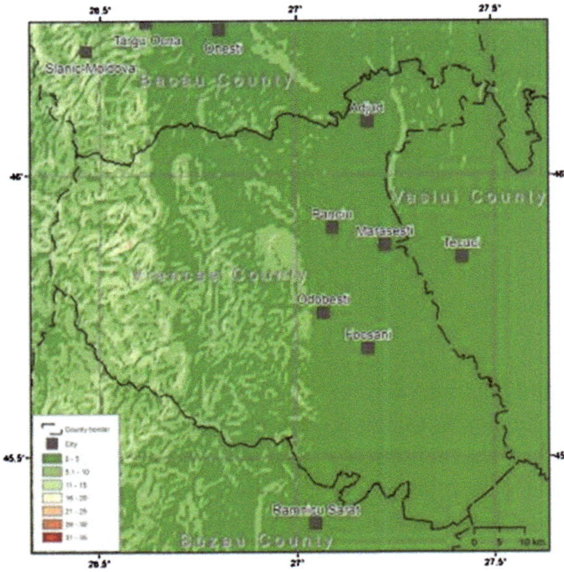

Fig. 9 Reclassification values for the slope angle raster

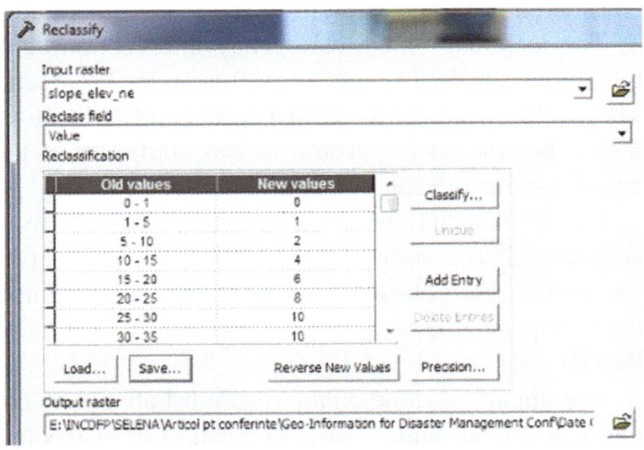

why, for the analysis, we used the isoseismal map, providing MSK intensities between IX 1/2 and VII ½ (Fig. 10). Considering the effects associated with these high intensities, multiplication values were defined (×2 for I = VII½, ×3 for I = VIII or VIII½ and ×4 for I = IX and more) and later added in multiplying risk elements that might be triggered by the earthquake, such as slope angle (causing landslides or soil cracks), bridges or industrial sites failure.

Because road networks also depend on the adjacent land cover they pass through, GIS data from the CORINE 2006 mission (http://www.eea.europa.eu/data-and-maps/data/corine-land-cover-2006-raster) was also used. In this data, major river bodies are defined, so bridge locations can be identified; knowing the characteristics of the area, these crossings can receive a certain vulnerability value, multiplied with the value of intensity. As a further work in the field, bridges can be

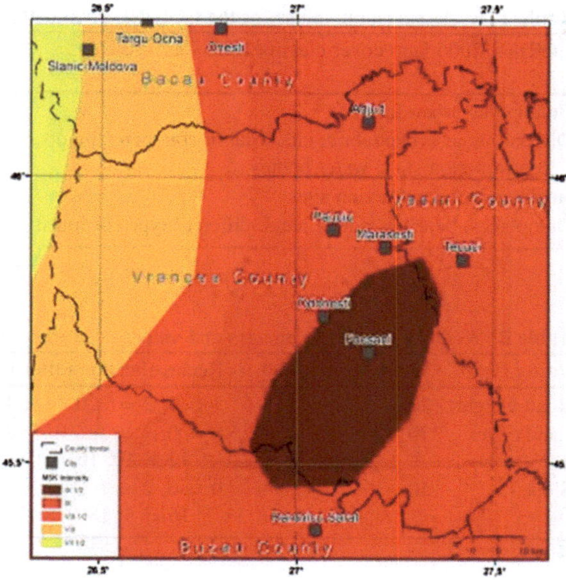

Fig. 10 Isoseismal map of the maximum credible Vrancea earthquake (Mw = 7.7) [3]

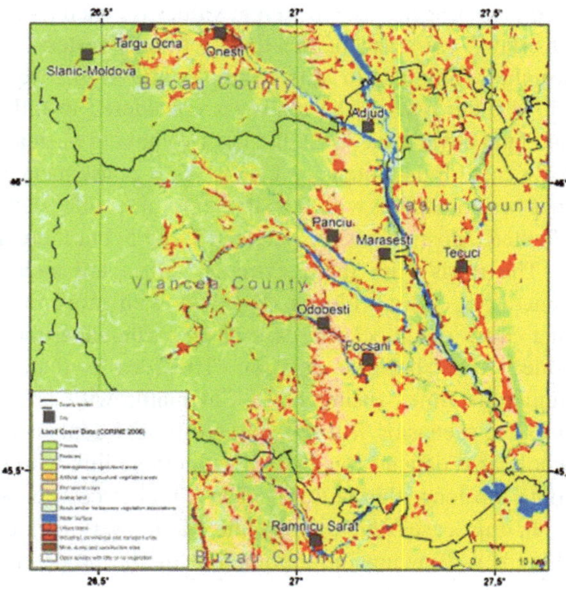

Fig. 11 CORINE 2006 data used for the Vrancea County

added to the analysis as structures with specific damage functions, possible to be also analyzed in real time by sensors. Beside bridges, CORINE data also includes possible hazardous areas, like industrial facilities, swamps, rock formations or types of forest (Fig. 11). Table 1 shows how the classification was carried out.

Table 1 Reclassification values for the CORINE 2006 data

CORINE 2006 land cover classification	Reclassification value
Continuous urban fabric	6
Discontinuous urban fabric, industrial or commercial units, construction sites, water bodies	4
Rocks, mineral extraction sites	3
Dump sites, beaches, dunes, sand, different types of forest	2

Table 2 Weight overlay percentages and values used for generating the basic cost raster

Raster type	Weight overlay (%)	Reclassification (after multiplication with MSK Intensity)
Road weight	20	1 = 1; 2 = 2; 3 = 4; 4 = 6
Land cover	30	0 = 0; 4 = 2; 6 = 5; 8 = 7; 9 = 8; 12 = 10; 16 = 10
Slope	50	0 = 0; 2 = 1; 3 = 1; 4 = 2; 6 = 4; 8 = 6; 12 = 6; 16 = 8; 18 = 8; 24 = 10; 30 = 10; 32 = 10

After all necessary data is converted into raster and reclassified, an important step is to properly assign weight values for each risk factor. By using the Weighted Overlay Tool, each of the final 3 impact rasters receive importance values and weight (Table 2). In this example, the slope was considered more important (50 %), the type of the road was considered to influence the vulnerability 20 %, and land cover data 30 %, the last two having been also multiplied according to the MSK maximum intensity factor. As a consequence, a preliminary vulnerability map was obtained, with 0 as the less vulnerable road cell and 10 the possible most affected. Most values are between 1 and 5, but there are also some extreme values.

The areas with these extreme values are later considered as very possible generators of road blocks. Together with more personal knowledge about areas that cause roadblocks in case of an earthquake, these most vulnerable points are added to a shapefile polygon. The idea is that, by assigning random values numerous times (using Field Calculator with arc rand expression), all various possible scenarios will reflect the different dependencies in the road network; if a whole area depends on a single vulnerable link with the origin points or the detour route too long, this will reflect in a higher cost-distance.

In this case study, a specific model was created in ArcGIS Model Builder (Fig. 12), allowing the automation of the process. 100 different scenarios were randomly computed, each one showing a particular situation that might occur, with more or less hot points determining road blocks. All the cost-distance values in each cell were then mediated, and a final vulnerability map was obtained (Fig. 13). Comparing it with the initial vulnerability map, an increase in the vulnerability of some more isolated areas is revealed.

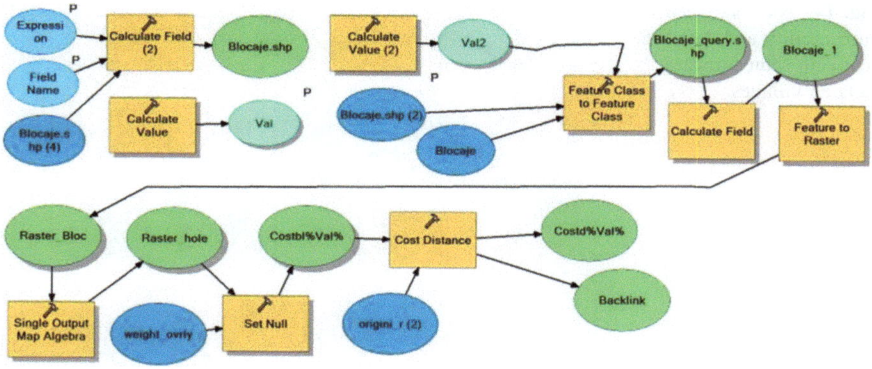

Fig. 12 Scheme of the model built in order to create random simulations, for the cost-distance method

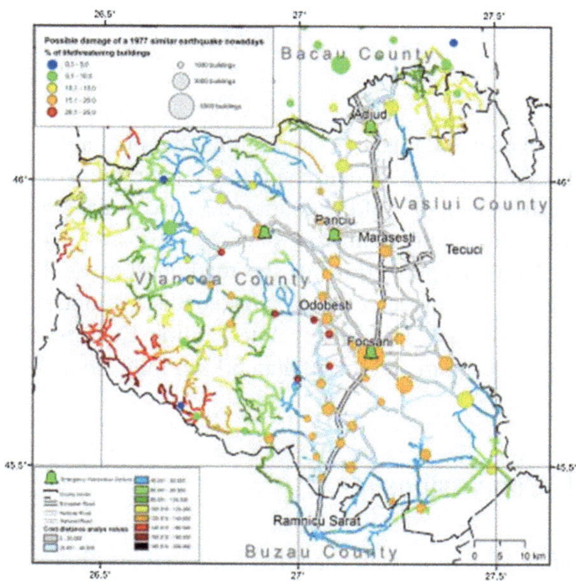

Fig. 13 The road network vulnerability and the building loss estimates during a 1977 similar earthquake

3.3 Result Analysis

The final results are more relevant showed on maps, which is in fact the purpose of using a GIS. The distance raster can be symbolized in order to show different vulnerability degrees, and various overlays can be made, in order to better understand the implications and the solutions to the questions in the introduction.

In Fig. 13 is shown the road network vulnerability, together with building damage estimates obtained from a simulation for the 1977 earthquake, with SE-LENA Software [8]. It can be noticed that although the most vulnerable roads are in the western part of the county, there are not a lot of settlements there, and also

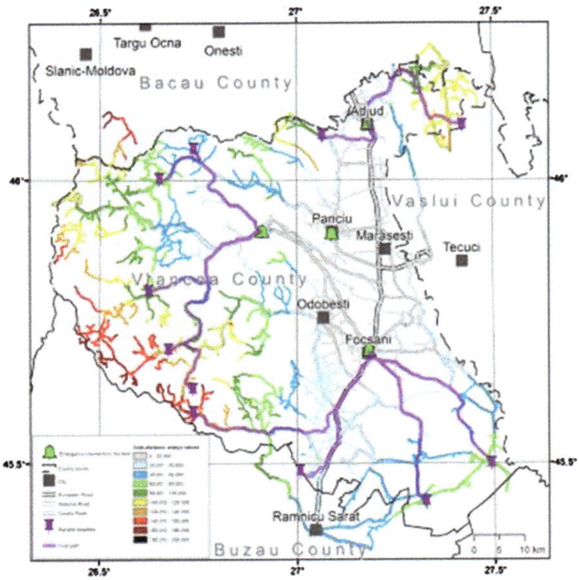

Fig. 14 The road network vulnerability and the safest cost path routes toward isolated villages

the expected damage is smaller than the one in the central and eastern parts, where the network is more dense and less vulnerable. A potential vulnerable area is in the north-eastern part of the county, because of the single link in the network and the 5 communes with significant damage estimates.

Another application for the methodology is the use of Cost Path Method. In Fig. 14 are presented the safest paths (computed automatically) from isolated villages to the Emergency Intervention Centers. Some of them (the ones in the south-west part) are more than 50 km long, so a closer center must be established.

4 Conclusions

After providing these detailed explanations about the methodology, some important advantages can be extracted:

- First of all, the answers aimed to be provided refer to real and important situations; they try to clarify the vulnerability of a certain type of network, in reaction to natural disasters, and that is most certainly an analysis that can save lives and decrease other losses.
- The easy integration within a GIS software like ArcGIS allows it to be flexible and easily operable, providing also the means for a further geographic analysis.
- Apparently, the chain of methods is simple, but the straight forward approach allows for a quick understanding of the methodology and various possibilities of adapting it to the specific purpose and level of detail intended. From real time

network monitoring to planning purposes, the methodology can assess the behavior of the network and its vulnerability.
- The dependency for a start-up complex GIS database is reduced.
- Maps can be created very easily, giving that all the output is in GIS format.
- The Cost-Path Method can be applied in order to find the best routes.

As this paper marks the first steps in using the mentioned approach, if scientifically approved, further studies will take place, in order to find better ways of reclassifying the risk factors rasters, creating a complete module for ArcGIS or experimenting with more complex data and network models.

References

1. D.M. Atkinson, P. Deadman, D. Dudycha, S. Traynor, Multi-criteria evaluation and least cost path analysis for an arctic all-weather road. Appl. Geogr. **25**(4), 287–307 (2005)
2. M.A. LaRue, C.K. Nielsen, Modelling potential dispersal corridors for cougars in midwestern North America using least-cost path methods. Ecol. Model. **212**, 372–381 (2008)
3. G. Marmureanu, C.O. Cioflan, A. Marmureanu, Intensity seismic hazard map of Romania by probabilistic and (neo)deterministic approaches, linear and nonlinear analyses. Rom. Rep. Phys. **63**(1), 226–239 (2011)
4. N. Mandrescu (ed.), *The Large Vrancea Intermediate Depth Earthquakes Occurred in the XXth Century and Their Effects on the Romanian Territory; Photographic Testimonies* (Academiei Romane, Bucharest, 2008)
5. National Institute for Earth Physics (NIEP), Romplus earthquake catalogue 2012, Romania
6. Pho Thanh Tung, Road vulnerability assessment for earthquakes. Ph.D. Dissertation, International Institute for Geo-information Science and Earth Observation, Enschede, 2004
7. P.E. Pinto, F. Cavalieri, P. Franchin, A. Lupoi, Definition of system components and the formulation of system functions to evaluate the performance of transportation infrastructures. SYNER-G Project No. 244061, Deliverable no. D2.6, June 2011
8. D. Toma-Danila, *A Seismic Risk Analysis for Focsani and Bacau Counties, Using SELENA in Near Real-Time and GIS Tools, Moldavian Risks—from Global to Local Scale* (Bacau, Romania, 2012)

Part III
System Requirements and Analysis

Effectiveness of Net-Centric Support Tools for Traffic Incident Management

Results of a Field Experiment

John Steenbruggen, Maarten Krieckaert, Piet Rietveld, Henk Scholten and Maarten van der Vlist

Abstract In the last two decades traffic Incident Management (IM) has become an important tool to reduce and prevent congestion on the road network, especially in urban areas. IM involves the coordinated interactions of many public and private actors. To support their tasks in an effective way, information systems are becoming increasingly important. In particular, information and system quality and Situational Awareness (SA) have been identified as major hurdles for effective emergency response. This paper reports the results of an empirical analysis of the effectiveness of net-centric information systems to improve the cooperation between public and private IM organizations. A set of controlled experiments were conducted with 16 participants. Data on the responses of the participants were collected through questionnaires and observer notes. The analysis focused on: a comparison of the tools tested, in terms of the appreciation of information and system quality, a comparison of the communication and coordination of a test group and a control group of emergency workers; the value of SA in the

J. Steenbruggen (✉) · M. Krieckaert · P. Rietveld · H. Scholten
Department of Spatial Economics, VU University, De Boelelaan 1105, 1081 HV Amsterdam, The Netherlands
e-mail: j.g.m.steenbruggen@vu.nl

M. Krieckaert
e-mail: m.n.m.krieckaert@vu.nl

P. Rietveld
e-mail: p.rietveld@vu.nl

H. Scholten
e-mail: hscholten@vu.nl

M. van der Vlist
Ministry Infrastructure and Environment, Zuiderwagenplein 2 8224 AD Lelystad, The Netherlands
e-mail: maarten.vander.vlist@rws.nl

S. Zlatanova et al. (eds.), *Intelligent Systems for Crisis Management*, Lecture Notes in Geoinformation and Cartography, DOI: 10.1007/978-3-642-33218-0_16,

performance of the decision-making process; and, how scenario complexity can affect the design principles of net-centric systems.

Keywords Traffic incident management · Net-centric information systems · Common operational picture · Situational awareness

1 Introduction

The early and reliable detection and verification of incidents, together with integrated traffic management strategies, are important contributions to improve the efficiency of the incident response. In the Netherlands, several studies have analysed the relationship between Incident Management (IM) measures and the consequences of incidents [35, 44, 45, 53, 80, 86]. IM emerges as one of the most important instruments of traffic management in the Netherlands, as it serves to mitigate and reduce incident effects, congestion and traffic jams, and eventually it may considerably contribute to reducing the number of casualties on the roads. All studies conclude that investing in IM measures is a very cost-effective strategy in road traffic management.

IM measures affect multiple phases of the incident-handling process. Classical IM strategies are aimed at minimizing the negative effects of congestion caused by an incident. Because of the quadratic relationship between the duration of an incident and the response time of the emergency services, response time (speed of emergency aid) plays an important role in determining the overall incident effects. For instance, the number of lost vehicle-hours as the result of an incident depends on the time required to clear the road for traffic following an accident, the road capacity, and the extent to which the road capacity is filled [33]. The basic idea is that fast clearance of the incident scene can help to reduce the incident-related congestion. Improving Situational Awareness (SA) for emergency services is crucial for the quick clearance of the incident scene. Klein [42] present four reasons why SA is important: (a) SA appears to be linked to performance; (b) limitations in SA may results in errors; (c) SA may be related to expertise; and (d) SA is the basis for decision making.

Breton and Rousseau [67] state that SA measurement can be seen as a process where three questions need to be answered: (1) Why measureng SA?; (2) What type of SA is measured?; and (3) How it can be measured. In the literature there are many definitions of SA. However various papers address the difficulty in the development of SA measurement techniques [18, 24]. Stanton et al. [77] identified over 30 different approaches to measure SA. Salmon et al. [62] categorize these into different types of SA measure. Models for SA that currently dominate the literature [see [76]] are individually oriented theories, including Endsley's three-level model [17]; the perceptual cycle model [75]; and the activity theory model [3]. Of the individual oriented SA theories, Endsley's information processing based three-level model is the most popular [17]. Its counterpart measurement approach, the Situation Awareness Global Assessment Technique [SAGAT: [17]] is the most commonly used procedure for measuring SA, despite questions regarding its validity as an SA measure [62]. However in the literature there is no general model that can be applied to traffic IM.

To measure SA it is crucial to include the concept of quality of information systems. The term 'information systems' covers collecting, processing, distributing, and using data by organizational processes or people [78]. The concept of 'quality' is often defined in terms of *'fitness for use'* [41], *'fitness for intended use'* [40] or *'fitness for purpose'* which has been a widely used approach by quality agencies that is usually based on the ability of an institution to fulfil its mission [27]. In the information systems literature, quality itself is relatively "ill-defined" [59]. The concept of quality is usually bound to the specific object, such as a specific product, process, service, or information system. Different authors identify information and system quality as the key factor for information system success [6, 13, 68]. However, various authors have concluded that information quality and system quality are the major hurdles for efficient and effective multi-agency emergency services, and are crucial for information systems success [48]. Information Quality (IQ) and System Quality (SQ) form an important requisite to achieve SA. There is a wealth of literature on information system success in profit-oriented business environments research regarding information quality dimensions. However, the literature on the public sector emergency services and traffic IM regarding information-sharing across different agencies and the quality of information-sharing is scarce, and empirical support is almost non-existent. Our research will address the following questions:

- Which constructs are relevant to measure information quality and system quality for traffic IM literature review
- How was the new information system appreciated by end-users in terms of IQ and SQ (questionnaires)?
- Is there a difference in terms of communication and coordination between the two groups?
- How has SA improved the performance in the decision-making process (outcomes);
- What are the main issues using net-centric systems as experienced by end-users?
- What are the effects of scenario complexity on the benefits of net-centric systems.

2 Assessing the Effectiveness of Net-Centric Information Systems

2.1 *Situation Awareness*

Endsley [56] defines SA as a product comprising "the perception of the elements in the environment within a volume of time and space, the comprehension of their meaning, and the projection of their status in the near future". A Common

Operational Picture (COP) is the basis to create SA for the emergency services to support traffic IM. A COP is established and maintained by gathering, collating, synthesizing, and disseminating incident information to all appropriate parties [30]. Achieving a COP allows on-scene and off-scene personnel to have the same information about the incident, including the availability and location of resources and the status of assistance requests. This can be achieved by the introduction of net-centric information systems which provide the capability to acquire, generate, distribute, manipulate, and utilize information. The relationship between these new information concepts for traffic IM is extensively described by Steenbruggen et al. [36].

SA consist of the terms 'Situation' and 'Awareness' which are both relevant in terms of measuring the value added services to improve existing information systems. 'Awareness' systems can be broadly defined as those systems that help people (emergency workers) to construct and maintain awareness of each others identity, location, activities (tasks), context or status [61]. A general definition of 'Situation' can be found, for example in Pew [69] who defines it as the surrounding environment (spatial awareness), mission goals, system availability, and physical human resources, and each 'crew member' must know the current activity of other crew members. Wickens [85], for example, defines SA components as geographical awareness, system awareness, and task awareness. There are some common elements in these definitions such as organization, system- and environment-related variables. However, in the literature there is no general model that can be applied to traffic IM. Therefore, we use three elements of information which define 'Situation' in order to support traffic IM awareness: incident information; information specifically related to the environment of the incident; and information about the emergency organizations involved in dealing with the incident.

Hone et al. [23] has stated that, although Endsley's definition [56] of SA is very good, it has some major problems. The definition cannot be operationalized. The three levels are treated as sequential and are often called perception, comprehension, and prediction. In the real world, *perceptual inputs* are both sequential and parallel. *Comprehension* starts with the first perceptual input. *Prediction* can start before comprehension is completed. So, Level 1 and Level 2 cannot be separated, and it may be hard to distinguish Level 3. Hone et al. [23] reformulates Endsley's definition of SA: (1). 'a person's perception of elements in the environment within a volume of time and space' to 'Who is where'?; (2). 'the comprehension of their meaning' to 'What are they doing'? and; (3) 'the projection of their near future' to 'What will they do'? Hone et al. [23] made a good step for SA to be operationalized. However, he only talks about people (organizations), but for traffic IM there are also other elements that are relevant. In Table 1, we give an overview of some examples of traffic IM information components and what to measure at the different levels of SA.

Table 1 Measuring levels for SA and traffic IM information components

SA components	Incident	Environment (surrounding of the incident)	Organization (emergency services)
Level 1 Perception of elements in the environment within a volume of time and space (Who or what is where?)	What is the incident location What type of incident What is the nature of the incident	Where is the congestion located Where are the traffic jams Where are the road users	Which emergency organizations are involved Where are the managers/emergency services
Level 2 The comprehension of their meaning (What are they doing or What does it mean)	What causes the incident What is the number of injuries Is there release of dangerous goods	What causes the congestion: incident, events, weather What is the site accessibility for emergency services How many people are in the area	How should we respond Which traffic management strategies do I have at my disposal
Level 3 The projection of their near future (What will they do or Which impact it will have)	When is the road cleaned	How far the consequences of an incident reverberate on the road network What risks are there for the surrounding area (e.g. chemical releases)	What will they do (activity) At what time will they be (t)here

Note Adopted from [17, 23]

2.2 *Information Quality*

The management of information is essential for the coordination of emergency response [70]. Information (or data) quality (IQ) can be seen as an important requisite for improving cooperation between IM emergency responders. IQ must be in line with the requirements of end-users [83]. IQ is difficult to observe, capture or measure [63], and can be considered a confusing concept [39]. In the literature a great deal of attention has been paid to the attributes of IQ. This refers to attributes that are important for end-users and IQ has multiple dimensions by which we can measure it [54]. Data quality is established during three procedures within the information manufacturing cycle, which evolves through a sequence of stages: data collection, organization, presentation, and application [78]. Many studies have confirmed that IQ is a multi-dimensional concept [2, 32, 82, 83]. Several researchers have identified different dimensions of IQ. However, until now, a uniform list of the IQ attributes (constructs) does not exist. For example, Strong et al. [78] group the IQ dimensions into four categories. These categories

Table 2 Most relevant information quality dimensions identified in the literature

IQ groups	Information quality constructs
Intrinsic	Accuracy, Objectivity, Believability, Reputation
Accessibility	Accessibility, Security
Contextual	Relevancy, Value added, Timeliness, Completeness, Quantity (information overload)
Representation	Interpretability, Concise, Consistency, Comprehensive
Others	Correctness, Currency, Precision, Format, Availability, Reliability (validation), Personalization

capture different dimensions with a similar degree of information quality. The categories are intrinsic; accessibility; contextual; and representation. These categories are widely acceptable in the literature [49] and form the only framework that has been involved and has been refined over the years, and proposes empirically tested items for IQ measurements [47]. However, in the literature many papers have their own classification. We analysed 15 papers from the literature to see which IQ dimensions are most used. We made a distinction between generic information quality dimensions [20, 34, 47, 51, 54, 78, 83, 87] and specific applied IQ dimensions for emergency services [4, 57, 63, 65]. We looked at which dimensions—within the five sidentified IQ groups—are most relevant for the emergency services (Table 2).

Intrinsic data quality indicates that information has quality in its own right, that is inherent to the data, and which consists of context-independent dimensions. *Accessibility of data quality* focuses on the role of information systems that store, process, and deliver data to the end-user, and, in particular it refers to the ease with which available information can be accessed and/or is easily and quickly retrievable (extracted) from the system and very relevant for the emergency services. Relatively few researchers have paid attention to conceptual definitions [43]. The definition of *accessibility* is framework-dependent. Some frameworks do not even consider it as a dimension of IQ [16]. In some papers, *security* is also seen as an important dimension [64]. In ECIS [16] the increasing importance and relationship between *accessibility* and *security* is analysed in detail. Emergency services are information intensive processes [12] and their effectiveness is largely dependent on the availability of the necessary information [10]. There is also an ongoing debate about the relation of *accessibility* to IQ and System Quality (SQ). Some see *accessibility* more as a system quality dimension. *Contextual data quality* highlights that data quality must be considered within the context of the task concerned. The three most used contextual quality dimensions for emergency services are: *timeliness* [5, 11, 31, 64, 66, 79], *relevancy* [64], and *completeness* [21, 73]. One of the main problems is information overload [19]. Simply put, information overload is the notion of receiving too much information. It is widely agreed that more data does not mean better information. In an information rich

Table 3 Overview of the selected information quality dimensions

Construct	Definitions [adapted from Perry et al. [65]]
Timeliness (currency)	The extent to which the currency of information is suitable to its use Applicable and helpful for the task at hand [64]
Correctness	The extent to which information is consistent with ground truth
Completeness	The extent to which information relevant to ground truth is collected
Relevance	The proportion of information collected that is related to the task concerned
Consistency	The extent to which information is in agreement with related or prior information
Quantity (overload)	Information overload occurs when the information-processing requirements (information needed to complete a task) exceed the information-processing capacity (the quantity of information one can integrate into the decision-making process) [55]
Reliability (verification)	The extent to which information is correct and that one can trust it [83]

environment, users can be easily overloaded [19]. This must be in line with the concept of '*bounded rationality*' [25]. Therefore, *quantity* is also identified as an important construct. Eppler and Mengis [55] provide a framework for information overload. *Representational data quality* looks at aspects related to the format of the information and its meaning. It concerns whether the information is presented in an easily interpretable, understandable, concise, and consistent way. The most used representational quality dimensions for emergency services is *consistency* [63, 65, 78]. For example, if several organizations identify an inconsistency in a different incident location, this delays decision making [22]. Inconsistent information from multiple sources sometimes points to different answers. It is difficult to determine which information is correct, and which is false. Besides the four categories of Strong et al. [78], there are also other IQ dimensions that are relevant for the emergency services. *Correctness* is mentioned as relevant in several studies and has a strong relation with the contextual data quality *completeness*. Validation of data is also mentioned as an important dimension [60, 63] which is strongly related to *correctness* and *reliability*. Reliable information is needs to be correct and is based on data that you can trust on [83]. Two relatively new dimensions are *personalization* and *context awareness*, which both have strong relationships with the contextual data quality dimension *quantity*. *Personalization* is related to context-aware computing whose primary goal is to make interaction with computers easier and more supportive for human activity. This can be done in several ways, one of the most important being the filtering of the information flow from application to user to avoid receiving irrelevant information and thus preventing the problem of information overload [Schmidt et al. 1999]. In other words, it is crucial to get the right information at the right moment in the right context. Table 3 contains an overview of the IQ constructs we used for our field exercise.

2.3 System Quality

Although the study of System Quality (SQ) has a long history [for an extensive historical overview see [13, 81], it has received less attention than information quality in the literature [4, 48]. SQ is a concept used to measure and evaluate the multiple dimensions of the information-processing system itself [13]. SQ is related more to the characteristics of the information-processing system, and closely related to service quality and ease of use than to its resulting product. However, they are not the same. Ease of use can be seen as a consequence of SQ. *Ease of use* is more an overall indicator of perceived user satisfaction. *Usability* is how the system supports the primary tasks of the end-user. IQ constructs are more related to the *output* of an information system. SQ reflects the information processing system required to produce that output [59]. *Accessibility* and *system reliability* are seen more as system-related SQ dimensions. They represent defined properties that are largely independent of usage. *Accessibility* has been suggested as an important dimension in emergency response [5, 11, 31, 46, 66]. It defines the role of information systems, that store, process, and deliver data to the end-user, and, in particular, it refers to the ease with which information (data) is available, can be accessed and or easily and quickly retrievable. *System reliability* is the technical availability of the system. *Response time, integration, memory, format and situational awareness* (SA) are more task-related SQ dimensions. In the literature, format has also been identified as an IQ construct [47, 63, 87]. It can be defined as the meaning of the information, and concerns whether the information is presented in an easily interpretable, understandable, concise, and consistent way. Most of the time this is presented in a predefined format. Therefore, we include this construct in SQ. Table 4 contains an overview of which SQ constructs are the most important to support a net-centric traffic IM system for our field exercise.

2.4 Impact on the Decision Process

Hone et al. [23] stated that in the real world the perceptual inputs for cognitive processes for SA are both sequential and parallel. This means that the level and quality of SA need to be combined with the time duration of the incident. The duration is defined as the period of time in which traffic flow is disrupted due to an incident. The amount of delay and the number of impacts that result from the incident depend on the duration of the different distinct phases. The following phases (or time periods) can be identified [based on Zwaneveld et al. 2000]: detection, verification and warning time; response, driving, and arrival time; site management operation or action time; and normalization and flow-recovery time.

The 3D model in Fig. 1 is the basis to create a COP to support personalized SA related to the different user-perspectives of the emergency services involved. The term "picture" in a COP refers not so much to a graphical representation, but

Table 4 Overview of selected system quality constructs

Construct	Definitions	Category
Accessibility	The degree to which a system and the systems related information it contains can be accessed with relatively low effort [59]	System-related
System reliability	The degree to which a system is dependable (e.g.technically available) over time [59]	
System response time	The degree to which a system offers quick (or timely) responses to requests for information or actions [59]	
Integration	The degree to which a system facilitates the combination of information from various sources to support business decisions [59]	Task-related
Memory	The degree to which the information (flow) and, tacit and explicit knowledge can be stored and organized in the system for reuse	
Format	The degree to which information is presented in an easily interpretable, understandable, concise and consistent way	
Situational awareness	The degree to which the system supports knowing what is going on around you [1, 15, 18].	
Ease of use	Satisfaction of user-interface [59]	Perceived operational satisfaction
Usability	'*fitness for use*' [41] or '*fitness for purpose*' ,which is based on the ability of an institution to fulfil its mission [27]	

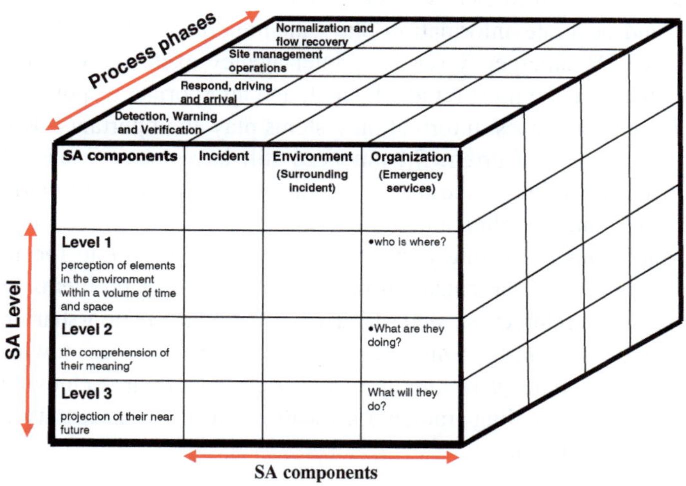

Fig. 1 3D model for measuring situational awareness for traffic IM

rather to the data used to define the operational situation. As such, "*the creation and dissemination of the COP is as much an information management challenge as it is a visualization challenge*" [71]. Both aspects are relevant to support SA. This means that for each IM phase, information needs to be available and shared in the right place and time, and presented in a way that a task-technology fit is accomplished for different end-users so information overload is avoided [19]. A way to achieve this goal is net-centric working (see [36]).

3 Design of the Experiment

3.1 Supporting Traffic IM in the Netherlands

Incident Management (IM) is, in general, the policy that, through a set of measures, aims to reduce both the negative effects on the traffic flow conditions and the effects on safety, by shortening the period needed to clear the road after an accident has happened. It can also be seen as a process to detect, respond, and clean-up traffic incidents, and to restore traffic capacity. There are many private and public organizations involved in the daily handling of IM. The public IM emergency services are the Road Authority, Police, Fire Brigade, and the Ambulance services. In the Netherlands, the Rijkswaterstaat has, in the context of the law called the Rijkswaterstaat works Management Act [1996], the public responsibility for an efficient and safe use of the main road network. Towing, repair, and insurance services are the main tasks of private IM parties. Together, the different parties have set up new guidelines and protocols in order to shorten the time that is needed to clear the road after incidents.

Timely and accurate information plays an important role in the information chain between IM emergency services. Inter-agency exchange of information is the key to obtaining the most rapid, efficient, and appropriate response to highway incidents from all agencies. Information systems play an important role within and between organizations. Current IM practices still have many issues which have regularly been identified in the daily operations in terms of information-sharing communication, and coordination.

IM organizations are strongly related and need to collaborate for an effective incident response. Each organization has the same kind of problems in terms of system diversity, architecture, and standards used. information-sharing between traffic management control centres, emergency control centres, towing services and insurance companies is becoming increasingly important. Achieving identical SA and a common handling framework for effective IM is necessary for further improvement of cooperation.

3.2 Hypotheses to be Tested

Four hypotheses are tested to evaluate the effectiveness of the net-centric information systems:

1. Net-centric systems improve the appreciation of IQ and SQ by end-users.
2. SA improve the performance in the decision making process of the emergency organizations.
3. Net-centric systems improve the communication and coordination of emergency organizations.
4. Scenario complexity affects the design principles of net-centric systems.

3.3 Set-up of the Experiment

In the field exercise we introduce new information concepts such as net-centric working, and a COP to improve information and system quality. This is the basis for an improved SA, which leads to better decisions, better actions, and thus better effects. To test these concepts we set up a field exercise. On 25 May 2012, a national IM test took place in the city of Eindhoven (the Netherlands). Five incident scenarios were simulated to measure the value added by a net-centric system. Participants were randomly assigned to one of two groups. The test group used the net-centric systems while the control group used traditional tools. The net-centric system provided the possibility to exchange pictures of the incident (see Table 6), to see where other parties were, and send text-messages to all parties at once. The control group used a system that had similar capabilities as their daily practice systems and communicated via telephone. The communication between the different actors was recorded by a group of students, based on shadowing. Computer loggings of the communication that took place in the system were also recorded. Questionnaires were handed out to all actors participating on the tests. At the end of each scenario, respondents filled out a questionnaire on IQ. After all scenarios were considered, questions on SQ were answered.

3.3.1 Participants

The demographics of the participants in the field exercise are shown in Table 5, which shows: number of participants, average age, gender, condition, organization, work experience, GRIP [74] experience and education. Due the limited number of participants in the field experiment, these attributes are not used for statistics but simply for information.

Table 5 Demographics of participants field exercise

Participants	16
Average age	44.1
Gender	n
Male	14
Female	2
Condition	n
Test group	9
Control group	7
Organization	n
Regional traffic management centre	2
Road district Rijkswaterstaat	4
Emergency room ANWB	3
Towing services	3
Towing emergency room CMV (trucks)	3
Towing emergency room LCM (cars)	1
Experience (years)	n
0–1	0
1–5	4
5–10	5
10–20	5
20–30	1
More than 30 years	1
Experience Grip 2 incidents or higher	n
0 times	8
1–5 times	7
More than 5 times	0
Education	n
Lager onderwijs	1
LBO, LAVO, MAVO, MULO	1
MBO, VMBO, HAVO	9
MMS, HBS, VWO	0
HBO, Universiteit	5

3.3.2 Net-Centric Software Tool

The introduction of new data sets and information concepts can be helpful in solving the identified problems and in reducing the time interval between the detection of the incident and the re-establishment of traffic flows in a significant way. This is particularly the case when information systems are linked to the information needs of actors involved in the IM process. The report '*Successful Response Starts with a Map*' [NSF, [58]] concludes, on the basis of various workshops and interviews, that during major disasters in the United States, there was a lack of correct information. The report indicates that the information needed for disaster response consists primarily of specific location information. This information is also referred to as spatial information. The report also recommends

Table 6 Scenario descriptions

Scenario 1a, Truck next to highway (GRIP-0) *Thursday 15.30 on the A67, 42.1* km

This is an incident without victims, only material damage and the congestion builds up. There is a truck stranded next to the highway. The incident is reported to the police by passing road users mobile calls to the emergency number (E112). The incident notifications differ in accuracy. The 112 emergency room sent the notification to the allocated emergency centre (Police, Fire Brigade, GHOR). They sent a police car. In the net-centric environment the regional traffic centre is able to view this notification. The focus of this scenario is a confusion of the exact incident location that was caused on purpose by the researchers. The first emergency vehicle heading for the incident notices this mistake. In normal situations, the other involved actors would not receive this information, but hopefully now they will, and this confusion will not cause delay for the other involved actors

Goal of scenario: *This scenario is based on the fact that the emergency centre (Police, Fire Brigade, GHOR) sometimes does not know the exact location of the incident. They don't have access to camera images from the traffic management centre. Therefore, they do not have a good overview of the incident scene. This causes different problems, such as sending emergency cars to the wrong location and not having detailed information about which measures need to be taken. Therefore, they sometimes allocate inappropriate resources to the incident scene. If they have access to real-time cameras they have better situational awareness of the incident*

Scenario 1b, Broken-down car (GRIP-0) *Tuesday 8.00 on A2 ring road Eindhoven busy but no traffic jams*

There is a broken-down vehicle on the ring road of Eindhoven. The driver of the vehicle contacts the ANWB. The ANWB dispatch centre immediately sends a service car to help the driver

Goal of scenario: *The importance of this scenario is a detection of the incident with wrong location information of the incident. The first emergency service arrives at the wrong location. In normal situations other emergency services will also drive to the wrong location. Net-centric information sharing assumes that this wrong information is detected more quickly. Other emergency centres will communicate this information directly to their own field workers. This will avoid a waste of valuable process time in the handling the incident*

Scenario 2, Truck loaded with iron scrap and several victims (GRIP-1) *A2 Right, 171.1* km

This scenario is a GRIP 1 scenario. Several victims are involved. A truck driver loses control, slips through the crash barrier and hits a pillar supporting a flyover on the right side of the road. The truck is loaded with iron scrap. The driver and his companion are severely injured and the fire department needs to cut them loose from the cabin. The cargo is scattered over the road. The pillar supporting the fly over is severely damaged. A large traffic jam starts to build up behind the incident

(continued)

Table 6 (continued)

Goal of scenario: *This is a large incident with severe traffic problems and congestion. The incident escalates on a national scale with the involvement of all emergency organizations. The aim of this scenario is to show that, with quick information sharing between all emergency services, the incident can be cleared more rapidly. To complicate the scenario, there is a secondary incident in the tail of the traffic jam. Therefore, the resources of the emergency services need to be managed over the different incidents. There are several casualties, there more vehicles involved, and the road is blocked due to lost cargo of the truck*

Scenario 3, Truck loaded with iron scrap and several victims (GRIP-1) *A2 Right, 171.1* km

A truck catches fire. The driver panics and makes an emergency stop, causing the truck then slip and eventually stop horizontally across the road, blocking all the traffic. Driving behind the slipping truck, there is another truck loaded with meat, which is also forced to brake hard. In doing so, it also slips on the tarmac and loses some of the cargo. Behind the two slipping trucks enormous damage and congestion builds up in which several trucks and private cars are involved. Because of the many (emergency) resources involved in this incident, the route to the incident gets blocked. An alternative route is needed to reduce traffic jams

Goal of scenario: *This scenario is created to demonstrate that an early shared common operational picture between the emergency services involved can improve the decision making process so that the necessary actions can be taken more quickly. Due to the great chaos on and around the incident scene, many emergency services struggle to get a good overview of the impact of the incident. There are many issues such as applying appropriate traffic and safety measures. Apart from that, it is also very difficult for the emergency services to arrive at the incident location owing to blocked roads and traffic jams*

Scenario 4, Hazardous cargo and fire, fatal casualties (GRIP-3) *A2 East 159,3*

A collision between a truck (transporting a tank containing isobutene) and a passenger car. The truck has tipped over and the car has caught fire. A second car gets involved in this fire, followed by two more cars. The fire causes black smoke that can be seen from a great distance. Many people call into report the incident. The truck was transporting hazardous cargo (isobutane) and hit the casing of the electrical infrastructure, while the traffic management systems in the immediate area breaks down. As soon as the Fire Departement arrives, they confirm that this incident concerns the transport of isobutene, which is highly flammable and explosive (risk of explosion from heating of the tank containing liquid gas). The situation is scaled up to GRIP 3 and everyone is evacuated within a radius of 500 metres. The driver of the truck and the drivers of the two cars that caught fire did not survive the accident

Goal of scenario: *This is a full scenario with all emergency services and the safety region. This scenario shows that sharing information on the environment of the incident scene helps the traffic management centre to better coordinate the incident, and helps to apply effective traffic management measures* that preparations for future disasters must always be based on this spatial information. Attention to the spatial information also remained limited in the Netherlands until recently, but people are becoming increasingly aware that the spatial component is crucial, not only for the realization of the information but also for communication of the information [37].

Scholten et al. [28] drew up a conceptual diagram on how to work with this spatial information, which was based on four frameworks that are integrated using technology, and which can realise the information in question. Firstly, there is the *Organisational framework*, in which the boundary conditions are established, such as standards, legal conditions, security, etc. The *Data framework* contains a collection of all the necessary basic data, both static and dynamic. The *Analytical framework* describes the way in which the processes that play a role in a disaster can be analysed and modelled. The most important models pertain to floods, forest fires, evacuations, and the spread of hazardous materials. Finally, there is the *Visualization and Communication framework*, in which descriptions are given of how the spatial information is displayed and communicated, using maps, images and audio as well as texts.

The technology (GIS) enables the frameworks to be integrated and information systems to be built. These systems also have various forms, such as the part for the crisis control centre (single or multiple), the part for the drivers of the vehicles, the part for the mobile crisis control centre, and the part for the mobile users in the field. Communication between these users is crucial and must take place seamlessly. Each of them has the same common picture, and supplements this picture with specific information from the appropriate field of knowledge. This means that the information is not shared in a hierarchical manner, in which a central point of information usually does the sharing (Client–Server Model). Instead, each organization involved is both a source and a recipient. This model is referred to as 'peer-to-peer technology'. Technical details for such a peer-to-peer model for disaster response is provided in Scholten et al. [29]. This form of communication improves the speed of information exchange, and makes the network more robust. Further detailing of such a net-centric approach to provide spatial information for disaster response has the following functionalities:

- Information comes from various sources and various areas of knowledge, and also goes back to them;
- Information exchange takes place between the experts, without the intervention of the management hierarchy;
- The information is Geo-information, because the location aspect (location awareness) is essential;
- Ultimately, decision making takes place within the management hierarchy;
- Decision making requires complex information, sitreps and sitplots.

It is assumed that better and faster sharing of information in this network will result in a better deployment, resulting in increased efficiency during disaster response. The bases for this are: correct information: the right people: the right place, and the right time. The starting points for the Traffic IM System (TIMS) are as follows:

- The TIMS must be seen as a basis facility which can be expanded if necessary to include additional facilities, functionalities, data, and participants;
- The TIMS consists of a Geographical Information System, a Text System, a Logging System, a voice system, and a Security System. All these components are integrated;
- The participating actors are connected to the TIMS, which gives them access to all the information that is being shared; and
- The TIMS supports the disaster-response decision making, in terms of both operations and policy.

The functionalities of the TIMS include a text application for writing and sending messages and instructions to participants. Symbols are used to check whether the messages on the user's tab have been read and acted on. The functionalities of TIMS also include a Geographical Information System (GIS) for sharing, combining, analysing, and visualizing data and information. The GIS makes it possible to clarify the current and future disaster situation in a single map image. The question we are asking ourselves is an obvious one. Will the use of such a TIMS system also result in an improved IM response?

The support system used for sharing textual information was developed in MS-Groove and is known as 'sitekst'. The system works with tabs, and each participating organization has its own tab. The tabs are primarily intended to indicate the information position of the various departments; other actors can view the tabs. All messages sent and received are automatically placed and stored on the tab. This also makes each tab a logbook of the information exchange. The user- functionalities for sharing spatial information in TIMS were designed on the basis of a location-driven approach, so that, with the help of the sitplot application, it is possible to gain insight into where the incident is, what the context of the environment of the incident is, and which measures have been taken. Various analyses can also be carried out on the basis of the available data.

All the functionalities are targeted at achieving a complete, current and common picture of the situation as quickly as possible, and at anticipating future developments on site. This COP, with the sitplot as information product, is built up by all the plotters in the various emergency centres. The common situation picture is visible on each individual PC on which the sitplot application is running. The plotters in the various organizations can build up their situation picture separately. Active users are shown on the user-interface by means of different colours. If a user has added data to a sitplot, or amended data in a sitplot, a notification message is generated. By clicking on a user, the map layers of the user are added to the list of map layers.

3.3.3 Scenario Descriptions

To create realistic use cases that reflect the daily IM work activities, the scenarios were built upon several IM reports, which describe in detail the IM process phases and the role of the different emergency organizations [14] and contain input from well-trained emergency workers. By the design of the scenarios, we include different operational user-perspectives (e.g. centralist, road inspectors, road users), as well as the specific goals of strategic management operations and policy makers. In the field exercise five different simulated IM scenarios were played. They varied in complexity from the breakdown of vehicles and small collisions (GRIP 1) up to complex incidents with dangerous goods, serious impact on the environment, multiple involvement of cars and trucks, severe casualties, and complex traffic management measures (GRIP 3), which involve complex organization and coordination measures from multiple emergency organizations.

Each scenario is based on logs of real incidents and covers all existing work processes such as applying safety measures, avoiding congestion, traffic management for closing and redirecting traffic flows, towing activities, cleaning roads, and repairing the damage of infrastructural works. The test group and the control group played the scenarios simultaneously. The situation in the 'field' was simulated by a maquette. Table 6 provides a brief description of the five scenarios including the specific aim of each one.

3.3.4 The Experiment

This study is based on realistic traffic IM scenarios that cover a wide range of different types of incidents in terms of vehicles involved, casualties and complexity. The field exercise was set up with two groups of participants: a test group that used the specially developed net-centric systems and a control group that used traditional systems. Both groups were able to use telephone communication. Each group consisted of emergency centralists and emergency fieldworkers. The actual incident scene was simulated by a maquette (see Fig. 2b). For each group the scenarios were facilitated by an exercise staff (experiment organization). They initiated text messages to create the starting point for each played scenario. After each scenario a central evaluation of the participant experience was carried out. These discussions were input for the exercise staff to improve the next scenario (see Fig. 3).

We used different research methods to be able to analyse the results (see Fig. 4). Data on individual perceptions regarding the tools were acquired from the participants' responses by questionnaires. After each scenario, both groups had to fill in a questionnaire on IQ. After all scenarios were considered, both groups filled in a questionnaire on SQ. All the scenarios were 'shadowed' by students. Shadowing is a useful method for observing participant behaviour [52]. The shadowing of all participants was done by a group of students who had been instructed to use a predefined form. All text messages with the net-centric system were logged (Fig. 2a).

3.3.5 Limitations

Even though this net-centric field exercise is based on realistic scenarios, the present empirical approach has some limitations. We chose to play the scenarios

Fig. 2 **a** Centralist in action. **b** Fieldworkers on the scene

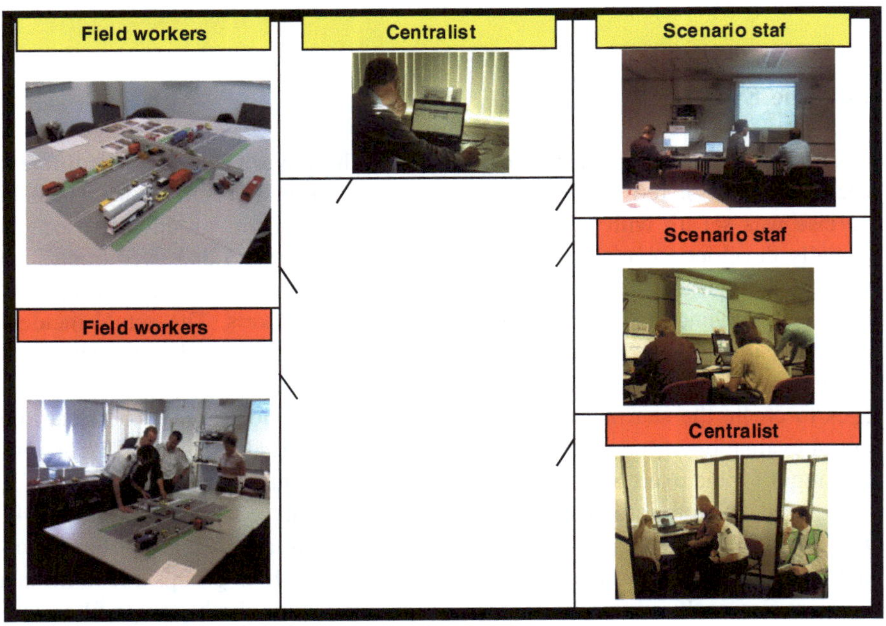

Fig. 3 Arrangements of participants

with well-trained emergency workers. An important consequence of this decision is that we needed to ask operational organizations to provide us with the necessary resources; these organizations had to plan these activities in a busy operational environment. This proved to be extremely difficult. Furthermore, we worked with a test group and a control group. That meant we had to double the necessary capacity. Given these constraints, the field test was limited to 16 observations This relatively small sample size might have influenced the precision of the results.

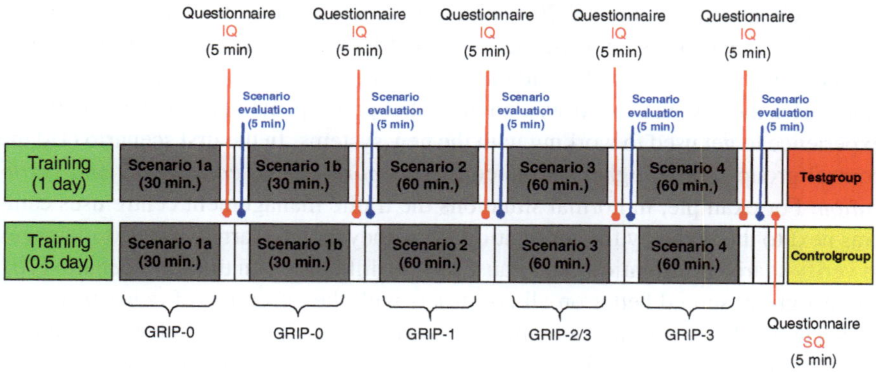

Fig. 4 Timelines representing the experiment protocol

A larger sample size would also have provided results with more statistical significance. A larger sample size proved to be impossible to realize due to limitations of time and the emergency workers availability.

We used real stakeholders who are all well-trained and skilled emergency workers. However, they had different backgrounds. Their work experience varied between 1 and roughly 30 years. The participants also had different educational backgrounds and did not have hands-on experience with net-centric information systems. To overcome this limitation, we provided all participants with one full day of training to acquire some background knowledge on the information concepts, and to train them in use of the system. Next to that, the participants for the test group and control group were randomly chosen. To avoid the control group being influenced, only the test group was trained in the net-centric system. This excludes the effect of background knowledge and experience.

Another limitation is that not all organizations were involved. Police, fire brigade and GHOR did not participate as centralist; they only had a role as the field-workers with the maquette. The research staff simulated the role of the centralist. Loggings were predefined as input for the other emergency services.

4 Results of the Experiment

4.1 IQ and SQ Questionnaires

The *first* hypothesis was: "Net-centric information systems improve the appreciation of IQ and SQ for traffic IM". Testing this hypothesis involved comparing the perceived IQ and SQ of each scenario between the test group and the control group. To measure IQ we use seven constructs. To validate each construct we asked two or three questions after each scenario. The most common measure of reliability of scores for a sample is the coefficient of internal consistency, or Cronbach's alpha

[50]. A Cronbach's alpha of 0.70 or higher is generally considered as an acceptable value of internal consistency. Scenarios 1a and 1b were used as cases on GRIP level 1. These are relatively small incidents with vehicles. Following evaluation discussions after these scenarios, it was obvious that participants needed some hands-on experience to get used to working with the new systems. In the first scenario (1a) the control group scored higher on the constructs *correctness*, *consistency* and *verification*. For example, in normal situations the traffic management centre uses cameras next to the highway for verification. Now they had to learn to use text messages to verify shared information. This caused some difficulties. In the next scenario (1b), the test group scored better on all constructs with the exception of *consistency*.

Scenario 2 was the first complex scenario. This scenario is characterized by many phone calls and sharing text messages in the net-centric system. As well as that, there were multiple incidents to complicate the decision-making process in the scenario. This causes many problems in communication between the emergency service and the coordination of activities. The control group scored better on all constructs with exception of *timelines* and *verification*. In the *t* test, timelines scored significantly better then the test group. This is in line with the loggings of the shadow observations. The incident notification information and the arrival of the emergency services to the incident scene was significantly faster in the test group with exception of the Fire Brigade and the Ambulance services. They arrived 3 min later. Only with the environmental information such as traffic management measures the control group perform better. The main issue in the test group was information *overload*.

In Scenario 3 there was clearly a learning effect from the second scenario. The test group scored better on all constructs with the exception of *overload* and *verification*. Overload was still the main issue. They had difficulties in using the predefined tools to filter relevant information. This also made it hard to verify shared information. However, the test group performed significantly faster than the control group. In Scenario 4 it was clearly visible that the test group were starting to have hands-on experience using the net-centric system. They scored better on all constructs with the exception of *relevance*. This is mainly because the filters for personalization of information system were still too complicated to use.

The outcomes of perceptions on information quality IQ in the various scenarios are presented in Tables 7, 8, 9, 10, 11. We find that the internal consistency of the various items to measure IQ dimensions is, on average, satisfactory (Cronbach's alpha is larger than 0.7 in a clear majority of the cases). With the exception of Scenario 2, the test group reports higher information quality dimensions than the control group. However, given the small number of participants, the differences are in most cases not significant. Timeliness is the dimension with the best score in the comparison between test group and control group.

To measure SQ we used nine constructs. To validate each construct we asked two or three questions at the end of all scenarios. *Accessibility* was the only system related construct that scored significantly higher in the test group. *Response time* and *system reliability* scored higher in the control group. This is mainly because we used an Internet version of the net-centric application. The system had some trouble in the

Table 7 IQ for scenario 1a

Scale*	Items	Average Test group N= 8	Average Control group N=6	Indication Reliability (Cronb. α)*	Test value	*p*
Timeliness1a	3	3.9	3.4	0.84	1.57	0.173
Correctness1a	3	3.4	3.6	0.71	-0.55	0.594
Completeness1a	3	3.3	3.2	0.88	0.26	0.798
Relevance1a	2	3.8	3.4	0.75	1.12	0.286
Consistency1a	3	3.4	3.6	0.76	-0.45	0.569
Overload1a	3	3.2	3.2	0.34	0.13	0.900
Verification1a	3	3.2	3.6	0.61	-1.05	0.316
Total 1a	20	3.4	3.4		0.06	0.950

Table 8 IQ for scenario 1b

Scale*	Items	Average Test group N = 8	Average Control group N = 4	Indication Reliability (Cronb. α)*	Test value	*p*
Timeliness1b	3	3.0	2.8	0.52	0.47	0.651
Correctness1b	2	2.8	2.4	0.75	0.85	0.418
Completeness1b	3	2.7	2.3	0.81	0.97	0.356
Relevance1b	3	3.5	3.1	0.79	1.02	0.333
Consistency1b	2	2.5	2.8	0.74	-0.45	0.663
Overload1b	3	3.4	3.3	0.23	0.29	0.775
Verification1b	2	3.5	3.0	0.73	0.82	0.433
Total 1b	18	3.1	2.8		1.46	0.176

Table 9 IQ for scenario 2

Scale*	Items	Average Test group N = 9	Average Control group N= 7	Indication Reliability (Cronb. α)*	Test value	*p*
Timeliness2	3	3.4	2.9	0.43	2.12 sign.	0.053
Correctness2	2	3.4	3.5	0.74	-0.41	0.686
Completeness2	3	2.1	2.8	0.81	-1.78 sign.	0.096
Relevance2	2	3.1	3.4	0.77	-0.77	0.454
Consistency2	3	3.5	3.7	0.65	-0.55	0.592
Overload2	3	3.3	3.4	0.54	-0.34	0.738
Verification2	3	3.5	3.2	0.58	0.80	0.438
Total2	19	3.2	3.2		-0.32	0.751

Table 10 IQ for scenario 3

Scale*	Items	Average Test group N = 7	Average Control group N = 6	Indication Reliability (Cronb. α)*	Test value	*p*
Timeliness3	3	3.6	2.9	0.82	1.73	0.113
Correctness3	3	3.3	2.9	0.92	0.83	0.426
Completeness3	2	3.0	2.9	0.78	0.40	0.701
Relevance3	3	3.5	3.2	0.16	1.02	0.328
Consistency3	3	3.3	3.2	0.93	0.14	0.888
Overload3	2	3.3	3.4	0.86	0.10	0.923
Verification3	2	3.1	3.6	0.87	-1.05	0.361
Total3	18	3.3	3.1		0.77	0.463

Table 11 IQ for scenario 4

Scale*	Items	Average Test group N= 6	Average Control group N= 4	Indication Reliability (Cronb. α)*	Test value	*p*
Timeliness4	3	3.9	3.3	0.85	1.45	0.185
Correctness4	3	3.8	3.6	0.87	0.52	0.618
Completeness4	3	3.7	3.3	0.54	1.15	0.284
Relevance4	2	3.4	4.3	0.83	-1.57	0.156
Consistency4	3	3.9	3.6	0.83	0.77	0.461
Overload4	3	3.5	3.3	0.65	0.39	0.706
Verification4	3	3.4	3.3	0.71	0.22	0.834
Total	20	3.9	3.7		0.72	0.490

Note Green, test group performed better

performance. This is a technical issue which can be easily solved. For the task-related SQ constructs the test group scored significantly higher on *integration*, *memory* and *SA*. Only the construct *format* scored significantly lower. This is strongly related to IQ constructs *overload* and *verification*. Personalization seemed to be a key issue in an information-rich environment. For perceived operational satisfaction we measured two constructs. The test group found the system complicated to use. The SQ construct *ease of use* scored significantly lower in the test group. However, a learning effect was visible. The test group started to perform relatively better after each scenario. *Usability* was scored significantly better in the test group. Here we can conclude that the test group recognized the value-added service of a net-centric system, but that they still perceived it as complex to use.

An important other issue is that, although the field-workers in the test group had access to the net-centric system, and had 1 day training before the field exercise, they did not use the systems. They only used the phone with their own centralist.

Table 12 SQ for all scenarios

Scale	Items	Average Test group N = 7	Average Control group N = 4	Indication Reliability (Cronb. α)*	Test value	*p*
System related						
Accessibility	2	3.5	2.3	0.88	4.03	0.003 **
Reliability	3	2.4	2.9	0.70	-1.17	0.274
Response time	2	2.5	3.2	0.76	-1.19	0.266
Task related						
Integration	5	3.5	2.6	0.82	3.71	0.005**
Memory	2	3.1	2.1	0.62	3.83	0.004**
Format	3	2.0	2.6	0.58	-2.35	0.044*
Sit. awareness	3	3.4	2.2	0.80	5.64	0.000**
Perceived operational satisfaction						
Ease of use	3	2.5	3.4	0.84	-2.25	0.052*
Usability	3	3.3	2.5	0.83	2.53	0.032*
TotalSQ	27	2.9	2.7			

* significant at a level of 0.05
** significant at a level of 0.01

This situation is similar with the daily IM handling. We may conclude there is clearly more need to integrate such systems in their work processes. This confirms that net-centric systems for traffic IM need to be introduced in different stages, as described by Steenbruggen et al. [36]. The introduction of these concepts are extremely difficult, and short-term strategies are doomed to fail [26]. This means that, for a successful adoption, these concepts need to be carefully introduced. A logical choice is the user-perspective. It makes sense to start with those persons who are controlling and coordinating the response and recovery processes. They are those who will attain and maintain an accurate, shared COP and SA, as stated by Harrald and Jefferson [26]. For traffic IM this means the traffic management centre and the centralist of the emergency rooms.

System quality assessments are reported in Table 12. The Cronbach alpha results show that the internal consistency of the items is reasonable to good. The differences in the assessments of the test group and the control group is rather large and significant in most cases. According to five of the SQ dimensions, net-centric working has clear advantages above current routines: accessibility, integration, memory, situational awareness, and usability. For two dimensions disadvantages are reported: format and ease of use.

4.2 Shadowing and System Logs

The main findings of each individual scenario based on the detailed observations recorded by the 'shadowing' evaluation process and the logs created by the participants are summarized in Table 13. Each participant was shadowed by an

Table 13 Data on group performance were collected from observer notes (Shadowing) and system logs

	Scenario 1a		Scenario 1b		Scenario 2		Scenario 3		Scenario 4	
	Test group	Contr. group	Test group	Contr. group	Test group	Contr. group	Test group	Contr. group	Test group	Contr. group
Incident notification information										
First incident notification	0	0	0	1	0. (2e 34)	0. (2e 36)	0	0	0	5
Incident location known	1	3	5	9	2	4	0 (12)	0 (9)	2	11
Type of incident known	1	3	0	1	2	4	23	32	7	11
Number of vehicles known	1	4	0	1	2	4	23	32	6	24
Number of victims known	1	3			7	12	31	31	24	-
Involvement of dangerous goods known	6 (14)	9 (-)			3	12	12	45	7	12
Environmental information										
Environmental consequence	12	6	14	-	8	7	Changed	Changed	8	11
Safety measures applied (500 m.)	8	Not relevant							18 (25)	11 (14)
First decision lanes closed	12	7			8	8	5	8	8	9
Decision close underlying network									18	14
Decision close entire road					15 unnecessary	9	16	25	14	14
Decision alternative routes					15 unnecessary	8	4	20	25	34
Organisation and coordination information										
Road inspector at incident the location	11	6	11	-	8	10	7	11	12	12
RWS Officer of Duty at the incident location	16	10			11	14	16	16	12	18
Police at the incident location	15	6			6	17	8	11	6	12
Fire brigade at the incident location	14	18 unnecessary			12	9	8	11	6	11
Ambulance at the incident location					12	9	No ambulance	18 order	12	12
Towing car at the incident location			8	-	37 (2e incident.)	29 unnecessary	10	27 2e 28	26	30
Towing truck at the incident location	14 (2e 21)	24 (2e 26)			20	28	33	46	-	-
ANWB at the incident location			8 (15)	10 wrong loc.						
Trauma helicopter					29 ordered	17 ordered	28 ordered 41	No helicopter		
Demand additional transport					42	47				
Demand environmental expert	9	23								
Demand for STI expert					14	10				
First COPI meeting					No COPI	21	34 Motorkap	38 COPI	14	22
COPI escalating GRIP 1					22	33	21 GRIP 1	23 GRIP2		
COPI escalating GRIP 2/3							28 GRIP 2		17	22
COPI conclude safety guaranteed									19	28
COPI conclude no treat for explosion									34	31
ROT operational									25	27
Cause known no camera images available									2	4
Camera images helicopter available									15	-
Insufficient water for fire brigade									7	25
Rest time incident	0	9	16	16	20	17	-	-	22	34

Note Green faster results test group, *Yellow* equal results both groups, *Red* faster results control group, *Orange* different interpretation scenario, *Grey* not relevant

individual observer. This information was used to reconstruct a detailed overall process description. The table is divided into three main groups: incident notification; surrounding environment consequences of the incident, and; organization and coordination activities of the emergency services involved.

4.3 Scenario Evaluation

In this section we describe the main findings concerning the differences between the test group and the control group for each scenario.

Scenario 1a (GRIP 0)

The activities of the two groups had some significant differences. In general, the incident notification was available a few minutes earlier in the test group. The test group recognize that there was the possibility of an incident involving dangerous goods and decided to keep a safe distance of 500 metres till the Fire Brigade decided that the cargo of the truck was safe. The Fire Brigade arrived after 15 min and cverified that there was no direct danger. In the control group, after 10 min it was established that there were no dangerous goods involved. The source of this information remained unclear. However, at that moment there has already been a fire brigade under way to the incident. The road inspector reports this to the traffic management centre. However, after 18 min the fire brigade of the control group arrives unnecessary at the incident scene, the fire brigade wasn't needed. In a shared information environment this would have been known by the GMK. In the test group all organizations take measures and decisions based on shared information. They follow the procedure for trucks with dangerous goods. The different log and shadow information confirms this picture. However, in the control group the information on dangerous goods was shared by 1 on 1 communication. This led to confusion, wrong conclusion and unnecessary measures and activities. In the test group the CMV towing services ask after 9 min with the netcentric system for a environmental expert because the truck might have dangerous goods. In the control group, CMV towing service makes after 23 min the same request. The test group has access to more information and so makes this request 13 min earlier. In the test group the towing service arrives 10 min earlier at the incident than the control group. The request for a second towing vehicle is 5 min earlier in the test group

Scenario 1b (GRIP 0)

This is a simple scenario about a breakdown vehicle. Initially, the wrong location of the incident scene is communicated between the driver and the ANWB dispatch centre. The traffic management centre sees a traffic jam on the other side of the road and contacts ANWB, LCM towing services and the road inspector. There are no cameras available but based on the detection loop data and the traffic management information system the conclude that the location of the incident happened on the other side of the road. The correct location of the incident is detected 4 min earlier by the test group. Due to technical problems with the system this scenario has to be stopped half way.

Scenario 2 (GRIP 1)

The detection and driving phase is almost identical between the two groups in the first couple of minutes of the incident. The information about involved victims is available 5 min earlier in the test group. The CMV towing service reports about possible dangerous goods after 3 min. This stays unclear for a long time in the test group. After 12 min there is an indication about dangerous goods, however, during the process there is no communication anymore about this subject. All emergency services arrive earlier at the incident scene with exception of the fire brigade and ambulance. They arrive only a few minutes later. The test group escalates to GRIP1 12 min later. During the handling process there is a second traffic incident which causes much confusion. The towing service for the second incident is informed after 14 min, however with the wrong location information. After 27 min the test group is informed about the second incident, but with the good location. Ten minutes later the towing service of the test group arrive at the incident. The towing service of the control group is unable to find the incident and returns with a false incident notification. The safety screen is after 13 min at the incident location. In the control group the safety screen is moved to the second incident. They confuse information about the first incident with the second incident. Decisions on guided transport after the second incident is 5 min earlier in the test group. After 52 min the test group takes measures on alternative driving routes. The control group made no traffic management measures.

The main issue in the test group was information overload. This was the first complex scenario. They had many difficulties in using the system. They used the telephone to verify text messages in the net-centric system. Next to that, the text message were too long and contained many specific terminology that was only known by specialist of some organizations. This also causes much confusion in communication. After the evaluation of this scenario participants improved the text messages for the next scenarios. This was the main lesson learned form this scenario. Also, both groups had major difficulties handling more incidents in the same scenario. However, in the overall results of the decisions and outcomes, the test group performed slightly better.

Scenario 3 (GRIP 2/3)

There were some major differences in how the both groups handled the incident. In the test group all organizations are informed about the best driving route (since regular routes are blocked), in the control group this is only communicated after 21 min. In the control group there is hardly any communication about a truck fire and the status of the fire. Within the test group there is frequent communication about this subject. In the control group the severity of the fire was never communicated. This was requested several times by the truck towing organization (CMV). The escalating to GRIP2 is 3 min earlier in the test group. The fire was under control 11 min earlier in the test group. Because roads were blocked, the test group requested a trauma helicopter after 28 min, which arrived at the incident scene after 13 min. The control group requested an ambulance after 18 min. However, they were not aware of the difficulty to arrive at the incident scene. In the test group a picture of the incident situation is shared. This helps emergency services to get a good overview of the incident. For example, there is detailed information available about lost cargo on the road. This is never shared within the control group. Clearly, the test group learned from the previous scenario. There was less telephonic communication, text message were more compact and only contained the relevant information. Also, pictures were shared to communicate about the impact of the incident.

Scenario 4 (GRIP 3)

This is a full scenario were all emergency services and the safety region are involved. The benefits of a net-centric system are clearly visible in this scenario. Also the experience of previous scenarios helped the test group to improve their performances. The incident detection information is a couple of minutes earlier available in the test group. Especially information about number of involved victims and dangerous goods caused some troubles. Also, the exact incident location was known 5 min earlier in the test group. Besides that, the control group assumed that the incident was on the wrong side of the highway.

Information about the impact of the environment also had some difficulties. The test group had a quicker overview, however, the 500 m safety measures were quicker applied in the control group. They were also 4 min earlier to close the underlying road network. The emergency service in the test group arrived equally or some minutes quicker on the incident scene. In almost all cases, the coordination activities of the emergency services were better in the test group. The first COPI meeting was 8 min earlier. They detected, for example, 18 min earlier that there was not enough water on the incident scene for the fire brigade. They also confirm 9 min earlier that the incident scene was safe. The overall conclusion of this scenario is that the test group performed significantly better on almost all aspects.

5 Discussions

Applying net-centric information concepts is a promising solution for improving cooperation between private and public emergency services. They may provide useful tools in the daily handling of traffic IM. The main goal is to improve SA which contributes to faster and effective collaborative decision making. However, the research assessing about the effectiveness of these decision support tools is still ongoing. To date, there are no concrete guidelines and design principles in the literature of net-centric systems for traffic IM. Net-centric information systems find its roots in the military domain. In recent studies these concepts have also been applied in disaster management. Traffic IM can be seen as a special form of disaster management. However, literature in the public sector on emergency services and traffic IM regarding information sharing across different agencies and the quality of information sharing is scarce and empirical support is almost nonexistent. This study takes a step forward by evaluating the effectiveness of net centric systems between two groups. This evaluation is based on a framework that includes tests of usefulness of the tools, information quality and system quality. In drawing conclusions, this section discusses the results on the basis of three aspects: first, the comparison between communication and coordination of emergency organizations; second, the value of SA in the performance of the decision making process; and, third, how scenario complexity can effect design principles of netcentric systems.

5.1 Communication, Coordination and Performance Decision Making Process

The *second* hypothesis was: “SA improve the performance in the decision making process of emergency organizations”. Testing this hypothesis involved comparing the observed outcomes (shadowing) of each scenario between the test group and the control group. To validate this hypothesis we focus on the speed of completeness of incident notification information and how fast emergency services arrive at the incident location. Table 14 provides an overview of the sum of minutes gained in the test group. Only the ambulance was 3 min later at the incident location in scenario 3. In scenario 4 the test group used a trauma helicopter. In scenario 5 the ambulance arrived at the same time. We can conclude that the net-centric group in general performed better in the scenarios.

Table 14 Sum of minutes gained in information sharing and coordination for all scenarios

Incident notification information (min)	Coordination and performance (min)
First notification : 7	WIS: 1
Location : 17	OvD: 3
Type : 16	Police: 11
Vehicles : 33	Fire brigade: 9
Victems : 7	Ambulance: −3
dangerous goods : 49	Towing cars 13
	Towing trucks: 41
	ANWB: 2

Note 1 Not all information categories are relevant in the each scenario
Note 2 Not all emergency services play a role in each scenario

Table 15 Current identified problems for communication and coordination for traffic IM

	Accuracy	Availabilit	Complete	Consisten	Correctne	Format	Personalis	Relevancy	Reliability	Timeliness
Communication and coordination issues										
E112 informs different centers which starts separate uncoordinated processes				■						
police sometimes have no capacity after been informed by TMC										■
communication about opening closed or blocked lanes (Police – TMC)		■								
no (time) information available when emergency services arrive at incident		■								
knowing status and real-time location emergency services		■								
resource information not always available for towing services for RWS			■							
relatively many unnecessary towing trips (false incident notification)									■	
different centralists do not communicate with each other										
central police communicate with regional police by E112 control room										■
sometimes fire brigade is informed too late										■
information about incident is sometimes communicated too late to RWS										■
information to TMC is sometimes wrong, incomplete and unclear	■		■		■					
incident detection via (0900-8844) is not always known by TMC		■								
no uniform incident definition and registration				■						
sometimes no registration, RIS needs to explain situation multiple times to TMC		■								
same incident registered independently by all involved organizations				■						
communicate only relevant information to emergency services						■	■			
communication between TMC and RIS not optimal due capacity problems										■
information between TMC and RIS sometimes incorrect and own interpretation	■							■		
information notification provided by the police is often very summier			■			■	■			
more and better information sharing during driving phase										■
communication only by phones causes misinterpretation					■	■			■	
mobile phones sometimes fail by system problems (coverage/capacity/accu)		■								■
webcam /video images could provide useful information for all actors		■	■							
sometimes the first safety measure are not appropriately applied										
direct involvement of TMC helps to ensure safety incident location										■

Source Ministry of Water Management and Transportation [2012]

The *third* hypothesis was: Net-centric systems improve the communication and coordination of emergency organizations. The current identified problems for communication and coordination for traffic IM are summarized in Table 15. These were collected during 10 regional evaluation meetings with Rijkswaterstaat (road inspector, road traffic coordinator, traffic officer of duty) and personnel from the police, fire brigade and ambulance service. Together with an evaluation team, they replayed past incidents step by step. Each incident was evaluated in great detail and then recommendations for improvement were clustered. For the purpose of this paper the categories which focus specifically on improvements for information, communication and coordination are used.

In the first two scenarios there were hardly any complex communication and coordination issues identified. Main difficulties were related to identify environmental consequences such as traffic management measures. Here the control group slightly performed better in some aspects. In the more complex scenarios hands-on experience helped the test group to perform better. In the more complex scenarios there was a strong need for sharing information. As observed in the first scenarios, the participants still use mainly telephone communication. Especially in scenario 2, they had strong difficulties combining the many text messages with telephone communication. This made coordination activities complicated. However, in the last two scenarios participants start to have experience writing compact messages. Also, we observed that the frequency of telephone communication decreases and participants start to rely more on the net-centric system. For scenarios 3 and 4, it is clearly visible in Table 13, that the test group performed significantly better. We can conclude that the net-centric system clearly improved communication and coordination activities of the test group.

5.2 Design Principles of Net-Centric Systems

The *fourth* hypothesis was: “Scenario complexity affect the design principles of net-centric systems”. The main goal of net-centric working is to improve SA by a Common Operational Picture. The criteria to design information systems which fits the needs and benefits of end users are more than just technology. Systems must be designed that insure information needs of centralist and provide tools that support the cognitive and psychological capabilities, especially in an information rich and dynamic environment. Several factors will influence the accuracy and completeness of SA. Humans are limited by working memory and attention. New information from multiple sources must be integrated with other knowledge. How people direct their attention in acquiring new information has a fundamental impact on which elements are incorporate in their SA. Jones and Endsley [38] found that the most caused error (35 % of all SA errors) was that al information was present but was not noted by the operator. The limits of working memory also cause constraints on SA [56]. Net-centric tools must be designed to support work memory and attention. This is strongly related to information overload. Most detected problems in our field exercise measuring IQ in the more simple GRIP0 scenarios, was related to *consistency* of information. This means that only a small amount of information was shared. The work memory of the participants could handle the information flow and could easily judge the (in) consistency of the data. Telephone communication plays still an important role here. In the more complex scenarios (2, 3 and 4) participants had most problems with *relevance*, *overload* and *verification* of information. This is directly related to system quality constructs. Participants in the test group were pleased that the system supports the *accessibility* and *integration* of many data. They also scored higher in the task related construct *memory*. However, the construct *format* was clearly not used and designed to avoid information overload, help their work memory and support their attention. This is partly due the lack of little or now experience of the participants with net-centric systems; however we observed a learning effect during the scenarios. Clearly, more complex incidents need to have appropriate formats which are specially designed for different types of incidents (GRIP 0) and the more higher GRIP incidents (GRIP 1–4). Supporting long term memory can be achieved by creating memory functionality for later data retrieval. Formats need to be more personalized to the specific goals and tasks of each organization and the different roles within the organizations. Related to format, the nature of the information and its presentation also causes problems for end users. Creating SA is more than just simply reading 'dots' on maps [7]. It is about understanding the significance of those information in a operational context and decision making process. The traditional COP does not support these aspects of situational awareness, but leaves this cognitive load on the user [72, 84]. A more effective approach to shared situation awareness for NCW is to be able to push and pull the story behind the data, and not just the underlying data [8, 9]. These are the main reasons that the system is perceived as complex. The IQ construct *times lines* and SQ constructs *situational awareness* and *usability* scored higher in the test group. This means that net-centric system is perceived as *useful* but clearly there is a need to improve some technical system functionality to support IQ for daily use.

References

1. M. Adams, Y. Tenney, R. Pew, Situation awareness and the cognitive management of complex systems. Hum. Factors **37**(1), 85–104 (1995)
2. D.P. Ballou, H.L. Pazer, Modeling completeness versus consistency tradeoffs in information decision contexts. IEEE Trans. Knowl. Data Eng. **15**(1), 240–243 (2003)
3. G. Bedny, D. Meister, Theory of activity and situation awareness. Int. J. Cogn. Ergon. **3**(1), 63–72 (1999)
4. N. Bharosa, B. Van Zanten, A. Zuurmond, J. Appelman, (2009). Identifying and confirming information and system quality requirements for multi-agency disaster management. in ed. by J. Landgren, U. Nulden, B. Van De Walle Proceedings of the 6th International Conference on Information Systems for Crisis Response and Management (ISCRAM)
5. C. Christopher, B. Robert, *Disaster: Hurricane Katrina and the Failure of Homeland Security* (Macmillan Publishers, London, 2006)
6. C.E. Shannon, W. Weaver, *The Mathematical Theory of Communication* (University of Illinois Press, Urbana, 1949)

7. D. Lambert, J. Scholz, A dialectic for network centric warfare. in *Proceedings of the 10th International Command and Control Research and Technology Symposium (ICCRTS)*, MacLean, VA, 13–16 June 2005
8. D.A. Lambert, Grand Challenges of Information Fusion. in *Proceedings of the 6th International Conference on Information Fusion* Cairns, Australia, 2003
9. D.A. Lambert, Situations for situation awareness. in *Proceedings of the 4th International Conference on Information Fusion.* Montreal, Canada, 2001
10. T.H. Davenport, L. Prusak, *Working Knowledge.How Organizations Manage What They Know* (Harvard Business School Press, Boston, 1998)
11. S. Dawes, A. Creswell, B. Cahan, Learning from crisis: lessons in human and information infrastructure from the World Trade Center response. Soc. Sci. Comput. Rev. **22**(1), 52–66 (2004)
12. H. de Bruijn, One fight, one team: the 9/11 commission report on intelligence, fragmentation and information. Public Adm. **84**(2), 267–287 (2006)
13. W. Delone, E. McLean, Information systems success: the quest for the dependent variable. Inf. Syst. Res. **3**(1), 60–95 (1992)
14. Dutch Ministry of Transportation and Water Management, Red blue booklet, the roles of Emergency services in Incident Management in the Netherlands. ISBN 90-369-0097-2 (2005)
15. E.C. Adam, Fighter cockpits of the future. in *Proceedings of 12th IEEE/AIAA digital avionics systems conference (DASC),* pp. 318–323 1993
16. ECIS, Analyzing the effect of security on information, quality dimensions. 17th European conference on information systems, Verona, research paper, ECIS2009-0688.R1, (2009)
17. M.R. Endsley, Toward a theory of situation awareness in dynamic systems. Hum. Factors **37**(1), 32–64 (1995)
18. M.R. Endsley, D.G. Garland, *Situation Awareness. Analysis and Measurement* (Lawrence Erlbaum Associates, Publishers, Mahway, 2000)
19. M.R. Endsley, E.O. Kiris, The out-of-the-loop performance problem and level of control in automation. Hum. Factors **37**(2), 381–394 (1995)
20. M. Eppler, *Managing Information Quality. Increasing the Value of Information in knowledge-intensive Products and Processes* (Springer, Berlin, 2003)
21. F. Townsend, P. Rapuano, J.B. Bagnal, M.L. Malvesti, K.M. Nielsen, et al, The Federal Response to Hurricane Katrina: Lessons Learned. Retrieved from http://www.whitehouse.gov/reports/katrina-lessons-learned.pdf (2006)
22. C.W. Fisher, B.R. Kingma, Criticality of data quality as exemplified in two disasters. Inf. Manag. **39**(2), 109–116 (2001)
23. G. Hone, L. Martin, R. Ayres, Awareness—does the acronym "SA" still have any value? in *Proceedings of the 11th International Command and Control Research and Technology Symposium,* Coalition Command and Control in the Networked Era. held at Cambridge UK. Washington, 2006
24. J. Gibson, J. Orasanu, E. Villeda, T.E. Nygren, (1997). Loss of situation awareness: causes and consequences. in ed. by R.S. Jenssen, R.L.A. Proceedings of the eight international; symposium on aviation psychology, Colombus Ohio State University, pp. 1417–1421
25. H.A. Simon, Theories of bounded rationality. in ed. by C.B. McGuire, Roy Radner, *Decision and Organization*, (North Holland Publishing Company, New York, 1972) pp. 161–176
26. J. Harrald, T. Jefferson, (2007). Shared Situational Awareness in Emergency Management Mitigation and Response. in *Proceedings of the 40th Hawaii International Conference on System Sciences*, 1530-1605/07, 2007 IEEE
27. L. Harvey, D. Green, Defining quality. Assess. Eval. High. Educ. **18**(1), 9–34 (1993)
28. H.J. Scholten, R.J. van de Velde, N. van Manen, *The role of Geo-ICT and Spatial Approaches in Science* (Springer, Dordrecht, 2009)
29. H.J. Scholten, S. Fruijter, A. Dilo, E. Van Borkulo, Spatial data infrastructures for emergency response in the Netherlands. in ed. by S. Nayak, S. Zlatanova, *Remote sensing and GIS technologies for monitoring en prediction of disasters* (Springer, Berlin, 2008) pp. 179–197
30. Homeland Security, *National Incident Management System*, Dec 2008

31. T. Horan, B. Schooley, Time-critical information services. Commun. ACM **50**(3), 73–78 (2007)
32. K.T. Huang, Y.W. Lee, R.Y. Wang, *Quality Information and Knowledge Management* (Prentice Hall, Upper Saddle River, 1999)
33. L.H. Immers, Guide to professional Incident Management, TNO-rapport 2007-D-R1242/B. Wegwijzer naar professioneel Incident Management, Nov 2007
34. W.H. Delone, E.R. McLean, Information systems success: a ten-year update. J. Manag. Inf. Syst. /Spring **19**(4), 9–30 (2003) (E-Commerce Success Metrics)
35. J. Schrijver, B. Immers, M. Snelder, R. de Jong, Effecten van de landelijke invoering van incident management maatregelen op de voertuigverliestijd in het netwerk, TNO Mobility and Logistics (in Dutch, 2006)
36. J. Steenbruggen, P. Nijkamp, J. Smits, M. Grothe, Traffic incident management, a common operational picture to support situational awareness of sustainable mobility. Int. J. Transp. Econ. **XXXIX**(1), 131–170 (2012)
37. J.M.M. Neuvel, H.J. Scholten, A. van den Brink, From Spatial Data to Synchronised Actions: The Network-centric Organisation of Spatial Decision Support for Risk and Emergency Management, Applied Spatial Analysis and Policy, 2011
38. D.G. Jones, M.R. Endsley, Sources in situation awareness errors in aviation. Aviat. Space Environ. Med. **67**(6), 507–512 (1996)
39. J.R. Evans, W.M. Lindsay, *The management and control of quality*. 6th edn. (South-Western/ Thomson Learning, Cincinnati, 2005)
40. J.M. Juran, A.B. Godfry, *Juran's quality handbook*, 5th edn. (McGraw-Hill, NewYork, 1999)
41. J.M. Juran, F.M.J. Gryna, R.S. Bingham, *Quality Control Handbook*, 3rd edn. (McGraw-Hill, New York, 1974)
42. G. Klein (2000) Analysis of situation awareness from critical incident reports. in: ed. by M.R. Endsley, D.J. Garland, *Situation awareness analysis and measurement* (Lawrence Erlbaum Associates, Mahwah, 2000) pp. 51–71
43. S. Knight, J. Burn, Developing a framework for assessing information quality on the World Wide Web. Inf. Sci. **8**, 159–172 (2005)
44. V.L. Knoop, Road Incidents and Network Dynamics Effects on driving behaviour and traffic congestion. Dissertation 2 December 2009, Technische Universiteit Delft, 2009
45. M. Kouwenhoven, H. Siemonsma, R. van Grol, Voertuigverliesuren door incidenten, report PM-2066-AVV (in Dutch, 2006)
46. L. Comfort, N. Kapucu, Inter-organizational coordination in extreme events: the World Trade Center attacks, September 11, 2001. Nat. Hazards, **39**(2), 309–327 2006
47. Y.W. Lee, D.M. Strong, B.K. Kahn, R.Y. Wang, AIMQ: a methodology for information quality assessment. Inf. Manag. **40**(2), 133–146 (2002)
48. Y. Lee, N. Bharosa, J. Yang, M. Janssen, H. Rao, Group value and intention to use—a study of multi-agency disaster management information systems for public safety. Decis. Support Syst. **50**(1), 404–414 (2011)
49. N.Y. Li, K.C. Tan, M. Xie, Factor analysis of service quality dimension shifts in the information age. Managerial Auditing J. **18**(4), 297–302 (2003)
50. L.J. Cronbach, Coefficient alpha and the internal structure of tests. Psychometrika **16**(3), 297e334 (1951)
51. M.B. Parker, V. Moleshe, R. De la Harpe, G.B. Wills, An evaluation of Information quality frameworks for the World Wide Web. In: *8th Annual Conference on WWW Applications*, Bloemfontein, Free State Province, South Africa, 6–8th Sept 2006
52. C. McDaniel, R. Gates, *Contemporary marketing research* (Sams & Co., New York, 1998)
53. McKinsey and Company, *File arm wegbeheer (Eng: Tailback free traffic management)* (Ministerie van Verkeer en Vervoer, The Hague, 1995)
54. H. Miller, The multiple dimensions of information quality. Inf. Syst. Manag. **13**(2), 79–82 (1996)

55. M.J. Eppler, J. Mngis, A framework for information overload research in organizations. Insights from organization science, accounting, marketing, MIS, and related disciplines. paper # 1/2003, Sept 2003
56. M.R. Endsley, Design and evaluation for situation awareness enhancement. in *Proceedings of the Human Factors Society 32nd Annual Meeting*. Human Factors Society, Santa Monica, (1988) pp. 97–101
57. N. Bharosa, Net-centric information orchestration, assuring information and system quality in public safety networks. Dissertation, Technische Universiteit Delft, 2011, ISBN 978-90-8891-231-3
58. National Research Council, *Successful Response Starts With a Map. Improving Geospatial Support for Disaster Management* (The National Academies Press, Washington, 2007)
59. R.R. Nelson, P.A. Todd, B.H. Wixom, Antecedents of information and system quality: an empirical examination within the context of data warehousing. J. Manag. Inf. Syst. **21**(4), 199–236 (2005)
60. M. O'Leary, *Measuring Disaster Preparedness: A Practical Guide to Indicator Development and Application* (iUniverse, Lincoln, 2004)
61. P. Markopoulos, B. De Ruyter, W. Mackay, *Advances in Theory, Methodology and Design Series: Human–Computer Interaction Series,* Springer, 1st edn., vol. XVI, ISBN 978-1-84882-476-8, 2009, p. 470
62. P. Salmon, N. Stanton, G. Walker, D. Green, *Situation Awareness Measurement: A review of applicability for C4i environments.* (Defence Technology Centre for Human Factors Integration (DTC-HFI) Brunel University, 2006)
63. P. Singh, P. Singh, I. Park, J. Lee, H.R. Rao, Information Sharing: A study of information attributes and their relative significance during catastrophic events, (2007)
64. P. Singh, P. Singh, I. Park, J. Lee, H.R. Rao, Information sharing: a study of information attributes and their relative significance during catastrophic events, in ed. by J.K. Kenneth, Cyber Security and Global Information Assurance: Threat Analysis and Response Solutions IGI Global, 2009
65. W. Perry, D. Signori, J. Boon, *Exploring Information Superiority: A Methodology for Measuring the Quality of Information and its Impact on Shared Awareness* (The RAND Corporation, Santa Monica, 2004). ISBN 0-8330-3489-8
66. E.L. Quarantelli, Problematical aspects of the information/communication revolution for disaster planning and research: ten non-technical issues and questions. Disaster Prev. Manag. **6**(2), 94–106 (1997)
67. R. Breton, R. Rousseau, Situation awareness: a review of the concepts and its measurement. Defence Research and Development Canada—Valcartier, Technical Report. DRDC Valcartier, 2003, TR-2001-220. 2003-02-05
68. R.O. Mason, Measuring information systems output: a communication systems approach. Inf. Manag. **1**(5) (1978), pp. 219–234
69. R.W. Pew, The State of Situation Awareness Measurement: Heading Toward the Next Century, in ed. by M. R. Endsley, D.J. Garland, *Situation Awareness Analysis and Measurement* (Lawrence Erlbaum, Mahwah, 2000), pp. 33–47
70. J. Ryoo, Y.B. Choi, A comparison and classification framework for disaster information management systems. Int. J. Emergency Manage. **3**(4), 264–279 (2006)
71. S. Mulgund, S. Landsman, User defined operational pictures for tailored situation awareness. in *Proceedings of the 12th international command and control research and technology symposium—Adapting C2 to the 21st century,* Newport, RI, 19–21 June 2007
72. S. Wark, D.A. Lambert, Presenting The Story Behind The Data: Enhancing Situational Awareness Using Multimedia Narrative. in MILCOM 2007, IEEE, Orlando, 2007
73. R. Samarajiva, Mobilizing information and communications technologies for effective disaster warning: Lessons from the 2004 tsunami. N. Media Soc. **7**(6), 731–747 (2005)
74. GRIP, Scaling up reference framework (GRIP). 2006. Circular, Ministry of the Interior. http://www.minbzk.nl/actueel?ActItmIdt=99130. Accessed 11 Sept 2006

75. K. Smith, P.A. Hancock, Situation awareness is adaptive, externally directed consciousness. Hum. Factors **37**(1), 137–148 (1995)
76. N.A. Stanton, P.R.G. Chambers, J. Piggott, Situational awareness and safety. Saf. Sci. **39**(3), 189–204 (2001)
77. N.A. Stanton, P.M. Salmon, G.H. Walker, C. Baber, D. Jenkins, *Human factors methods: a practical guide for engineering and design* (Ashgate, Aldershot, 2005)
78. D.M. Strong, Y.W. Lee, R.Y. Wang, Data quality on context. Commun. ACM **40**(5), 103–110 (1997)
79. B. van de Walle, M. Turoff, Emergency response information systems: emerging trends and technologies. Commun. ACM **50**(3), 29–31 (2007)
80. R. Van Reisen, Incidentele files, de kenmerken, de kosten en het beleid, SEO Discussion paper No. 50 (in Dutch, 2006)
81. W. Delone, E. McLean, The DeLone and McLean model of information systems success: a ten-year update. J. Manag. Inf. Syst. /Spring **19**(4), 9–30 (2003)
82. Y. Wand, R.Y. Wang, Anchoring data quality dimensions in ontological foundations. Commun. ACM **39**(11), 86–95 (1996)
83. R.Y. Wang, D.M. Strong, Beyond accuracy: what data quality means to data consumers. J. Manag. Inf. Syst. **12**(4), 5–34 (1996)
84. S. Wark, D. Lambert, M. Nowina-Krowicki, A. Zschorn, D. Pang, Situational Awareness: Beyond Dots On Maps To Virtually Anywhere (2009)
85. C.D. Wickens, Situation Awareness and Workload in Aviation". Curr. Directions Psychol. Sci. **11**(4), 128–133 (2002)
86. I.R. Wilmink, L.H. Immers, Deriving Incident Management Measures Using Incident Probability Models and Simulation, Trans. Res. Rec, **1554**,184 (1996)
87. B.H. Wixom, P.A. Todd, A theoretical integration of user satisfaction and technology acceptance. Inf. Syst. Res. **16**(1), 85–102 (2005)

A Customizable Maturity Model for Assessing Collaboration in Disaster Management

Jaana Mäkelä and Kirsi Virrantaus

Abstract A maturity model can be used as a tool to assess the current state of competence, to set a roadmap for improvement and to assess the effects of development of collaboration in disaster management. Customized maturity models were developed for two disaster management exercises: a search and rescue exercise and an ICT exercise. The contents of the models were based on a literature review on maturity models, on the goals of the exercises, and on the interviews of the participating organizations. In the exercises, human agents performed a quick maturity assessment where they assessed the maturity level of the key capabilities of collaboration such as the sharing of critical information. Critical information such as geographic information is a prerequisite for shared situational awareness. The results of the maturity assessments were visualized on radar charts which facilitate the display and comparison of multivariate data from different exercises. They highlighted the strengths of collaboration and capabilities that need further development. The process itself: the development of the maturity model, the quick maturity assessment, and the presentation of the results is the main result of the research.

Keywords Disaster management · Collaboration · Shared situational awareness · Geographic information · Critical information · Maturity assessment

J. Mäkelä (✉) · K. Virrantaus
Department of Real Estate, Planning and Geoinformatics, Aalto University School of Engineering, PO Box 15800, FI-00076 Aalto, Finland
e-mail: jaana.makela@aalto.fi

K. Virrantaus
e-mail: kirsi.virrantaus@aalto.fi

S. Zlatanova et al. (eds.), *Intelligent Systems for Crisis Management*, Lecture Notes in Geoinformation and Cartography, DOI: 10.1007/978-3-642-33218-0_17,

1 Introduction

The probability of major accidents on land, at sea, or in air traffic is increasing continuously in our insecure world. Because of extreme natural phenomena serious disturbances in the power supply, public utilities, or transport logistics are increasing as well. Societies have to be prepared for cyber threats or serious disturbances in their telecommunication and information systems. The fast and efficient management of these disasters to reduce property losses and to minimize damage to critical infrastructure requires good interagency collaboration between both the public and the private sectors.

A vital prerequisite for successful collaboration is the shared situational awareness (SSA), which comprises the critical factors affecting a situation that the human agents of participating organizations must develop [1]. An SSA between all actors is necessary for the rapid estimation of consequences, decision making, management of resources, and communications [1, 2]. Seppänen et al. [2] have defined critical information as the minimum information that is needed for an adequate SSA. Geographic information and spatial methods can play an essential role because they enable critical information to be integrated, analyzed, and visualized [3–5]. As an example we can mention topographic map and terrain information that support the accurate positioning of the accident as well as the optimum route finding to the place. Therefore, in the rest of our paper the concept of information embodies geographic information as well.

In Finland preparedness for disaster management and collaboration capabilities are trained constantly in exercises such as search and rescue (SAR) exercises and information and communications technology (ICT) exercises. The evaluation of the results is often a qualitative report that describes verbally and generally the successes of the collaboration and aspects that should be improved in the future. The problem is that the existing evaluation methods do not clearly identify the challenges and they do not produce quantitative data for the evaluation and the development of collaboration. Therefore, more quantitative and measurable methods are needed [6]. The importance of a constant assessment of the progress of a comprehensive approach to both national and multinational crisis management has been highlighted as well [7].

Some formal methods exist, such as the comprehensive CTEF 2.0 model for measuring the effectiveness of a multi-team system [8] and maturity models for assessing and developing collaboration in network-centric environments and strategic alliances [9–13]. A maturity model is a tool to assess the current stage of competence, to set a roadmap for improvement, and to assess the effects of development. A maturity assessment identifies the strengths and weaknesses of the key capabilities of an organization and produces quantitative values as a result [14].

1.1 Maturity Models

The focus in maturity models is on the capabilities that are required at each level of collaboration. Maturity models have been developed for multi-national purposes such as the NATO Network Enabled Command and Control Maturity Model (N2C2M2) [9] for civilian-military peace-keeping operations and the Emergency Management Capability Maturity Model (EM-CMM) [12] for global emergency management operations. Kuusisto [13] combined his information exchange meta-model and N2C2M2 in order to enable information exchange processes to be developed both in national and multi-national collaboration situations. Griffin et al. [11] emphasize spatially enabled emergency management and they have developed a tool for evaluating interoperability maturity in a national network-centric environment. The Strategic Alliance Formative Assessment Rubric (SAFAR) [10] helps to develop collaboration maturity in national Safe School/Healthy School Initiatives.

The structure of a maturity model is typically organized into key areas and their sub-areas over five maturity levels [14]. The above-mentioned maturity models include five maturity levels that describe the different stages of collaboration. In the maturity models of [9, 13] the collaboration develops from the lowest maturity level, "Conflicted", to the highest level, "Edge or Agile", through "De-conflicted", "Coordinated", and "Collaborative" actions. An agile collaboration is responsive, flexible, innovative, and adaptive. Peer-to-peer interactions are rich and the degree of shared situational awareness and understanding is high. In the EM-CMM model [11] the management of an emergency situation can mature from ad hoc to standardized process management and finally to optimized change management. In the SAFAR [10] collaboration develops even further and at the fourth level the resources of organizations are merged and at the fifth level organizations are unified to form a single structure.

The actual contents of maturity models describe key areas and capabilities that are required for successful collaboration. People—that is, positive personal relations and commitment [9, 10, 12], common processes [9, 11, 12], enabling technologies [11, 12], and the sharing of information [9, 13] are emphasized. According to Kuusisto [15], human agents would like to release different information, such as their own decisions and interpretations, from the information they require from others. The N2C2M2 [9] describes well how the maturing of the structure of a coalition enhances the key capabilities of collaboration. At the "Conflicted" level organizations do not plan or train operations together, and nor do they share information. At the "De-conflicted" level organizations operate in silos and they share very limited and sharply focused information. At this stage, a coalition can manage only complicated situations, where cause-and-effect relationships are generally well understood but not complex situations, which involve changes and behaviors that cannot be predicted in detail. At the "Coordinated" level some task-specific groups that have members from different organizations are possible. Coordination processes and linked plans exist as well. However, the sharing of information is still limited and covers information about coordinated areas and functions. At the "Collaborative" level trust enables the interdependency and collaborative processes of organizations to take place. Information sharing is significantly broad. At the "Agile" level interactions are rich and continuous and information is exchanged widely. This is supported by a secure, ubiquitous, and interoperable information structure.

In a dynamic and complex disaster operation or exercise it is not possible to perform a detailed and laborious maturity assessment that needs time and resources. Therefore, a quick maturity assessment and a maturity model where each maturity level is described by a few key definitions could be useful. According to Mäkelä [16], the results of a quick maturity assessment were close to the actual detailed maturity assessments of organizations. Although the maturity values from quick assessments were slightly higher, they nonetheless indicated the right level of maturity.

1.2 Objectives of This Research

Our opinion is that it is not possible to develop one general maturity model that would be suitable for all cases. The objectives of national security strategies and disaster management exercises and organizational goals should be considered as well. Instead of a generic model, we suggest a customized maturity model that is based both on the generally identified key areas and on the specific goals.

The objective of our research was to answer the following research questions: (1) How can the content of a customizable maturity model be created? (2) How and when can maturity in an exercise be assessed? (3) How can the results be presented in such a way that the maturity assessment supports the development of interagency collaboration in disaster management in the best possible way? For this purpose two maturity models were developed, related to two cases. The first model was developed for a SAR exercise and the second model for an ICT exercise, executed in Finland in 2010 and in 2011. These two exercises and our research methods—the developments of the maturity models, the maturity assessments in the exercises, and the reporting of the results—are described in Sect. 2. In Sect. 3, we will briefly both present and discuss some observations from the results of maturity assessments focused on the development of the SSA and availability of information. Discussion about the results and conclusions from the research are provided in Sect. 4.

2 Materials and Methods

In this Chapter the two case studies are introduced: the SAR and the ICT exercises and the methods that were used in the development of the maturity models, in the maturity assessments, and in the presenting of results.

In the first case study, in the SAR exercise a passenger airplane crashed into an airport. A temporary SAR organization was established to manage the emergency operation. The organization had a pre-defined structure and it consisted of several actors, such as rescue authorities, the police, medical care services, the airline, the airport, the Red Cross, and the local authorities. The Command Body of the Area of Operation (CBAO) was responsible for the actual search and rescue actions. A Command Center of Regional Rescue Services (CCRRS) was responsible for the coordination and operational planning of the operation. A Command Body of Supporting Services (CBSS) was formed for support activities for patients and their relatives. The law obliges similar SAR exercises to be planned and carried out annually by the same group of organizations. Therefore, the systematic development of collaboration and common processes for emergency management is possible and worth doing.

The second case study was a large national ICT exercise where preparedness for disturbances in information systems and networks was trained. Organizations from the local, regional, and national administrative levels of several ministries and from the private sector took part in the exercise. Corresponding large exercises for managing possible threats that are directed towards the information society are organized regularly. However, the scenarios and organizations change in each exercise. The main principle in the management of disturbances is that "the divisions of duties, as well as operational models customary under normal conditions, are retained as long as possible" [17]. However, intersectoral cooperation bodies such as the meeting of Heads of Preparedness can be used to support the management of a disturbance.

The contents of the maturity model that was developed for the SAR exercise were based on the objectives of the exercise, on a literature review on maturity models and collaboration in disaster and crisis management, and on feedback from previous exercises, such as factors that hamper an SAR organization in achieving an adequate SSA [2]. In regular SAR exercises the predetermined roles and structures, processes, and practices were emphasized, and, in addition, common ground and tools, communication and interaction, competence, and a general goal were identified as important areas. They formed the six key areas of the maturity model. Each key area includes several sub-areas. The sub-areas of the key area *common ground and tools* are presented in Fig. 1, together with the results of the maturity assessment.

For a quick maturity assessment both the names and the descriptions of maturity levels should enunciate how the interagency collaboration develops from separate actions into comprehensive collaboration. Table 1 shows the five maturity levels and their descriptions that were used in the SAR exercise. The descriptions

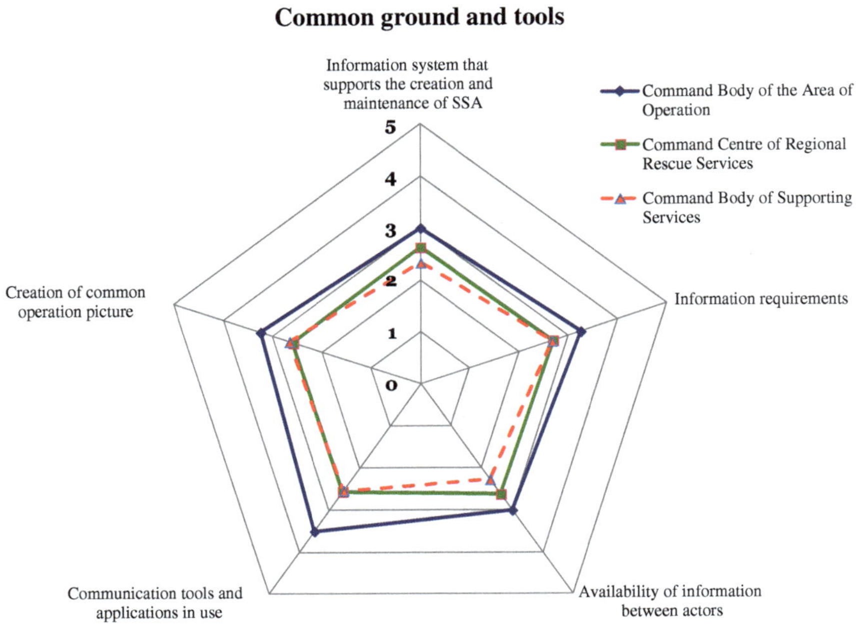

Fig. 1 The radar chart showing the maturity values on a scale of one to five given by the three Command Bodies in the SAR exercise

Table 1 The five maturity levels of collaboration in the SAR exercise

Maturity levels	Description
1 Separate	Everyone acts on the basis of his own needs. Solo actions
2 Learnable	A little collaboration on the basis of common SAR processes. Requirements for collaboration have been taken into account to some extent in the preparedness plans of organizations
3 Established	Operations of organizations have been developed on the basis of common exercises. Actors value both the training and the practicing of collaboration. The active planning, development, and measurement of collaboration and common SAR processes are progressing
4 Proactive	The actors perceive both the whole and the different phases of a SAR operation. The cooperation is systematic. The actors execute the most important goals of collaboration through common SAR processes
5 Optimized	Collaboration produces the desired results without distractions. The functioning collaboration is an ideal for other corresponding organizations. Complete collaboration

of the maturity levels emphasize the common SAR processes [18] that outline the SAR operation as a whole, the culmination points of collaboration, and the main duties of each agency.

The data for the maturity assessment were collected at the end of the one-day exercise. Totally 35 human agents: 16 from the CCRRS, 15 from the CBSS and four from the CBAO answered a questionnaire where they assessed the maturity level of each sub-area. For data analysis, each respondent filled out a form dealing with background information such as the field of operation, the command body during the exercise, and his or her role in the command body.

For the ICT exercise the maturity model was further developed in order to meet better the viewpoints of the participating organizations. The purpose of the exercise was to train preparedness on the basis of the Security Strategy for Society [17], which provides the common basis for preparedness and crisis management for all the actors in the society. Therefore, the goals come from the Strategy, in which a comprehensive and intersectoral approach is underlined. Organizations from both the public and the private sectors that had the main roles in the ICT exercise were interviewed before the exercise. The interviewees gave feedback on the first draft of the maturity model and highlighted issues that, from their points of view, were critical in the intersectoral cooperation. They emphasized capabilities such as the sharing of information, interaction between organizations, and the agility of organizations as critical success factors for collaboration. These viewpoints were taken into account in the final descriptions of the maturity levels (Table 2). The maturity model includes four key areas: management of disturbances, situation awareness and situation picture, the support provided by legislation, and communications in crisis situations and disturbances. The sub-areas of

Table 2 The five maturity levels of collaboration in the ICT preparedness exercise

Maturity levels	Description
1 Separate actions	The sharing of information and interaction between organizations is very limited. The need for common processes of cooperation has been identified
2 Starting cooperation	Internal cooperation in administrative sectors. The ensemble does not work yet. Formal interaction and sharing of information. Low intersectoral shared situational awareness
3 Coordinating	Adequate instruction enables networked cooperation to take place. Both formal and informal interaction and exchange of information. Shared situational awareness available only at the top management level
4 Adaptable, capable of changes	Organizations are capable of reacting together in a fast and agile manner to new situations. Dynamic information exchange ensures good shared situational awareness and understanding. Intersectoral cooperation between authorities and private companies at all operational levels
5 Appropriate cooperation	Cooperation produces the desired results without distractions. Availability of critical information that is needed for actions is excellent. Optimal sharing of resources and responsibilities

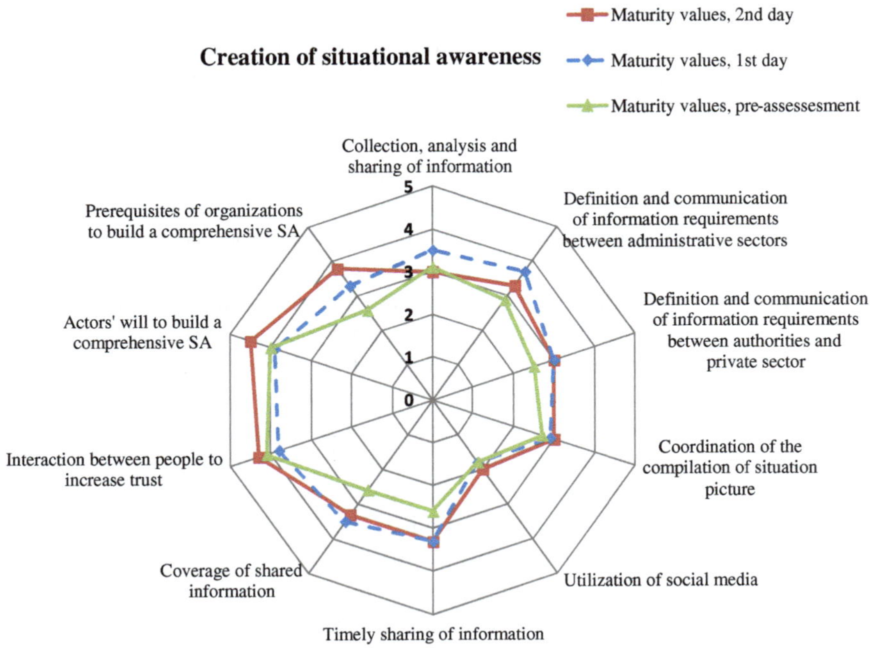

Fig. 2 The radar chart showing the maturity values on a scale of one to five from the pre-assessment and the first- and the second-day maturity assessments in the ICT exercise

the key area *creation of situational awareness* are presented in Fig. 2, together with the results of the maturity assessment.

In the two-day ICT exercise the steering and evaluation group conducted a maturity pre-assessment before the exercise and the participants assessed the maturity of cooperation twice: late during the first day and at the end of the second day. In the multiple evaluation moments of all the administrative levels of one ministry that are shown in Fig. 2 the number of respondents varied between four and ten. For each sub-area participants chose the maturity level which, from their perspective, was reflected in what happened in the exercise. The respondents filled in the forms of background information such as their organizations and their roles in the exercise: top management, middle management, or expert as well. Thus the data that were collected could be analyzed on the basis of organizations, administrative areas of ministries, or roles.

In both of the exercises, the data from the questionnaires were exported for the analysis into Excel spreadsheet application. Calculation and graphing tools were used in order to produce the final maturity values and to create the radar charts. Visualization provides much better communication of information than verbal documents [19]. Good data visualization enables data to be viewed easily and provides an effective way to compare values and to do analysis. A radar chart, also known as a star plot [20], was used as a graphical method for displaying the results of maturity assessments. It has a star-shaped figure in which each ray represents one variable. The length of a ray is proportional to the value of the variable. A radar chart is a convenient method of visualizing and analyzing series of multivariate attribute data like the maturity values from different measurements.

3 Results

The main results are, first, the presentation of the results of the maturity assessments and their analysis and, second, the development process itself. Figures 1 and 2 display the maturity values as a radar chart where each sub-area represents one variable. Figure 1 shows the results from the SAR exercise. For each sub-area a mean maturity value was calculated from the single maturity values given by the human agents in the three Command Bodies.

According to the human agents in the CBAO, the total maturity of the key area was on level three or "*Established*". However, the human agents in both the CCRRS and the CBSS assessed the key area as being on maturity level two or "*Learnable*". The biggest differences in the maturity values are in two sub-areas: an information system that supports the creation and maintenance of the SSA and the availability of information between actors. The low maturity values may indicate that a correlation between these two sub-areas exists. The lack of information may partly arise from the non-integrated information systems of the organizations. Among the clear strengths of the CBAO are communication tools and applications, such as management of the situation picture. The sub-area that needs development is the availability of information in the CBSS. The results show that the maturity model reveals the known fact that the collaboration in the CBAO is more mature than in the two other bodies. This is due to the continuous collaboration of the rescue authorities, police, and medical care services in everyday emergency operations.

Figure 2 summarizes the results from the ICT exercise. The pre-assessment of the steering and evaluation group encompasses the overall maturity and concerns all participating organizations. The maturity values from the first- and the second-day assessments represent the opinions of human agents from all the administrative levels of one ministry.

The strengths in the collaboration are the human agents' willingness to collaborate and interact with each other. These sub-areas matured from *Coordinated* to *Adaptable* during the exercise. Therefore, trust increased and the building of a comprehensive SA was facilitated. All the human agents agreed that the low level of utilization of information from social media was a clear weakness. During the exercise, as the number of disturbances increased and they became more complicated, the organizations' capabilities to collect, analyze, and share information decreased. The communication of information requirements between administrative sectors decreased, whereas private sector and the authorities were able to maintain the same stage of communication, possibly on the basis of more active interaction. The maturity values of the pre-assessment show that the steering and evaluation group expected the capabilities of the organizations to be less mature, for example, in the sharing of information. However, the assessments during the exercise show that the organizations performed better than it was expected.

The data was also analyzed on the basis of the roles of all human agents that participated in the exercise. The results show that according to the top management the maturity of almost all sub-areas, especially in the definition and communication of information requirements, increased during the exercise. Similarly, the middle management assessed that the biggest development of the maturity occurred in the same sub-area. In contrast to the top and middle management, the experts saw that the maturity increased most in the interaction between people to increase trust.

Our approach during the exercises created as a result a process that could be used when customized maturity models for disaster management organizations are developed. The process has five steps:

1. *Selection*. A pre-study and a decision of the best suitable existing maturity model.
2. *Customization*. The identification of key areas and sub-areas and description of the maturity levels. These are based on the literature, on the goals of the exercise or operation, and on interviews of participating organizations.
3. *Assessment*. The maturity assessment in the exercise.
4. *Visualization*. The visualization of the results of the maturity assessment.
5. *Analysis*. The analysis of the results and utilization of the analysis, even during the exercise.

4 Discussion and Conclusion

Two customized maturity models were developed, maturity assessments were performed, and their results were visualized in an SAR exercise and in an ICT exercise. This approach produced both a process and quantitative data for the evaluation and the development of collaboration in disaster management. The contents of the customized maturity models reflect the national objectives, and the specific goals of the exercises. However, the key capabilities that have been emphasized in the existing maturity models were identified by the participating organizations as well, and they were taken into account in the new models. The SAR and ICT exercises train human agents to manage two very different disasters. The first maturity model was customized for a series of annual SAR exercises and the second model for a series of ICT exercises. Therefore, the contents of the maturity models are not comparable. Some of the human agents assessed the maturities on the basis of the maturity level descriptions, and some only on the basis of the names of the maturity levels. Therefore, descriptive names of the maturity levels are important and should not be any general ones. They should highlight the actual starting level of collaboration and the highest maturity level that is realistic to achieve. The names of the levels in the ICT maturity model seem to demonstrate better the achievable stages of collaboration than those in the SAR maturity model.

Quick maturity assessments were performed for the first time in the SAR and ICT exercises. As expected, the results showed the strengths, such as good collaboration in the CBAO in the SAR exercise, and weaknesses, such as the low level of utilization of social media in the ICT exercise. In the SAR exercise the maturity assessment was performed once after the exercise. The two-phased assessment during the ICT exercise gave valuable information about the agility of the collaboration that has been defined as a key capability in a complex disaster situation. The results from the first day highlighted capabilities, such as the prerequisites to build a comprehensive SA—including enabling technologies—that would have needed more attention and support when the disaster became more complicated.

Radar charts were used to display the results of the maturity assessments. They facilitate the visualization and comparison of the quantitative results more effectively than verbal descriptions. For the analysts, a radar chart is an informative tool that makes the visualization of data from different time series easier. We argue that concrete quantitative measures and graphical figures that reveal the weaknesses help the actors to realize that they have to change their behaviors in order to adopt new ways of working together. However, the clear presentations of the strengths and progress in the collaboration motivate the human agents to further improve their behavior.

According to the managements of both exercises, the results seemed to be reliable and useful. The process was formalized but the questionnaire and the data collection methods still need further development. Even if a proportion of human agents change in each exercise, the key organizations participate in the SAR and in the ICT exercises. Therefore, the methods and maturity models are usable in several exercises and worth developing. The content of a maturity model should, however, change in the course of time, and highlight the collaboration capabilities that are needed in complex disaster situations in the future.

The results of the quick maturity assessments highlight the opinions of all the human agents that participate in an exercise and not only the perspectives of the management. When the collected data are analyzed in a multifaceted way, this can reveal both the strengths and the weaknesses of collaboration, and discrepancies between organizations and human agents that have not been identified before. Thus, the targets and resources for further development can be better specified.

References

1. A. Nofi, *Defining and Measuring Shared Situational Awareness* (Center for Naval Analyses, Alexandria, 2000)
2. H. Seppänen, J. Mäkelä, P. Luokkala, K. Virrantaus, Developing shared situational awareness for emergency management. Submitted, unpublished paper, 2012

3. M. Konecny, S. Zlatanova, T.L. Bandrova, *Introduction in Geographic Information and Cartography for Risk and Crisis Management—Towards Better Solutions* (Springer, Berlin, 2010)
4. A.M. MacEachren, G. Cai, Supporting group work in crisis management: visually mediated human-GIS-human dialogue. Environ. Plan. B Plan. Design **33**, 435–456 (2006)
5. H. Seppänen, K. Virrantaus, The role of GI-supported methods in crisis management. Int. J. Digit. Earth **3**(4), 340–354 (2010)
6. V. Valtonen, *Collaboration of Security Actors—an Operational-Tactical Perspective (Finnish, English abstract)* (Ph.D. National Defence University, Helsinki, 2010)
7. K. Rintakoski, M. Autti, *Comprehensive Approach—Trends, Challenges and Possibilities for Cooperation in Crisis Prevention and Management* (Ministry of Defence, Helsinki, 2008)
8. T. Hof, L. de Koning, P. Essens, Measuring effectiveness of team and multi-team systems in operation. 15th international command and control research and technology symposium (2012), http://www.dodccrp.org/events/15th_iccrts_2010/papers/027.pdf. Accessed 15 Aug 2012
9. D. Alberts, R. Hayes, Planning complex endeavors. CCRP publication series (2007)
10. R. Gajda, Utilizing collaboration theory to evaluate strategic alliances. Am. J. Eval. **25**(1), 65–77 (2004)
11. G. Griffin, A. Rajabifard, D. Williams, The victorian emergency management continuum and the benefits of spatial enablement. GSDI 13 Conference (2012), http://www.gsdi.org/gsdiconf/gsdi13/papers/71.pdf. Accessed 28 June 2012
12. S.J. Krill, M. Dzirio-Ayvaz, An Innovative Capability Model for Global Emergency and Disaster Management. Booz Allen Hamilton (2008), http://www.slideshare.net/jmz2tsab/TIEMS2008AnnualConferenceProceedingsPaper. Accessed 2 July 2012
13. R. Kuusisto, Analyzing the command and control maturity levels of collaborating organizations. 13th international command and control research and technology symposium (2008), http://www.dodccrp.org/events/13th_iccrts_2008/CD/html/papers/028.pdf. Accessed 2 July 2012
14. M.C. Paulk, C.V. Weber, B. Curtis, M.B. Chrissis, A.W. Longman, *The Capability Maturity Model: Guidelines for Improving the Software Process* (Addison-Wesley, Reading, 1994)
15. R. Kuusisto, "SHIFT" The Theoretically-Practically Motivated Framework. Information Exchange viewpoint on developing collaboration support systems. Finnish Defence University, Department of Tactics and Operations Art, series 3, no. 1 (2008)
16. J. Mäkelä, Model for assessing GIS maturity of an organization, in *Spatially Enabling Government, Industry and Citizens*, ed. by A. Rajabifard, D. Coleman (GSDI Association Press, Needham, 2012)
17. Ministry of Defence, Security strategy for society, government resolution 16.12.2010 (2003), http://www.yhteiskunnanturvallisuus.fi/en/materials. Accessed 20 June 2012
18. H. Seppänen, V. Valtonen, SAR processes. National Defence University, Department of Tactics and Operations Art, vol. 1, no. 2 (2008) in Finnish only
19. M. Ward, G. Grinstein, D. Kleim, *Interactive Data Visualization: Foundations, Techniques, and Applications* (K Peters, Ltd., Natick, 2010)
20. M. Friendly, Statistical Graphics for Multivariate Data, SAS SUGI 16 Conference (1991)

Geographic Information for Command and Control Systems

Demonstration of Emergency Support System

Tomáš Řezník, Bronislava Horáková and Roman Szturc

Abstract This paper deals with the issue of integration of geographic information (GI) into (existing) command and control systems. The main motivations and benefits of such integration are described at the beginning. Open as well as proprietary solutions are discussed. The core of the paper lies in the definition of the Emergency Support System, including the user requirements, architecture, field tests and assessment. Beside others, conclusions depict issues of a system based on the Service Oriented Architecture that is integrated into a command and control system. Such issues consist of existing procedures and processes in crisis and emergency management, connection to other components of Spatial Data Infrastructures and future development.

Keywords Crisis management · User requirements · Service oriented architecture · Field test assessment · Inspire

1 Introduction

Over the last years, the international community has responded to an increasing number of various natural and man-made disasters. For instance, Konecny [6] emphasises the relative frequency of recorded disasters during decades in

T. Řezník (✉)
Department of Geography, Masaryk University, Kotlářská 2 611 37 Brno, The Czech Republic
e-mail: tom@mail.muni.cz

B. Horáková · R. Szturc
Prosek Point–budova A, Intergraph CS, Prosecká 851/64 190 00 Praha, The Czech Republic
e-mail: bronislava.horakova@intergraph.com

S. Zlatanova et al. (eds.), *Intelligent Systems for Crisis Management*, Lecture Notes in Geoinformation and Cartography, DOI: 10.1007/978-3-642-33218-0_18,

twentieth century: “about 100 per decade in the period 1900–1940, to 650 per decade in the 1960s, 2,000 per decade in the 1980s, and reached almost 2,800 per decade in the 1990s”.

At the same time, the emergency and crisis management has become increasingly complex due to the necessity of various institutions to cooperate during such disasters. Most of current activities in the emergency and crisis management domain aimed at developing more sophisticated tools for the production, processing and displaying information to end users. Abnormal events in recent years (like earthquakes in Chile, Haiti and Japan, flooding in Australia and/or the crises in the Middle East and North African regions) have confirmed the need for spatially based information—obtained from the Earth observation systems (like satellite and aerial images), Web-mapping technologies (for instance Web map services, clustering and processing Web technologies, etc.) as well as another Information and Communication Technologies (ICT) resources; for example starting with sensors and ending with the location based services based on information obtained from cell-phones. Many of these activities are stand-alone, based on the proprietary protocols, formats and methodologies and for that reason could be considered as the barrier of interoperability.

The Emergency Support System (ESS) as an FP7 European project, funded from 2009 to 2013, represents a suite of real-time (spatial) data-centric technologies. It aims to improve this situation (for more information about the project see ESS Project, [4]). The ESS consortium, consisting of 19 partners, is developing a crisis communication system that will reliably transmit filtered and re-organized information streams to crisis command systems, which will provide the relevant information that is actually needed to make critical decisions. The information streams in ESS are organized in such a way that they can be easily enhanced and combined with other available applications and databases. ESS provides an open Application Programming Interface (API) in order to allow any public authority, if needed, to add more applications customized to its particular needs. This API is thoroughly defined in the document referenced as ESS Consortium [3]. ESS data, functionalities and data flow are based on International Organization for Standardization (ISO), Open Geospatial Consortium (OGC), World Wide Web Consortium (W3C) and industrial standards. Therefore, each application which has adopted or will adopt these standards is able to connect to ESS—such as existing command and control systems. In order to validate the system, four field tests (defined in cooperation with end users of crisis systems) will be undertaken: a proof-of-concept field test, a fire in forested area, an abnormal event in a crowded stadium, and a toxic waste spillage accident. Operating ESS under different scenarios is necessary in order to test the system's capabilities in different kinds of crises using a variety of collection tools.

The following sections describe the concept of geographic information in command and control systems, user requirements in crisis and emergency management, overall architecture, and present a brief overview of ESS field tests and evaluation.

2 Geographic Information in Command and Control Systems

2.1 State-of-the-Art of Command and Control Systems

Command and control systems are designed to assist in decision-making and to control resources to successfully accomplish missions. Various authors have categorized different levels of these systems: one speaks about command and control systems (designated as C2 systems), the second about command, control and intelligence systems (C2I), the third about command, control, communication and intelligence (i.e. C3I), and the fourth about command, control, communication, computing and intelligence systems (with the abbreviation C4I). Please note that this is not an exhaustive list since there may be many different points of view. Such discussions are out of the scope of this scientific paper, therefore, in this paper the simplification of command and control systems is used as a designation for all similar systems; i.e. from C2 to C4I systems.

One important capability that command and control systems offer is situational awareness, i.e. information about the location and status of resources. Such information alone does not represent a decision-making process; it is still up to the commander to gather available information, to turn it into relevant knowledge and to combine it with his/her judgment in order to make appropriate decisions. The process, therefore, includes human behaviour (such as experience level, tiredness, stress level, etc.) that is hard to quantify and/or predict. The information on which decisions are based should be as accurate as possible to minimize the amount of uncertainty during the decision-making process. To achieve this, commanders are supported by tools, i.e. command and control systems, to enable and accelerate the planning and decision-making processes and to accomplish the missions. Since these systems cannot cover all data and functionality, we should assume that a command and control system is "a system of systems" in which other systems and sub-systems are used to solve specific functional needs in appropriate detail. This is exactly the reason why this paper deals also with the interfaces of command and control systems. The development of merging data from various data streams, fusion procedures as well as issues of transmissions and the pre-organisation of information to be prepared for effective decision-making have been a long-standing challenge for at least the past 20 years. We can observe this in several scientific papers and other scientific publications—for instance, in Llinas and Waltz [7], Curts [1], Hieb and Tolk [5], Tolk [10].

As described by Curts [1], clear definition of interfaces is a key for success in interoperability of all defence and/or command and control systems. Interoperability can thus be defined as the ability of systems, units, or forces to provide services to and accept services from other systems, units, or forces, and to use the services so exchanged to enable them to cooperate effectively. Interoperability depends on a consolidated, coordinated, organized architecture. It is stated in Curts [1] that the Office of the Assistant Secretary of Defence for Command, Control, Communications and Intelligence (OASD) commissioned a number of studies to

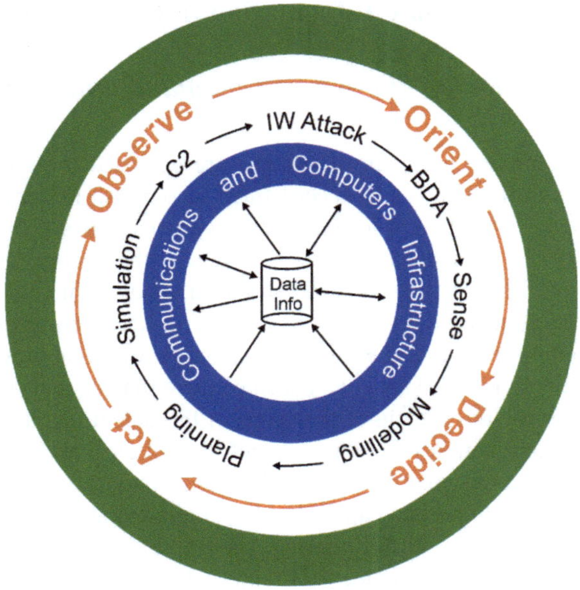

Fig. 1 The interoperability requirement diagram for command and control systems as defined in Curts [1], modified

review the status of architectures and interfaces within the U.S. Department of Defence. One of the most important findings is that architecture definitions have to start with interface points. Almost all architectures designate some peripheral node as the "connection to" the systems, structures, architectures of (usually) other agencies. The interoperability requirement for command and control systems has been proposed as a result of the report—which is depicted in Fig. 1. As you can see, (spatial) data are needed during all phases of crisis and/or emergency management.

According to Hieb and Tolk [5], the time when command and control systems must use the same commercial standards (i.e. commercial-off-the-shelf solutions) for fast and effective communication is over. Such commercial-off-the-shelf solutions are now seen as a barrier to the integration of different systems. Military systems like the North Atlantic Treaty Organisation (NATO) Standardization Agreements (STANAG) are composed of ISO, Institute of Electrical and Electronics Engineers Standards Association (IEEE), W3C or OGC standards. The use of the Unified Modelling Language (UML) is recommended in their development. This step has been carried out within many military systems so far. However, it has still not been implemented within the crisis and emergency management domain—including command and control systems.

Open standardized interfaces are the solution for all command and control systems, according to Curts [1], since no such system can guarantee that it will provide and support all the information and all functionality that will possibly be needed in the future. This stems from the fact that there may be new issues, new situations, new kinds of data, etc. On the other hand, there is always a system or structure which can at least provide a partial solution for this issue. Tasks analysed within command and control systems are complex and so, for that reason also, the architecture of these command and control systems is usually complex. Without open interfaces between systems and structures within this complex architecture we may close the possibility of answering the important questions of an incident commander in the future. Hieb and Tolk [5] state that open interfaces are not the only step towards interoperability. However, they are definitely the first one.

The ability to generate and move information has increased tremendously over the last 30 years. The services of crisis and emergency management organisations have all become much more reliant on information technology. Unfortunately, the current capability to generate information far exceeds our ability to control and use it effectively. To ensure information interoperability, system developers must comply with data and interface standards.

Michaels [8] states that data effort is uncoordinated and that there is currently no process to fix the problem. Many command and control systems are incapable of sharing and exchanging data, an interoperability problem that could result in the possible 'loss of life, equipment or supplies'. Correction of the problem requires both interoperable information architecture and a repository of systems' databases.

According to Curts [1], many command and control systems have not been certified for interoperability. The main goal for the near future is the utilization of an architecture that allows seamless interoperability between command and control systems and modelling/simulation systems (like map support systems, traffic control systems, systems for modelling a toxic plume evolution, etc.) Therefore, it is desired to allow the majority of command and control systems to "plug-and-play" into other supporting systems and exchange information without having to build unique interfaces. Also Tolk [10] declares the need for open interfaces with in existing command and control systems. This scientific paper (i.e. Tolk [10]), proposes the use of Web standards, like eXtensible Markup Language (XML) or Simple Object Access Protocol (SOAP), for open interfaces between command and control systems and other systems which may support these command and control systems. As stated in Tolk [10], there is a lack of interfaces for specific purposes–e.g. for an exchange of geographic information. In conclusion, definitions of open interfaces for exchange of spatial data as well as for pre-organized information based on this spatial data are desired to be developed.

2.2 *Benefits of GI Integration to Command and Control Systems*

We have discussed the several motivations of geographic information into the integration of command and control systems following the needs and principles described above. Summing up benefits arising from such integration, the main added value is:

- *Improved Command and Control decision making*: decisions that are based less on assumptions that need to be made in the absence of concrete information and more on information deduced through advanced data analysis.
- *Better coordination of emergency services in real time*: sharing of geographic information—force locations, reconnaissance results, risk assessments, etc.—during times of crisis.
- *Improved emergency force preparedness*: defining rules and delivering technologies for emergency or crisis response and recovery as well as providing historical data after the crisis for assessment and possible redefinition of crisis plans.
- *Coordinated and effective public alert systems*: rapid dissemination of alert messages to emergency authorities and to citizens. A GI-based system may send alerts through several channels like e-mail, Short Message Service (SMS) or Multimedia Message Service (MMS). Cell-phone based analyses enable to send locally determined alerts.
- *Better crisis recovery*: data gathered through the GI-based system aid in the targeted support of afflicted areas, their citizens and complex land recovery.

3 Emergency Support System

3.1 End User Requirements

End user requirements were needed to address issues depicted in previous section. At the same time, they are the foundation stone for the ESS architecture. End user requirements for the ESS were collected between 1 June 2009 and 15 December 2009, i.e. within the first 6 months of the ESS project funding. These requirements were recommended by various institutions involved in crisis and emergency management across Europe. Additional material resulting from e-mail and private conversations was used to obtain a more complete picture.

In total, the requirements model has incorporated 146 user requirements till 31 May 2012. The requirements model was divided into two main sub-catalogues, as depicted in the Fig. 2 bringing high-level overview.

The requirements model represents the main result of processing the user requirements. The ESS high-level architecture is based on those results and adds an ESS component model. The main purpose of the component model is to define the organisation and dependencies of the system components. External components are modelled as well to increase the understanding of the ESS system boundaries and potential interactions with external (command and control) systems. The software architecture is not a monolithic package, but a system consisting of components or subsystems. The components realize the functionality required from the overall system in a joint manner.

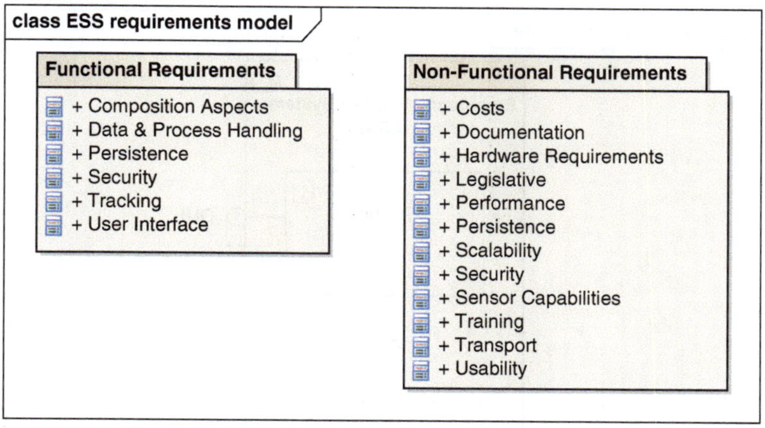

Fig. 2 ESS requirements model and its division into functional and non-functional requirements

4 ESS Architecture

As it was written above, the establishment of open interfaces is crucial for all levels of interoperability since it can be considered as the common starting point for open communication between existing command and control systems, on the one hand, and the Emergency Support System, on the other hand.

The ESS architecture is crucial to properly understand the definitions of the various open interfaces. The component model in Fig. 3 shows a simplified view of the high-level ESS architecture while a detailed one can be found in the ESS D2.4 document [2].
The Emergency Support System consists of three main subsystems:

- the Data Collection Tools (DCT);
- the Data Fusion and Mediation System (DFMS);
- the Portal.

A definition of the ESS open interfaces has been formulated on both levels—conceptual and implementation. The conceptual level depicts high-level information on how these interfaces are defined, which underlying standards are used, which functionality and data may be obtained through which interface as well as what is the added value for the existing command and control systems. Specifics on the technical level are briefly depicted to provide basic information on communication between the ESS and existing command and control systems. Thorough documentation of these interfaces is provided in the OGC implementation specifications as Web Map Service, Web Feature Service, Filter Encoding, or Catalogue Service for Web, etc.

The ESS offers more open interfaces than mentioned in the previous paragraph. For instance, interfaces from a family of standards called OGC Sensor Web Enablement are primarily aimed at communication with the back-end level. Such

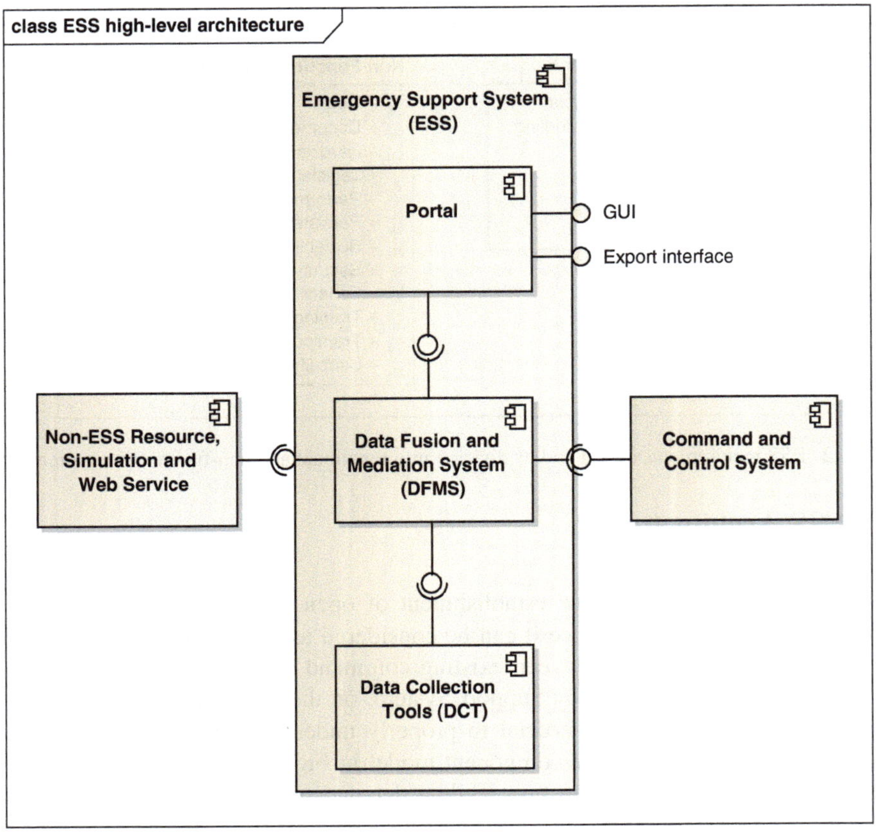

Fig. 3 High-level architecture of the Emergency support system expressed in the UML component (class) diagram

communication is between the core of the ESS, i.e. the Fusion and Mediation System, and the sensor layer where communication with basic ESS equipment, as depicted in Fig. 4, is carried out.

Geographic data in the ESS consist of so-called long-life data (topographic maps, satellite images, etc. from offline data stores as well as from Web services)

Fig. 4 Basic ESS equipment (from *left*): meteorological sensors with INCA modem, unmanned ground station (UGS), unmanned aerial vehicle (UAV), IMSI catcher

and real-time data (sensor measurements, video streams including corresponding telemetry, traffic data, etc.). Caching of Web services for long-life data is an irreplaceable technique in the ESS architecture since the ESS architecture is not fully dependent on any external system. For instance, we have established service caching in the form of the OGC WMS (Web Map Service) tiles in the ESS.

The ESS may be used in two ways:

- as a stand-alone supporting system (i.e. including the Portal);
- as an associated application of existing command and control systems (i.e. without the Portal).

The ESS is also prepared for simultaneous operations of both above mentioned possibilities—i.e. it may be used as a stand-alone system and as an associated application of existing command and control systems at the same time.

5 ESS Field Tests Realization and Assessment

Tests were foreseen to demonstrate the ESS success/functionality of innovations based on research and development undertaken during the Emergency Support system project. Laboratory as well as field tests are being conducted since 2010 till 2013.

The first field test, also called proof-of-concept (POC), demonstrated the applicability of the Emergency Support System. POC system and field test were devised by integrating essentially off-the-shelf components (COTS) provided by partners. Those COTS in a nutshell: an IMSI catcher enabled to discover people presence, intelligent video cameras were used to record motion of a person and/or to visually control the situation, thermal cameras detected sites of fire, UAV and air balloon assisted toxic plume and people presence analyses, wind and weather sensors were used to predict an evolution of a toxic plume, chemical poisons and chemical agents' sensors enabled to detect toxic chemical substances, GPS trackers added location to all sensors, Web map server published geographic data, etc. This was a challenging task due to the highly different technologies of the partners, the continuously evolving requirements of end users, and the relatively short time devoted to POC development. The POC system development took almost one year (from June 2009 to April 2010), while the field test was carried out at Bengener Heide air field in Germany between 7 and 10 June 2010.

Scenario for the first field test was following. A person at the front end calls 112 and reports burning containers containing possibly toxic materials, which have fallen from a van or truck. These burning containers are located at the Bengene-rHeide airfield, according to the first responders. An intelligence officer and fire fighters establish a mobile command post with the Emergency Support System to receive preliminary information about the incident. The incident commander classifies the incident as difficult to survey and potentially dangerous; therefore, (s)he orders mobilization of the ESS-vehicle.

After having checked the circumstances for intended use of ESS-equipment (see Fig. 4), (s)he chooses an appropriate configuration of ESS-vehicle with special sensors, cameras, Unmanned Aerial Vehicle (UAV), balloon and IMSI catcher. After the ESS-vehicle has arrived, the UAV is launched to get an overview and decide where to position the sensors and where to take measurements. The incident commander issues directives for positioning Unmanned Ground Stations (UGS) after having seen the videos from the UAV. The staff set up the UGS between hot and warm zones to take measurements and monitor the scene. The mobile International Mobile Subscriber Identity (IMSI) catcher is positioned to detect people with cell-phones in hot and warm zones. The ESS car is deployed to take various measurements in hot and warm zones. Its position can be tracked within the Portal.

The incident commander is notified via the Portal that cell-phones have been detected in the hot zone. The incident commander activates SMS alerts to the cell-phones detected in the hot zone instructing the users to immediately move away. The UGS video detects one person in a car on the runway. The incident commander checks data from the sensors. Chemical sensor data do not exceed threshold values; therefore, the fire brigade is told to start extinguishing the fire (which can be observed by the incident commander on the video). Suddenly an alarm is given that, according to values obtained from an ESS chemical sensor, a threshold value has been exceeded. Coloured smoke is seen on the UGS video. The incident commander instructs the fire fighters to stop fighting the fire and to treat the toxic leakage. The sensor values are below the threshold value again. The incident commander instructs the fire brigade to resume fire fighting. Special units dealing with chemical spillages take the first steps towards decontaminating the area. A final check via the ESS-equipment confirms that all values are within the normal range again.

The POC field test was attended by 24 independent observers representing various actors in Crisis and Emergency Management. Each observer completed a short questionnaire that contained 6 questions. Answers to these questions are described in the following paragraphs.

Only 1 person (4 %) did not get a general idea of ESS development, while 23 observers (96 %) obtained valuable information about ESS in general, details on ESS components, information about the components' integration framework, and details about their operational approach. One remark suggested that the system was not fully suitable for operational approach as it was considered more as a system test. This (operational) approach will be the subject of future scenarios, which will be focused on tests involving real crisis management situations.

All observers rated the POC field test highly in terms of innovation and user requirements needs. Some of the highly appreciated aspects of the system were its fusion capabilities, its scalability, its resilience, the level of visualization, the position plotting of video targets, the use of satellite phone links in the case of terrestrial links becoming blocked, and the scheduling of rescue workers. A weakness can be seen in its high ambitions with respect to toxic gas detection—according to the independent observers, this was the only issue requiring a separate research and development project. More than one-third of the observers were satisfied with the current version and would like to use ESS as is—i.e. without waiting for more than two years for a final version.

The third question related to the functionality that should be improved in terms of innovation and need. Our observers mentioned open interfaces for exchanging data with different command and control systems, simulation and risk assessment systems (Command and Control Systems, C3I, C4I, etc.) as the most crucial. These requirements were the subject of further development and were not intended to be features of the POC field test. Thus, it was expected that persons used to working in crisis and emergency management on a daily basis felt the lack of this functionality.

Extensions to the system in terms of technological innovation and operational need should—according to 75 % of our external users—include mainly the detection and repair of errors and gaps in sensor data. Further comment relates to access to archived datasets to compare the situation before and after the crisis. Debriefing historical data was not demonstrated in the POC field test; however, it was a part of ESS development.

The fifth question was on how the simulated scenario could be improved to test ESS in a more realistic event. In four cases, the low degree of complexity of the scenario was designated as the key limiting aspect, as real life situations would include the coordination of various types of rescue workers as well as integrating actors in the field, field base actors and decision maker stakeholders.

The last question was devoted to the potential risks or hindrances in using an emergency system like ESS. Only one person indicated the use of ESS as extremely risky for electronic communication (maps, e-mails, alerts, etc.) in comparison to phone calls. For instance, it was answered that it is less efficient to extract information from an ESS route map in comparison to express the same information through a phone call. All other observers evaluated ESS as applicable with minor potential risks—these risks were related to the special training needed to operate the system, purchase and maintenance costs, and disturbances of the cell-phone network during operation of the IMSI catcher. The last question revealed that legal restrictions and the obtaining of permission for operating various items of ESS-equipment seem to be one of the greatest obstacles. Also it was stated during the feedback session that an organization would share their data only with selected organizations and would need to know the sources of any information before trusting it.

The following tests have been planned for the second and third year after the Proof of Concept test. The tests are being aimed at demonstrating and validating the developments in the Emergency Support System since the initial test.

The first performance test was conducted in city of Nîmes in France on 24 March 2012. Performance of the ESS was tested against emergency management requirements related with the organization of crowd events (soccer match).

The first final test was conducted between 19 and 22 June 2012 in France. The scenario of an accident involving a truck containing chemical hazardous materials was planned in the area of Sisteron, a city about 100 km north of Marseille.

The second final test is planned for April 2013 (so far, the explicit date has not been set). An aircraft crash and forest fire has been selected as the scenario for the area on the French-Italian border between the villages of Sospel (on the French side) and Vescavo (on the Italian side).

The results from these tests were not available at the time of the writing of this scientific paper. All relevant information from each test will be published on the ESS project Website at http://www.ess-project.eu.

Please note, that there were provided only brief information about ESS (field) tests because of the limited extent of this paper. More information may be found in Řezník et al. [9].

6 Conclusions

This paper describes the Emergency Support System (ESS), a set of spatially-based tools aimed to support the decision making process during crisis and emergency management events. The ESS'spatially-based tools consist, beside others, of field sensor devices including the IMSI catcher, geographical databases, external Web services, metadata catalogue, business logic (like route planning, modelling of toxic plume, etc.), tasking interface, and event notification service. The basic premise of the whole system is not to affect any existing procedure and/or process in the crisis and emergency management domain. In other words, the ESS does not define how it should be operated since this always originates from the end users' purposes and procedures.

The paper reveals that the Service Oriented Architecture (SOA) brings significant advantages when applied to the development of a system for crisis and emergency management. The main advantages may be summarized as openness to other systems, the possibility to add any piece of functionality and/or data in the future, the lack of dependence on a software provider, and support for distributed solutions. This is in contrast to current practices. On the basis of the state-of-the-art analysis, it can be seen that most contemporary command and control systems are commercially driven proprietary desktop-based solutions. Interoperability among (spatial information) systems to support the decision making process is limited. SOA is desired by the end user community as well as by researchers; however, current systems remain desktop-based and blocked for any extension unless supported by a software developer. Similar barriers to existing command and control systems have been identified independently across many countries around the world.

The ESS supports up-to-date standards and best practices, including Service Oriented Architecture. For instance, the ESS can be connected to any service that is compliant with the Open Geospatial Consortium and/or INSPIRE Directive (2007/2/EC). As a result, more than 10000 service instances can be connected to the ESS to enhance its functionality and data. It can be operated with various inputs, from internal devices like sensors, IMSI catchers, UAVs, etc. to external

Web services. At the same time, ESS outputs are based on these standards and best practices as well. The ESS, or any service-based part of it, may therefore be integrated into a command and control system accepting these standards and best practices.

The ESS also offers frontier technologies and functionalities, such as to determine the number of people affected as well as their location in an affected area through an IMSI catcher, synchronisation of separate data streams (for instance for video and telemetry), automatic metadata creation from sensor data and their publication in the ESS catalogue, highlighting relevant information and suppressing the less relevant through Spatial Event Processing concept, etc. The basic limitation of ESS people presence detection lies in the fact that only people having a cell-phone may be localised. Evacuation of people without a cell-phone is realised trough remaining ESS equipment, like video sensors.

Future development is focused particularly on two field tests, which will bring further feedback from end users. Another development, not foreseen for a funded project phase, may lie in cost-benefit analyses, the enhancement of security principles, psychological (cognitive) testing and/or the training of potential ESS users.

This research has been supported by funding from FP7 Project No.217951, which is called ‘Emergency Support System’.

References

1. R.J. Curts, Architecture: the road to interoperability, in *Proceedings of Command & Control Research & Technology Symposium*. U.S. Naval War College, Rhode Island 1999
2. ESS Consortium, ESS Architecture. Deliverable D2.4 (internal project document). p. 251 (2010)
3. ESS Consortium, ESS integration to existing command and control systems. Deliverable D5.4 (internal project document). p. 57 (2011)
4. ESS Project, ESS project home page (online) (2012), http://www.ess-project.euAccessed 23 April 2012
5. M. R Hieb, A. Tolk, Building & integrating M&S components into C4ISR systems for supporting future military operations, in *Proceedings of International Conference on Grand Challenges for Modeling and Simulation*. ICGSMS, Orlando. Positioning paper (2003)
6. M. Konecny, Early warning and crisis management cartographic and geographic information researcher agenda, in *Proceedings of the 4th International Conference on Cartography & GIS*, Sofia ,2012, ed. by T. Bandrova, M. Konecny, G. Zhelezov. Bulgarian Cartographic Association, vol. 2, pp. 7–15. ISSN 1314-0604
7. J. Llinas, E. Waltz, *Multi Sensor Data Fusion* (Artech House, Boston London, 1990) p. 464. ISBN 0-89006-277-3
8. D. Michaels, *DoN Data Interoperability*, ed. by R.J. Curts,. Architecture: The Road to Interoperability. U.S. Naval War College, Rhode Island
9. T. Řezník, B. Horáková, D. Janiurek, Emergency support system: Actionable real-time intelligence with fusion capabilities and cartographic displays, in *Advances in Military Technology*. vol. 6, Issue no. 2 (University of Defense, Brno, 2011) pp. 83–97. ISSN 1802-2308
10. A. Tolk, *Composable Mission Spaces and M&S Repositories—Applicability of Open Standards*. Spring Simulation Interoperability Workshop, Washington, p. 13 (2004)

On the Roles of Geospatial Information for Risk Assessment of Land Subsidence in Urban Areas of Indonesia

Hasanuddin Z. Abidin, Heri Andreas, Irwan Gumilar, Teguh P. Sidiq and Yoichi Fukuda

Abstract Land subsidence is a silent hazard that may occurs in large urban areas, and usually caused by combination of excessive groundwater extraction, natural consolidation of alluvium soil, load of constructions and tectonic activities. Geospatial information is useful for studying the characteristics, causes, impacts and cost of land subsidence. This paper concentrates on the roles of geospatial information for risk assessment of land subsidence in three large cities in Indonesia, namely Jakarta, Bandung and Semarang. Geodetic based results show that land subsidence rates in all three cities generally have spatial and temporal variations, and their magnitude is in average about 5–10 cm/year and can reach up to about 20 cm/year at certain locations and times. The impact of land subsidence can be seen already in the field in forms of the buildings and infrastructure cracking, the wider expansion of (coastal) flooding areas, and increased inland sea water intrusion. Land subsidence has a strong linkage with urban development process. Urban development increases the built-up areas, population, economic and industrial activities, and also groundwater extraction, which can then lead to land subsidence.

Keywords GIS · Data integration · Data managemnet · Flooding · Spatial correlation

H. Z. Abidin (✉) · H. Andreas · I. Gumilar · T. P. Sidiq
Geodesy Research Group, Faculty of Earth Science and Technology,
Institute of Technology Bandung, Jl. Ganesha 10, Bandung, Indonesia
e-mail: hzabidin@gd.itb.ac.id

Y. Fukuda
Department of Geophysics, Graduate School of Science,
Kyoto University, Kyoto 606-8502, Japan

S. Zlatanova et al. (eds.), *Intelligent Systems for Crisis Management*, Lecture Notes in Geoinformation and Cartography, DOI: 10.1007/978-3-642-33218-0_19,

1 Introduction

Land subsidence is the downward displacement of the land surface relative to certain reference surface, such as mean sea level (MSL) or reference ellipsoid. It may occurs in active volcanic and tectonic areas, mining areas, oil and gas exploration areas, and large urban areas, and can be caused by natural and/or human activities.

Many large urban areas (cities) in the world have experienced land subsidence, such as Houston [10], Mexico City [11], Osaka [16], Tokyo [15], Shanghai [12], Taipei [13] and Bangkok [24]. In urban areas of Indonesia, this silent-type hazard have occurred in Jakarta [1, 2, 5, 7], Bandung [3, 4, 8], and Semarang [6, 9]. These urban land subsidences are mainly caused by the combination of excessive groundwater extraction, natural consolidation of alluvium soil, and load of constructions (i.e. settlement of high compressibility soil).

The impact of land subsidence in urban areas usually can be seen in several forms, such as cracking of buildings and infrastructure, the wider expansion of (coastal) flooding areas, and increased inland sea water intrusion. It will also badly influence the quality of living environment and life (e.g. health and sanitation condition) in the affected areas. In urban areas, comprehensive information on the characteristics of land subsidence is important to several important planning and risk assessment efforts (see Table 1), such as spatial-based groundwater extraction regulation, effective control of flood and seawater intrusion, conservation of environment, design and construction of infrastructures, and spatial development planning in general. Since data and information on land subsidence characteristics will be useful for many urban development and environmental aspects, systematic and continuous monitoring of land subsidence is obviously needed and critical to the welfare of the peoples.

Along with hydrogeological, geotechnical and socio-economic information, geospatial information is useful for studying the characteristics, causes and impacts (including economic losses) of land subsidence in urban areas. For example, results from geodetic measurement techniques, such as leveling surveys, global positioning system (GPS) surveys, interferometric synthetic aperture radar (INSAR), and microgravity surveys, provide magnitude of subsidence rates and their temporal and spatial variations. Satellite images, aerial photos and maps

Table 1 Importance of land subsidence information for various applications

Applications that should consider land subsidence information		
Groundwater extraction regulation	Spatial planning consideration	(Inland and Coastal) Flood Mitigation
Infrastructure design and construction	Environmental conservation	Seawater instrusion control
Sub-surface utility planning	Sewerage and drainage system design	

which provide information on geomorphology, land use and land cover in the urban areas also give more insights into the causes and impacts of land subsidence.

This paper presents and discusses the roles of geospatial information for risk assessment of land subsidence in three large cities in Indonesia, namely Jakarta, Bandung and Semarang.

2 Characteristics of Land Subsidence

Geospatial information for studying the characteristics of land subsidence can be in the forms of orthometric height differences (from Leveling surveys), ellipsoidal height differences (from GPS and InSAR methods), and gravity value differences (from microgravity surveys). Typical strength and limitation of those geodetic methods (i.e. Leveling surveys, GPS surveys, InSAR, and Microgravity surveys) in deriving land subsidence characteristics in urban area are given in Table 2.

Land subsidence in Jakarta has been studied using leveling surveys, GPS surveys, and InSAR techniques. The results obtained from leveling surveys, GPS surveys and InSAR technique over the period between 1982 and 2011 show that land subsidence in Jakarta has spatial and temporal variations, and in general the observed subsidence rates are about 1–15 cm/year, and can reach up to 20–28 cm/year at certain location and certain period. The summary of observed subsidence rates in Jakarta is given in Table 3. More detail information on the characteristics of land subsidence in Jakarta can be seen in Abidin et al. [1, 2, 5, 7] and Ng et al. [17].

In Bandung basin, land subsidence phenomenon has been studied using GPS surveys and InSAR techniques. Based on the results of these GPS surveys and InSAR, it was found that during the period between 2000 and 2011, several locations in the Bandung basin have experienced land subsidence, with an average rate of about 8 cm/year and can reach up to about 23 cm/year in certain location and time. The observed land subsidence rates have also spatial and temporal variations. More detail information on the characteristics of land subsidence in Bandung can be seen in Abidin et al. [3, 4, 8] and Sumantyo et al. [26].

In Semarang, land subsidence has been studied using Levelling surveys [23, 28], GPS surveys [6, 9]), Microgravity surveys [27, 14] and InSAR technique [21, 22]. Based on the estimation from those measurement methods, land subsidence with rates of up to about 19 cm/year were observed during the period of 1999–2011 (see Table 4). More detail information on the characteristics of land subsidence in Semarang can be seen in Abidin et al. [6, 9].

For land subsidence study in urban areas, geospatial information is also important for visualizing the subsidence rates as derived from various geodetic measurement techniques (see Fig. 1 as an example). It will give more insights into the possible causes of land subsidence and also the possible areas that will be affected by land subsidence.

Table 2 Typical strength and limitation of geodetic method for land subsidence (LS) study in urban areas

	Leveling surveys	GPS surveys	InSAR	Microgravity
LS information	Point-wise	Point-wise	Continuous	Point-wise
Spatial coverage	Local	Local to regional	Local to regional	Local
Temporal coverage	User dependent	User dependent	Images availability dependent	User dependent
Ground benchmark	Required	Required	Not required	Required
Data acquisition (survey)	Day time and weather dependent	Day and night, weather independent	Dependent on satellite passes in the region	Day and night, weather dependent
Typical limitation	Laborious and time consuming	Signal obstruction by buildings, infrastructures and trees	Poor image coherence due to land use and land cover dynamics	Requires stringent observation strategy and quite costly
Typical precision level of LS	mm (relative)	mm-cm (relative)	mm-cm (relative)	mm-cm (relative)

Table 3 Observed subsidence rates in Jakarta; after Abidin et al. [7]

No.	Method	Subsidence rates (cm/year)	Observation period
1	Leveling surveys	1–9	1982–1991
		1–25	1991–1997
2	GPS surveys	1–28	1997–2011
3	InSAR	1–12	2006–2007

Table 4 Land subsidence rates in Semarang as observed from several geodetic methods; after Abidin et al. [9, 21]

No.	Method	Subsidence rates (cm/year)	Observation period
1	Leveling surveys	1–17	1999–2003
2	GPS surveys	1–19	2008–2011
3	PS InSAR	1–10	2002–2006
4	Microgravity	1–15	2002–2005

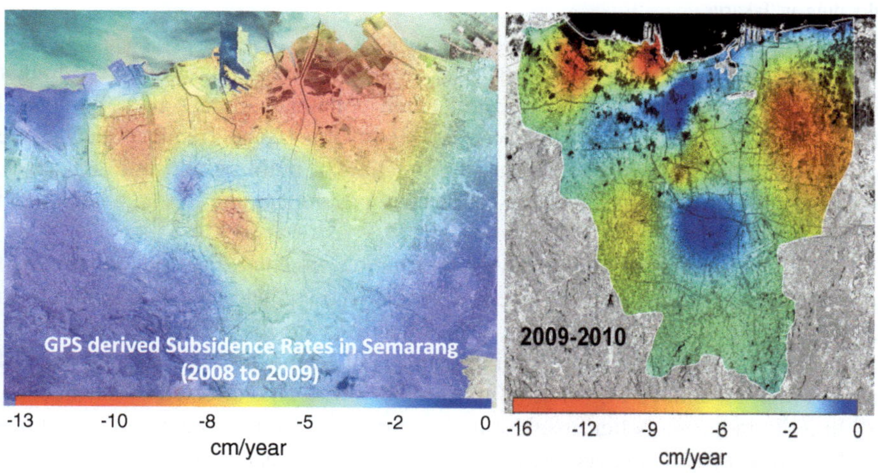

Fig. 1 Example of GPS-derived land subsidence rates in Semarang (*left*) and Jakarta (*right*). Kriging method was used for interpolating the data

3 Impacts of Land Subsidence

Geospatial information is also important for locating, mapping and studying the impacts of land subsidence in urban areas. In general, the impacts of land subsidence in urban areas can be seen in the forms of cracking of permanent constructions and roads, changes in river canal and drain flow systems, wider expansion of coastal and/or inland flooding areas, and malfunction of drainage system. Figure 2 shows some examples of land subsidence impacts at several locations in Jakarta, Bandung and Semarang which have relatively high observed subsidence rates.

Fig. 2 Examples of subsidence impacts in urban areas, after Abidin et al. [7–9]. **a** Subsided house in Bandung; **b** highway cracking in Bandung; **c** abandoned building in Semarang; **d** cracking in highway infrastructure and bridge in Semarang; **e** tilted building in Jakarta; **f** coastal flooding in Jakarta

In the case of Jakarta and Semarang, its coastal areas have relatively higher subsidence rates [5, 9]. This creates other collateral impact namely coastal flooding that its coverage tends to enlarge by times. The frequent and severe coastal flooding not just deteriorates the function of building and infrastructures, but also badly influences the quality of living environment and life (e.g. health and sanitation condition) in the affected areas.

Land subsidence will also have contribution in enlarging the coverage of inundated area during flooding event. Figure 3 shows that some flooded areas in 2010 are coinciding with areas of large subsidence in the period of 1999–2010. According to study done by Gumilar et al. [20], from the total flooded area of about 6388 ha, about 21 % or 1388 ha coincides with land affected by subsidence.

In order to have a better insight into the impacts of land subsidence, the features of land subsidence impacts should be located and mapped. In this case ground thruthing surveys should be conducted to find and locate the impact features and then plotted on the map. Figure 4 shows the distribution of land subsidence features in Bandung basin in 2011, overlapped with the GPS and InSAR derived subsidence from 1999 to 2010. From this figure it can be realized that the numbers of building, houses and other infrastructure affected by land subsidence phenomena are also numerous exhibiting either cracking, tilting, and general damage. The real number of land subsidence features maybe more than those shown in Fig. 4. This figure also shows that most damages occurred in the areas showing high subsidence, and also in the areas which spatially have differential subsidence.

The economic losses caused by inland and coastal subsidence are usually enormous, since many buildings and infrastructure are affected by land subsidence and its collateral coastal flooding disasters. Many houses, public utilities and a large population maybe also exposed to potential disaster. The corresponding maintenance cost is increasing by year. The related government and communities

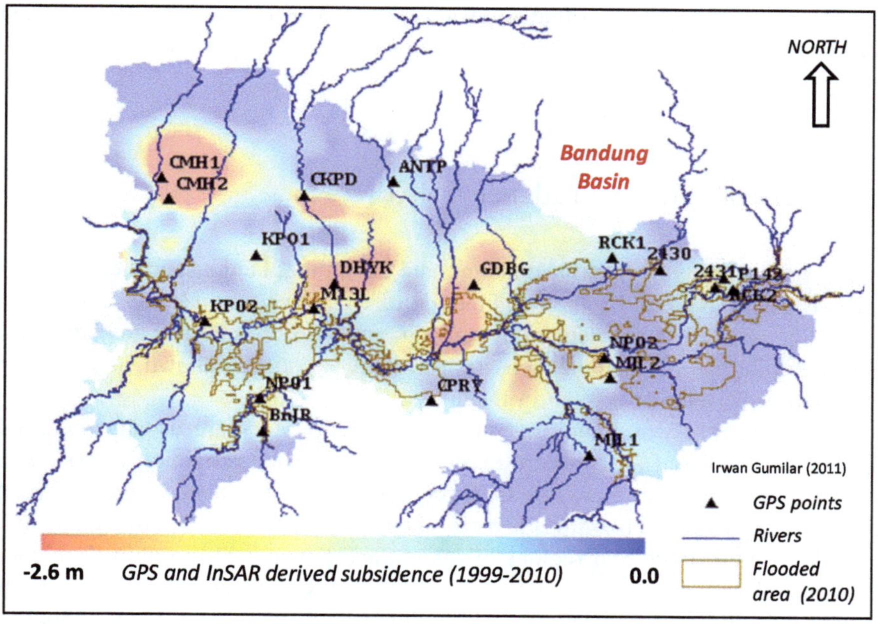

Fig. 3 Spatial correlation between land subsidence and flooding in Bandung basin; after Gumilar et al. [20]

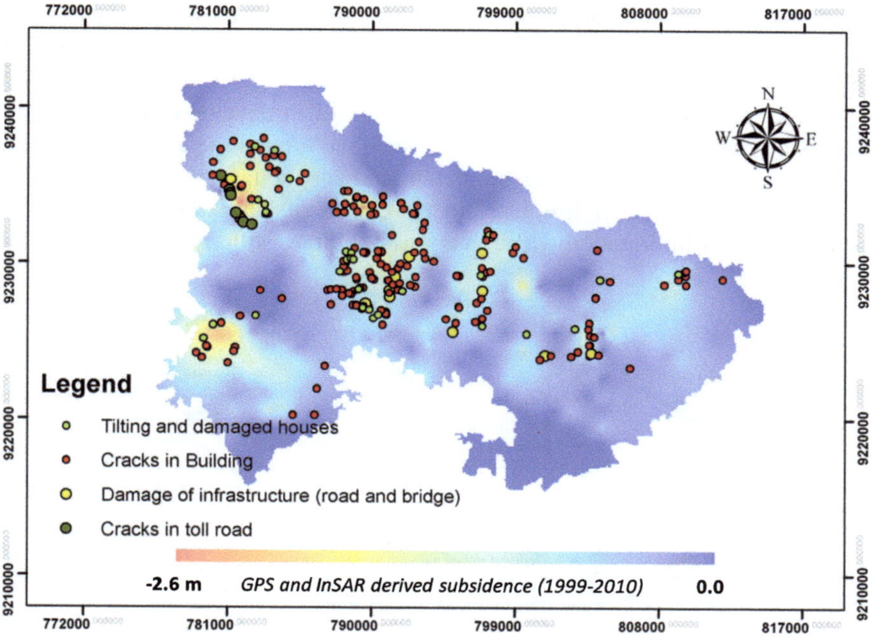

Fig. 4 Example of distribution of land subsidence impacts in Bandung, overlapped with the GPS and InSAR derived subsidence from 1999 to 2010; after Gumilar et al. [20]

are also usually required to frequently raise ground surface to protect roads and buildings. The living conditions, if the population affected by the coastal subsidence, are in general decreasing.

4 Causes of Land Subsidence

Based on our study in Jakarta, Bandung and Semarang, land subsidence in urban areas is mainly caused by the combination of excessive groundwater extraction, natural consolidation of young alluvium soil, and load of buildings and structure. The contribution of tectonic activity is relatively still unknown and can be expected to be relatively small in magnitude.

In this study it was found that higher subsidence rates are usually observed around the industrial areas where excessive groundwater extraction is expected. Correlation coefficient between GPS-observed land subsidence and observed groundwater level is also high, namely about 0.9 [4, 5]. It was also found that subsidence rates in coastal regions of Jakarta and Semarang are relatively higher than inland regions, indicating that there is also significant contribution from natural consolidation of relatively young alluvium deposits with high compressibility in those coastal regions.

Land subsidence also has a strong linkage with urban development process as has been shown by Abidin et al. [7]. Urban development in urban areas will usually increases built-up areas, population, economic and industrial activities, and also groundwater extraction, which can then lead to land subsidence phenomena. This on-going land subsidence in urban area will then affect the urban development process itself (see Fig. 5). For sustainable urban development, the characteristics of land subsidence phenomenon inside and around the urban area should be well understood for its proper mitigation and adaption.

In order to have better insights into the causes of land subsidence in urban areas, geospatial information will be very useful along with hydrogeological and

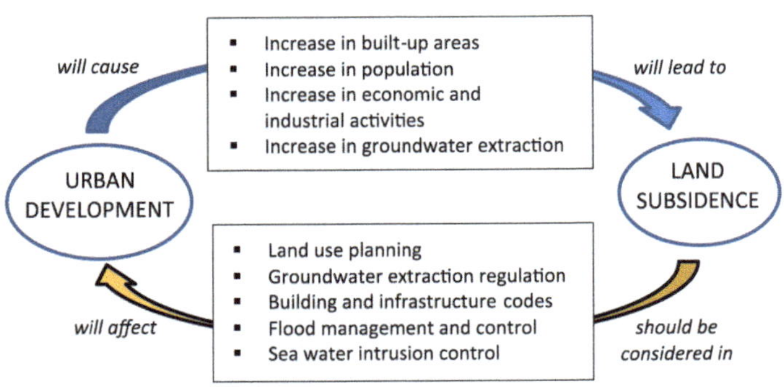

Fig. 5 Urban development and land subsidence relation; after Abidin et al. [7]

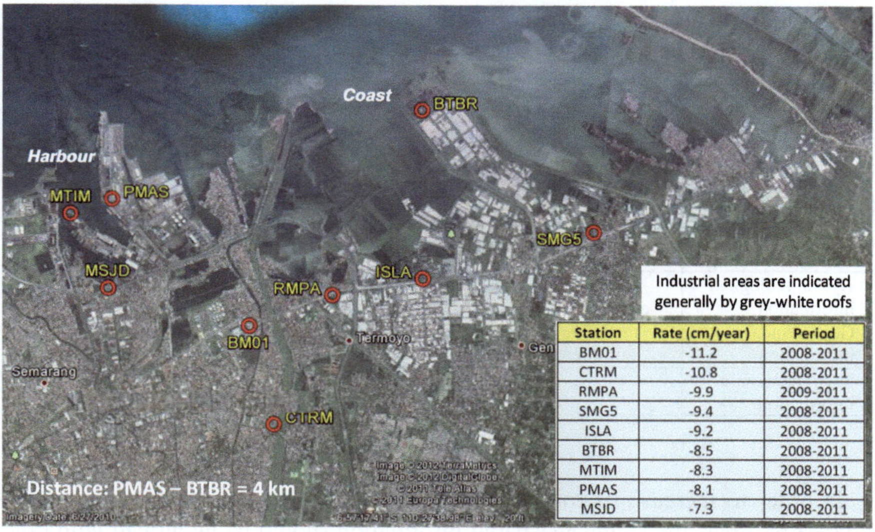

Station	Rate (cm/year)	Period
BM01	-11.2	2008-2011
CTRM	-10.8	2008-2011
RMPA	-9.9	2009-2011
SMG5	-9.4	2008-2011
ISLA	-9.2	2008-2011
BTBR	-8.5	2008-2011
MTIM	-8.3	2008-2011
PMAS	-8.1	2008-2011
MSJD	-7.3	2008-2011

Fig. 6 Location of several GPS stations and their estimated land subsidence rates in Semarang; after Abidin et al. [9]. Subsidence rates are based on four GPS surveys since 2008 up to 2001

geotechnical information. An example is given by Fig. 6 which is related to land subsidence in Semarang.

Geospatial information in Fig. 6 show that the relatively significant subsidence rates (i.e. 7–11 cm/year) were observed at stations in the coastal region of Semarang which its soil mainly composed by alluvial deposits, and closed to the industrial facilities which usually extract a lot of groundwater for their activities. This indicates the higher contribution of groundwater extraction and natural consolidation on the observed subsidence in this area. From land use and land coverage shown in Fig. 6 it can be expected that the load of buildings and constructions in the area can also have contribution in the observed subsidence rates. However, more data and further investigations are required to understand the intricacies of the relationship between land subsidence with natural consolidation, groundwater extraction and load of buildings and construction in this norther region of Semarang.

5 Closing Remarks

Geospatial information has important roles for risk assessment of land subsidence in urban areas (see Table 5). Its main role is in estimating the magnitudes and rates of land subsidence, and its variation both in spatial and temporal domain. Geospatial information will also needed for studying causes and impacts of land subsidence, and also for estimating economic losses due to land subsidence.

Table 5 Roles of geospatial information for risk assessment of subsidence in urban areas

<table>
<tr><th rowspan="2">GEOSPATIAL INFORMATION</th><th colspan="4">LAND SUBSIDENCE</th></tr>
<tr><th>Characteristics</th><th>Causes</th><th>Impacts</th><th>Cost</th></tr>
<tr><td>Leveling data</td><td rowspan="4">observed height differences in spatial and temporal domain</td><td></td><td></td><td></td></tr>
<tr><td>GPS data</td><td colspan="3">Location of land subsidence features in the field</td></tr>
<tr><td>InSAR data</td><td></td><td></td><td></td></tr>
<tr><td>Gravity data</td><td></td><td></td><td></td></tr>
<tr><td>Satellite images</td><td></td><td colspan="3" rowspan="3">spatial distribution of man-made and natural features (e.g. built-up areas, industries, buildings and infrastructures, forest, paddy fields, etc.)</td></tr>
<tr><td>Aerial photos</td><td></td></tr>
<tr><td>Topographic maps</td><td></td></tr>
<tr><td>Hydro-geological information</td><td rowspan="2">numerical estimation and modeling of subsidence</td><td colspan="2">spatial characteristics and dynamics of groundwater, aquifers and confining beds</td><td></td></tr>
<tr><td>Geotechnical information</td><td colspan="2">physical properties of soil and rock in spatial domain</td><td></td></tr>
<tr><td>Socio-economic information</td><td></td><td colspan="3">spatial distribution and characteristics of population and economic activities</td></tr>
</table>

In order to increase its role and contribution for risk assessment of land subsidence in urban areas, this geospatial information should be integrated with other spatio-temporal information, namely hydrogeological, geotechnical and also socio-economic information. It will be better if these information is managed by using GIS that is open and accessible by all related governmental institutions and if possible also by the public through web-based GIS.

References

1. H.Z. Abidin, R. Djaja, D. Darmawan, S. Hadi, A. Akbar, H. Rajiyowiryono, Y. Sudibyo, I. Meilano, M. A. Kusuma, J. Kahar, C. Subarya, Land subsidence of Jakarta (Indonesia) and its geodetic-based monitoring system. Nat. Hazards J. Int. Soc. Prevention Mitigation Nat. Hazards **23** (2/3), 365–387 (2001)
2. H.Z. Abidin, H. Andreas, R. Djaja, D. Darmawan, M. Gamal, Land subsidence characteristics of Jakarta between 1997 and 2005, as estimated using GPS surveys, *GPS Solutions* vol. 12, no. 1, (Springer, Berlin, 2008a), pp. 23–32
3. H.Z. Abidin, H. Andreas, M. Gamal, A. D. Wirakusumah, D. Darmawan, T. Deguchi, Y. Maruyama, Land subsidence characteristics of the Bandung Basin, Indonesia, as estimated from GPS and InSAR. J. Appl. Geodesy **2**(3), 167–177, de Gruyter, doi:10.1515/JAG.2008.019
4. H.Z. Abidin, H. Andreas, I. Gumilar, I. Wangsaatmaja, Y. Fukuda, T. Deguchi, Land subsidence and groundwater extraction in Bandung Basin (Indonesia), in *Trends and Sustainability of Groundwater in Highly Stressed Aquifers*, (IAHS Publication no. 329, 2009), ISSN: 0144-7815, pp. 145–156
5. H.Z. Abidin, H. Andreas, M. Gamal, I. Gumilar, M. Napitupulu, Y. Fukuda, T. Deguchi, Y. Maruyama, E. Riawan, Land subsidence characteristics of the Jakarta Basin (Indonesia) and its relation with groundwater extraction and sea level rise, in *Groundwater Response to Changing Climate*, IAH Selected Papers on Hydrogeology No. 16, ed. by M. Taniguchi, I.P. Holman, chap. 10 (CRC Press, London, 2010a), pp. 113–130, ISBN: 978-0-415-54493-1
6. H.Z. Abidin, H. Andreas, I. Gumilar, T.P. Sidiq, M. Gamal, D. Murdohardono, Supriyadi, Y. Fukuda, Studying land subsidence in Semarang (Indonesia) using geodetic methods, in

Proceedings of the FIG Congress 2010, FS 4D—Landslide and Subsidence Monitoring II, Sydney, Australia, 11–16 April 2010
7. H.Z. Abidin, H. Andreas, I. Gumilar, Y. Fukuda, Y.E. Pohan, T. Deguchi, Land subsidence of Jakarta (Indonesia) and its relation with urban development. Nat. Hazards **59**(3), 1753–1771 (2011)
8. H.Z. Abidin, I. Gumilar, H. Andreas, D. Murdohardono, Y. Fukuda, On causes and impacts of land subsidence in Bandung Basin, Indonesia. Environ. Earth Sci. (2012). doi:10.1007/s12665-012-1848-z
9. H.Z. Abidin, H. Andreas, I. Gumilar, T.P. Sidiq, Y. Fukuda, Land subsidence in coastal city of Semarang (Indonesia): characteristics, impacts and causes. J. Geomatics, Nat. Hazards Risk (2012). doi:10.1080/19475705.2012.692336
10. S.M. Buckley, P.A. Rosen, S. Hensley, B.D. Tapley, Land subsidence in Houston, Texas, measured by radar interferometry and constrained by extensometers. J. Geophys. Res. **108**(B11), 2542 (2003). doi:10.1029/2002JB001848
11. E. Cabral-Cano, T.H. Dixon, F. Miralles-Wilhelm, O. Díaz-Molina, O. Sánchez-Zamora, R.E. Carande, Space geodetic imaging of rapid ground subsidence in Mexico City. Geol. Soc. Am. Bull. **120**, 1556–1566 (2008)
12. J.C. Chai, S.L. Shen, H.H. Zhu, X.L. Zhang, Land subsidence due to groundwater draw down in Shanghai. Geotechnique **54**(2), 143–147 (2004)
13. C.T. Chen, J.C. Hu, C.Y. Lu, J.C. Lee, Y.C. Chan, Thirty-year land elevation change from subsidence to uplift following the termination of groundwater pumping and its geological implications in the Metropolitan Taipei Basin, Northern Taiwan. Eng. Geol. **95**, 30–47 (2007)
14. Y. Fukuda, T. Higashi, S. Miyazaki, T. Hasegawa, S. Yoshii, Y. Fukushima, J. Nishijima, M. Tanigushi, H.Z. Abidin, R.M. Delinom, Groundwater and land subsidence monitoring in 3 mega-cities, Indonesia, by means of integrated geodetic methods, Paper presented at the AGU Fall Meeting 2008, San Fransisco, 15–19 December 2008
15. M. Ishii, F. Kuramochi, T. Endo, Recent tendencies of the land subsidence in Tokyo, Publication no. 121 of the International Association of Hydrological Sciences, in *Proceedings of the Anaheim Symposium*, Dec, pp. 25–34 (1970)
16. S. Murayama, Land subsidence in Osaka, in *Proceedings of The Tokyo Symposium on Land Subsidence*, International Association of Scientific Hydrology and UNESCO, Tokyo , vol. 1, Sept 1969, pp. 105–130 (1970)
17. A.H.-M. Ng, L. Ge, X. Li, H.Z. Abidin, H. Andreas, K. Zhang, Mapping land subsidence in Jakarta, Indonesia using persistent scatterer interferometry (PSI) technique with ALOS PALSAR. Int. J. Appl. Earth Observ. Geoinf. **18**, 232–242 (2012)
18. G. Beutler, H. Bock, R. Dach, P. Fridez, A. Gade, U. Hugentobler, A. Jaggi, M. Meindl, L. Mervant, L. Prange, S. Schaer, T. Springer, C. Urschl, P. Walser, in *Bernese GPS software Version 5.0*, ed. by R. Dach, U. Hugentobler, P. Fridez, and M. Meindl (eds) Astronomical Institute, University of Berne, pp. 612 (2007)
19. M.A.C. Dam, P. Suparan, J.J. Nossin, R.P.G.A. Voskuil, GTL Group, A chronology for geomorphological developments in the greater Bandung area, West-Java, Indonesia". J. Southeast Asian Earth Sci. **14**(1–2), 101–115 (1996)
20. I. Gumilar, H.Z. Abidin, H. Andreas, T.P. Sidiq, M. Gamal, On mapping and evaluating the impacts of land subsidence in Bandung Basin, in *Proceedings of the FIG 2012 Working Week*, TS09E—Planning and Risk Management, Rome, Italy, 6–10 May 2012, Proceedings available at : http://www.fig.net/pub/fig2012/index.htm
21. F. Kuehn, D. Albiol, G. Cooksley, J. Duro, J. Granda, S. Haas, A. Hoffmann-Rothe, D. Murdohardono, Detection of land subsidence in Semarang, Indonesia, using stable points network (SPN) technique. Environ. Earth Sci. (2009). doi:10.1007/s12665-009-0227-x
22. A.M. Lubis, T. Sato, N. Tomiyama, N. Isezaki, T.Yamanokuchi, Ground subsidence in Semarang-Indonesia investigated by ALOS–PALSAR satellite SAR interferometry. J. Asian Earth Sci. **40**(5), 1079–1088
23. M.A. Marfai, L. King, Monitoring land subsidence in Semarang, Indonesia. *Environmental Geology*, vol. 53 (Springer, Verlag, 2007) pp. 651–659

24. N. Phien-wej, P.H. Giao, P. Nutalaya, Land subsidence in Bangkok, Thailand. Eng. Geol. **82**(4), 187–201 (2006)
25. D. Rismianto, W. Mak, Environmental aspects of groundwater extraction in DKI Jakarta: Changing views, in *Proceedings of the 22nd Annual Convention of the Indonesian Association of Geologists*, Bandung, 6–9 Dec 1993, pp. 327–345
26. J.T.S. Sumantyo, M. Shimada, P.P. Mathieu, H.Z. Abidin, Long-term consecutive DInSAR for volume change estimation of land deformation. IEEE Trans. Geosci. Remote Sensing **50**(1), 259–270 (2012)
27. Supriyadi, Separation of gravity anomaly caused subsidence and ground water level lowering of time lapse microgravity data using model based filter: case study Semarang Aluvial Plain (in Indonesian), in PhD Dissertation. Institute of Technology Bandung, September, p 146
28. H. Sutanta, A. Rachman, Sumaryo, Diyono, Predicting land use affected by land subsidence in semarang based on topographic map of scale 1:5.000 and leveling data. in *Proceedings of the Map Asia 2005 Conference* (in CDRom), Jakarta, 22–25 Aug 2005

The STIG: Framework for the Stress-Test for Infrastructures of Geographic Information

Bujar Nushi and Bastiaan Van Loenen

Abstract Spatial data infrastructure (SDI) facilitates the collection, maintenance, dissemination and use of spatial information. To stimulate SDI development effectively and efficiently, it is key to assess the progress and benefits of the SDI. SDI is difficult to assess because of its complex, dynamic, multi-faceted and constantly evolving nature. Several SDI assessment methods exist. However, these are still in an infancy stage and none of these appear to meet the requirements of practitioners. As a result, SDI decision makers are still without any guidance on the success of their SDI. In this paper we propose a new method for SDI assessment: The STIG, a Stress-test for Infrastructures of Geographical information. The development and application of the Stress-test methodology will provide new valuable information for decision-makers about the aspects of SDIs that need to be improved in order to take full advantage of the potential benefit of the SDIs, especially in the instance of disaster management.

Keywords SDI assessment · The STIG · Stress-test · Disaster management

1 Introduction

Information has grown to be of vital importance to the economic and social development of a country. Location-based information, in particular, is of increasing importance for the successful effecting of our daily private and public tasks.

B. Nushi (✉) · B. Van Loenen
TU Delft, Delft, The Netherlands
e-mail: b.nushi@tudelft.nl

B. Van Loenen
e-mail: B.vanLoenen@tudelft.nl

S. Zlatanova et al. (eds.), *Intelligent Systems for Crisis Management*, Lecture Notes in Geoinformation and Cartography, DOI: 10.1007/978-3-642-33218-0_20,

The specialty of geo-information has resulted in the emerging of Spatial Data Infrastructures (SDIs) or Geographic Information Infrastructures (GIIs) (see Masser [1], 2007). SDIs are network-based solutions, which enable easy, consistent and effective access to geo-information and services offered by public agencies and others (see [2]). SDIs enable users to save resources, time and effort when trying to acquire new data sets (Rajabifrad and Williamson 2002; Chan 2001). SDIs therefore play a crucial role in the management of geo-information and that associated to the administration of our societies. In the European Union, the INSPIRE Directive (2007/2/EC) aims to establish a European SDI promoting exchange, sharing, access and use of geo-information and services across the various levels of public authority and across different sectors.

Despite these many initiatives, widespread access and use of geographical information is still problematic in many instances. The existing activities are still fragmented and poorly coordinated in many countries and regions.

1.1 Why Assess the SDIs?

Large sums of money are and have been invested in SDI initiatives. Rhind (2000) estimated an expenditure of $10 billion for the US SDI and $2 billion for the SDI of the UK. The investment requirements for an Infrastructure for Spatial Information in the European Community (INSPIRE) was estimated to vary from €202 to €273 million each year for each European Member State [3].

Given this expenditure and society's interest in the effective and efficient use of public funds, it is imperative that these SDI services and initiatives should be assessed on their effectiveness and efficiency. Public sector administrators seek justification for providing public sources for SDIs, and SDI practitioners require a measure of success of their SDI strategy and implementation and guidance to move towards more useful, effective, and efficient SDIs. Since the value of geo-information comes from its use (Onsrud and Rushton 1995), one would expect that SDI assessment has put the use and user of geo-information central. However, currently developed and emerging assessment methods mostly ignore the user of geo-information.

2 Existing SDI Assessment Methods

Driven by different goals and interests, researchers of the last decade have tried several SDI assessment methods. Some authors focused on the description of the SDI itself (Onsrud 1998; [1]; Van Orshoven 2003–2004; Vandenbroucke and Janssen 2005–2006; Delgado-Fernandez and Crompvoets 2007), others paid more attention to the methodology (Steudler et al. 2004; [4, 5, 2]; Rodriguez-Pabon 2005; Grus 2006–2007). All these attempts either concentrate on one aspect of a

SDI, are restricted by one region, describe SDI development in only a few countries, or are still conceptual in nature.

Grus (2008) introduced the multi-view framework to assess SDIs. This framework acknowledges the difficult task of assessing SDIs due to their complex, multi-faceted, dynamic and constantly evolving nature, unclear objectives and poor knowledge about the implications of the current SDI-use and the current demands. It is argued that the main strengths of this assessment design lie in its wider scope, flexibility, multi-disciplinary and reduced bias in the assessment results. The main idea behind the framework is that it covers all three purposes of assessing SDIs—accountability, knowledge and development. It also acknowledges the multi-faceted character of SDIs. The core of the assessment framework is represented by the multiple assessment approaches that focus on different aspects of the SDI. Each approach treats SDIs from a different perspective. The essence of the multi view framework is that it accepts the multiple facets of an SDI and therefore accepts its complexity in terms of multiple definitions. Moreover, each assessment approach covers at least one of the three purposes of the assessment: accountability, knowledge and development.

Based on this overview of SDI assessment landscape, one can conclude that the SDI assessments are still in their infancy. Clearly there is a need for a new SDI assessment method, which is more extensive, comprehensive, user-oriented, demand-driven, diverse and closely tied to explicit targets.

3 Assessing an SDI from a User Perspective: The Stig?

"Stig exists for one reason alone—to wring every last drop of performance out of any car he drives around our demanding test track." J. Clarkson "BBC Top Gear" *(Stig = S-tress–Test for Cars. The STIG = Stress–Test for Infrastructures of Geo-information.)*

3.1 Multiple Disasters: A keystone for the SDI Assessment?

As indicated by Zlatanova et al. [6], SDIs can contribute to the effective management, the reduction and even the prevention of disasters. Recently, we have witnessed the increasing significance of disasters to society. In Haiti, a poorly 'mapped' country, the 2010 earthquake disaster response was primarily based on instantly, but ad hoc collected and on the fly generated maps with the help of software that had to be developed by international NGOs after the disaster (Clarke et al. 2010). The recent earthquake in Sendai, Japan, tragically followed by a tsunami and the Fukushima nuclear facility disaster, have taught us that even countries with a presumably well established SDI are experiencing enormous difficulties.

The success of disaster management does not only depend on well-defined policies and procedures, but also largely depends on the successful integration of related information to support well-informed decision-making during the disaster. Disaster managers have to deal with large amounts of existing and operational data under an unpredictable and time-critical manner. During the disaster, the availability and interoperability of the emergency services needs to be ensured as well as the availability and interoperability of the wide variety of required information. This would provide the appropriate information at the right place and at the right moment [6]. This is exactly the objective of many SDIs: to effectively and efficiently support decision-making by providing the needed information at the right time to the right place and/or people. The decision making includes decisions in disaster management and emergency response.

3.2 How may SDIs Contribute to Disaster Management?

One of the important benefits of well structured SDIs could be reduction and likely avoidance of disasters. Geo-information science in general, is very well suited to provide adequate information. Significant attention for disasters is not new, but we are seeing subtle change and developments in the perception, with the focus shifting away from the disaster itself and towards risk.

An earthquake may not only destroy buildings, but also destabilize dams or mountain slopes, resulting in the risk of a flood or a landslide. Hence an area is rarely exposed to a single hazard, in this case an earthquake, but also to secondary risks emanating from the same source. Multiple hazards may thus occur, each with their own area of impact and perhaps differing in times of recurrence, magnitude and nature of vulnerability. The impact of multiple hazards is determined by constructing single-hazard risk maps, subsequently summed up. However, summing up is often not the most appropriate mathematical operation, because hazards and vulnerabilities can compound and amplify each other.

Traditional SDI assessments have focused on the data access facility (Crompvoets 2006), readiness to embrace SDI in a specific country [4] or organizational issues [5]. But multiple disasters effects have a broader impact on SDI. However, risk theory has evolved to reflect a more complex and subtle reality, raising the question of whether current SDI assessment methods adequately reflect that theory. Risk theory distinguishes four main types of vulnerability: physical, social, environmental and economic, at times supplemented by technical, political, cultural, educational and institutional vulnerability.

3.3 A New SDI Assessment Methodology: Stress-Testing

Stress-Testing is an important risk management tool that is used by banks as part of their internal risk management. Stress-Testing is a form of testing that is used to determine the stability of a given system or entity. It involves testing beyond normal operational capacity, often to a breaking point, in order to observe the results (IMF [6]).

"Stress-testing" is a term that comes with much mystique. To put it more simply, a stress-test is a rough estimate of how the value of a portfolio changes when large changes are made to some of its risk factors (such as in case of disasters). A widespread example of a stress-test is from the financial sector. The object of a stress test is to understand the sensitivity of the portfolio to changes in various risk factors. The assumed changes in risk factors are usually made large enough to impose some "stress" on the portfolio. According to Illing and Liu [7], stress tests can be applied to both the asset and liability sides of a portfolio. More complex stress tests involving multiple risk scenarios or changes in the macroeconomic environment still amount to the same thing: revaluing a portfolio under a different set of assumptions. Stress tests usually produce a numerical estimate of the change in value of the portfolio.

The starting point for this research is the consideration that SDIs are organized similar to (large) financial institutions and it is therefore possible to apply the principles of 'Stress-Tests' to a user-oriented SDI. The SDI stress-tests can provide information on the behaviour of the SDI under exceptional, but plausible shocks (disasters), helping SDI policymakers to assess the significance of the system's vulnerabilities. Stress-tests cover a range of methodologies. Complexity can vary, ranging from simple sensitivity tests to complex stress-tests. Stress-tests may be performed at varying degrees of aggregation, from the level of an individual instrument up to the institutional level. The SDI Stress-Testing could alert SDIs organizations to adverse unexpected outcomes related to a variety of risks and provides an indication of how much useful data might be accessed in the times of large risk impact. While stress-tests provide an indication of the appropriate level of SDI necessary to endure deteriorating disaster conditions, alternatively an SDI structure may employ other actions to help mitigate increasing levels of risk.

4 The Stig Research Framework

In this paper we propose a new method for SDI assessment: The STIG, a Stress-test for Infrastructures of Geographical information. This paper is intended to provide guidance and answer some of the basic questions that may arise as part of the process of SDI stress-testing. This paper is an announcement of the research still to be carried out and does not focus on the technical aspects of SDI stress-testing but is intended as guidance at the beginning of the process. The development and application of the Stress-test methodology will provide new valuable information for decision-makers about the aspects of SDIs that need to be improved in order to fully utilise the potential benefit of the SDIs, especially in the instance of disaster management.

4.1 Research Objective

This research will develop a stress-test for infrastructures of geographic information (SDI) focusing on the geographic information needs of disaster managers and disaster rescue workers. The main objective of this research is: developing a Stress-Test for Infrastructures of Geographic information.

Based on the research objective the following general research questions can be formulated:

- To what extent are Infrastructures for Geographic Information supportive to managing disasters?
- What SDI capabilities does disaster management require?
- How to assess these required capabilities efficiently?
- How to communicate with/involve SDI practitioners so that they can utilise the results?
- What is the SDI stress-test result for several countries in Europe?

4.2 Method/Approach

The development of this kind of interactive method is usually designed through an iterative process that involves design, evaluation, and redesign. The research will be conducted in three major phases divided in 3 steps (Fig. 1): (1) Explore and Theorize, (2) Develop and Optimize and (3) Review and Evaluate.

The first three research question will be handled in the Explore and Theorize step (yellow) whiles other two subsequently in next two steps.

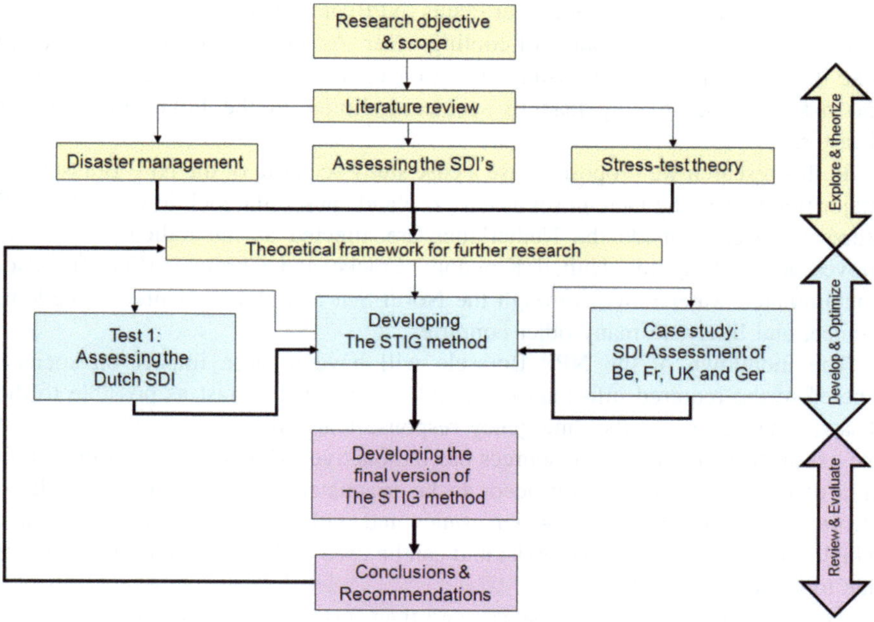

Fig. 1 Stages of The STIG research

4.2.1 Explore and Theorize

After defining the research objective and the scope of the research we will continue with the literature review. By critically reviewing existing literature on disaster management, different SDI assessment frameworks and stress-test assessment methods we attend to have a general theoretical overview of the relevant topics for this research. During this stage a formal definition of the users would be established so that their requirements of a practical stress-test assessment methodology could be identified. The needs of the users that are managing the disaster will also be assessed.

This information, along with the findings of detailed literature survey, will provide the basis for the development of The STIG methodology.

4.2.2 Develop and Optimize

To validate The STIG method on all aspects of the SDIs, the research uses a case study design. By generating the overlay of different hazard and risk layers from the ESPON Hazard project (European Spatial Planning Observation Network), we have generated an integrated vulnerability map, which will distinguish the case study area for this research.

In many countries Nuclear Power Plants (NPP) are often located on the coast, in order to provide a ready source of cooling water. As a consequence, NPPs need to take the risk of flooding and tsunamis into account. Failure to correctly calculate the risk of flooding may lead to accidents similar to the Fukushima nuclear disaster.

In this research we propose to use a case study scenario of the NPP Borssele. A major flooding in the Dutch province of Zeeland caused the meltdown of the NPP Borssele. Areas outside the Netherlands are affected as the radioactive plume moves across Belgium, United Kingdom, France and Germany (Fig. 2). Also, contaminated water moves through the North Sea to Belgium, United Kingdom, France, and likely, to many other countries.

The meltdown of the NPP Borssele will have a large impact on society. Therefore the required information should be provided as fast as possible to the disaster managers and the emergency response team. It concerns information of the extent to which natural resources are radioactive, who is present in the direct area of the NPP Borssele and needs to be evacuated, how these people will be informed and evacuated, via which routes and with what transportation means, which transportation networks exist and can be used and where the traffic jams are due to hysteric population.

The SDI stress-testing process is more than just applying a set of formulas to spreadsheets of numbers, but involves a series of judgments and assumptions that can be equally or perhaps even more, critical to producing meaningful results as the actual calculations themselves. Each assumption, aggregation, or analytical approximation made in the process can introduce wide margins of error to the results, and so much care should be taken in their estimation and interpretation.

The case study will be carefully selected so that it would allow us to assess the Dutch SDI as a first test of The STIG Assessment method. To predict similar results in means of a literal replication [7], we would implement The STIG method

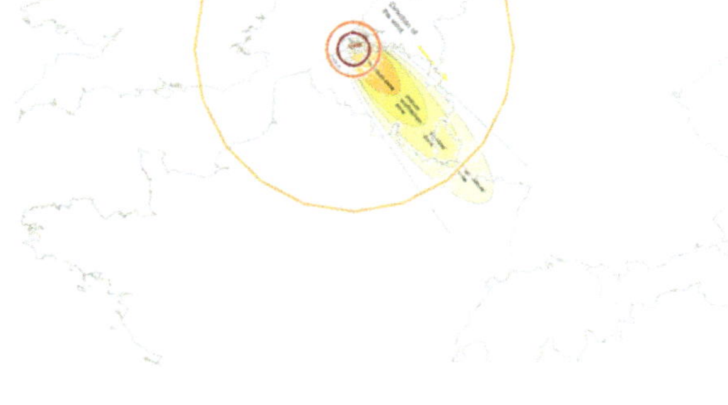

Fig. 2 Case study: 300 km zone around the fictive melt down of NPP Borssele

to assess the SDIs of Belgium, United Kingdom, France and Germany by using the same Borssele meltdown scenario. The methods that will be used include; hierarchical task analysis, informal interviews, questionnaires, structured discussion groups and user trials.

4.2.3 Review and Evaluate

Developing of the final prototype of The STIG method will be followed by a formal comparative usability evaluation of the method. The structure of The STIG method lifecycle will probably undergo a number of alterations as the research progresses.

In order to increase the valorisation of the research results, we plan to organize annual workshops, meetings and general conferences on The STIG SDI assessment method targeting SDI decision-makers and disaster managers and practitioners. During these events the key research results will be presented, along with practices justifying the results. Strategy documents, workshop presentations and publication in professional journals are other important means of increasing valorisation.

5 Discussion

This research aims at developing a stress-test assessment method for SDIs from a user perspective, especially for users from the field of disaster management. However, the research is still at an early stage and most of the work still needs to be done. The aim of the proposed The STIG assessment method is not only to assess SDI performance, but also to deepen our knowledge about SDI mechanisms and support SDI development.

Currently, there is no operational assessment method that allows for comprehensive SDI assessment supporting SDI practitioners in building a SDI. Our aim is to complement the existing array of SDI assessment with a new innovative comprehensive operational stress-test SDI assessment method which would be more efficient and effective for SDI practitioners.

Some obstacles and difficulties may be encountered when developing the STIG assessment method. The complexity of SDIs, the wide scope of disaster management, the high range of potential users and the wide-ranging set up of the study makes this research very demanding and innovative. The next consideration is the difference in data availability between various national SDIs involved in the case study.

The development and application of the stress-test methodology will provide new valuable information for decision-makers about the aspects of SDIs that need to be improved in order to fully exploit the potential benefit of the SDIs, especially in the instance of disaster management. Authors strongly believe that if The STIG methodology would be implemented the significance for theory and practice would be very significant.

References

1. I. Masser, All shapes and sizes: the first generation of national spatial data infrastructures. Int. J. Geogr. Inf. Sci. **3**(1), 67–84 (1999)
2. B. Van Loenen, Developing geographic information infrastructures, The role of information policies. Dissertation, 2006
3. INSPIRE (2003) (M. Craglia), (2003). Contribution to the extended impact assessment of INSPIRE. http://inspire.jrc.ec.europa.eu/reports/fds_report_sept2003.pdf
4. Delgado et al., Assessing an SDI Readiness Index. In FIG Working Week, 2005, and GSDI-8, From Pharaohs' to Geoinformatics, (Cairo, Egypt, 2005), 16–21 Apr 2005
5. B. Kok, B. van Loenen, How to assess the success of national spatial data infrastructures? Comput. Environ. Urban Syst. **29**, 699–717 (2005)
6. IMF (International Monetary Fund), (2006), "FSAP—Technical Note—Stress Testing Methodology and Results," IMF Country Report No. 06/216 (Washington: IMF), available at http://www.imf.org/external/pubs/ft/scr/2006/cr06216.pdf
7. M. Illing, Y. Liu, Measuring Financial Stress, Financial System Review, Bank of Canada, December (Ottawa, Bank of Canada), the Web at http://www.bankofcanada.ca/en/fsr/2003/fsr_1203.pdf

Part IV
Best Practices

Towards an Integrated Crowd Management Platform

Nico Van de Weghe, Rik Bellens, Tom De Jaeger, Sidharta Gautama, Roel Huybrechts, Beat Meier and Mathias Versichele

Abstract To date, numerous tracking technologies are being used in order to get insight in various situations where movements of objects are of major importance. Moving objects might be cars moving along the road network, animals running in the forest, people visiting mass events. Our research focuses on this last specific domain. Our aim is to build an integrated centralized system capable of acquiring, analyzing, and modeling the complex movement interactions occurring at mass-events in order to fully support the organizers in crowd management, security and possible evacuation. The integrated system will focus on two systems developed over the past years and successfully deployed at several mass events. The first

N. Van de Weghe (✉) · R. Huybrechts · M. Versichele
Department of Geography, Ghent University, Gent, Belgium
e-mail: nico.vandeweghe@ugent.be

R. Huybrechts
e-mail: roel.huybrechts@ugent.be

M. Versichele
e-mail: mathias.versichele@ugent.be

R. Bellens · S. Gautama
IOF Innovation Centre i-KNOW-UGent, Gent, Belgium
e-mail: rik.bellens@ugent.be

S. Gautama
e-mail: sidarta.gautama@ugent.be

T. De Jaeger
TERSEC, Gent, Belgium
e-mail: tdj@tersec.be

B. Meier
FAST Protect, Gent, Belgium
e-mail: beat.meier@fastprotect.net

S. Zlatanova et al. (eds.), *Intelligent Systems for Crisis Management*, Lecture Notes in Geoinformation and Cartography, DOI: 10.1007/978-3-642-33218-0_21,

system (*BlueMAP*) focuses on visitors, the second system (*Terra 3D Incident Management Platform*) focuses on the emergency services.

Keywords Mass event · Tracking · Crowd management · Security

1 BlueMAP

Given the ubiquity of Bluetooth-enabled electronic devices (e.g. mobile phones), Bluetooth technology is increasingly being put forward as a simple and low-cost alternative for the reconstruction of spatial behavior. Designed as a method of cable replacement between devices, Bluetooth constitutes a low-power, short-range radio technology. As users of portable applications of the technology very often leave their Bluetooth enabled as the default, scanners can trace the location of these devices—and by extension their users—over time by means of a unique Media Access Control (MAC) address. This way, the trajectory of an individual may be approximated by a spatiotemporal sequence of observations at different scanners in a study area. The MAC address, albeit unique, is not linked to a specific person through any type of central database, thus minimizing privacy concerns. As such, Bluetooth allows for non-participatory, unannounced and simultaneous tracking of a large number of anonymous individuals.

Despite its potential, however, the use of Bluetooth tracking to understand crowd dynamics at small-scale events with a large concentration of people has barely been examined to date. Furthermore, there are no generally accepted analysis tools that are specifically designed to derive useful spatiotemporal patterns from Bluetooth data. This is precisely one of the key research domains of the CartoGIS research group (http://geoweb.ugent.be/cartogis) of Ghent University. In the group, both fundamental and applied research on various aspects of cartography (e.g. map production, data quality, historical cartography) and geographical information science are conducted. Part of the group is specialized in the tracking of moving objects; visualizing, analyzing and modeling of spatiotemporal information; and cognition and linguistics of moving objects. The group gained a wide experience in setting up practical experiments in the area of Geographical Information Technology, more specifically with respect to the research of the movement behavior of persons at mass-events (such as The Ghent Festivities, a 10-day cultural and theatre festival taking place in the historic city center of Ghent in Belgium; and Rock Werchter, a major rock festival with about 300,000 people over 4 days) by use of the Bluetooth technology. These experiments resulted in several scientific publications (e.g. [1, 2]) and technical reports (see: http://geoweb.ugent.be/cartogis/research/tracking-of-moving-objects).

Typically the approach consists of a combination of data acquisition and data analysis. The data acquisition, based on Bluetooth scanners strategically located over the research area, is controlled by a central server system (See Fig. 1), and supports several types of visualizations and analyses (both in real-time (See Fig. 2)

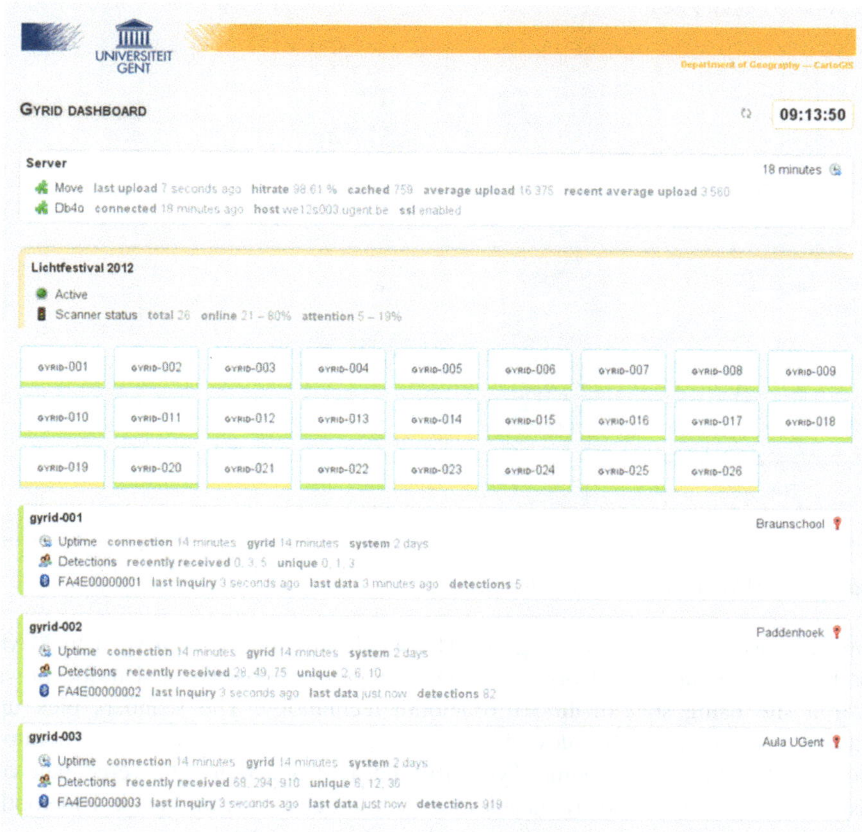

Fig. 1 BlueMAP: central server system

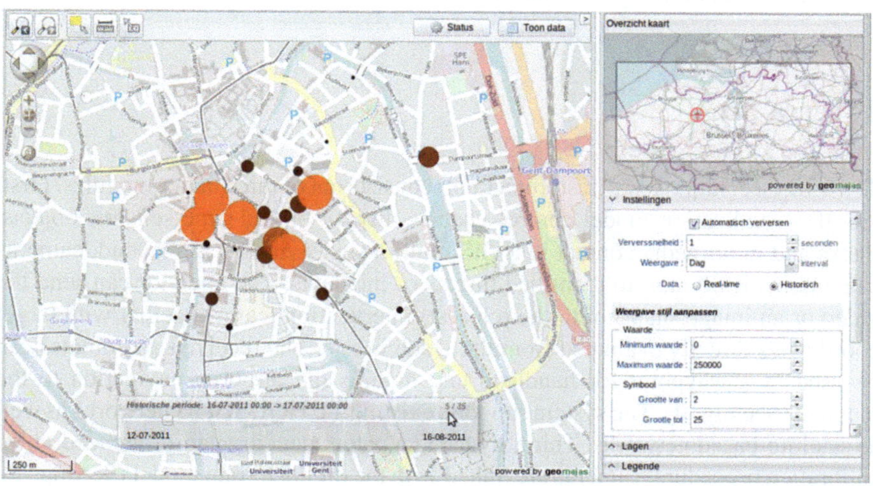

Fig. 2 BlueMAP: real-time interface

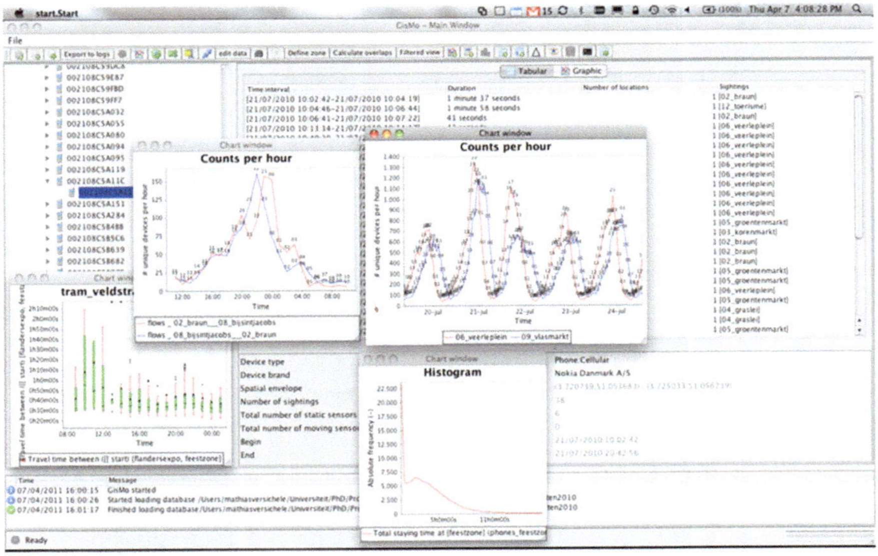

Fig. 3 BlueMAP: post-processing interface

and via post-processing (See Fig. 3). The whole is called BlueMAP. BlueMAP facilitates cities and event organizers in monitoring pedestrian and traffic flows over a site using state-of-the-art Bluetooth technology. The scanners pick up anonymous signals of BT devices like cell phones and car kits, and monitor movement from point to point. Typically, the detection ratio (the percentage of visitors that gets detected by means of a mobile device with a visible Bluetooth interface, with respect to the entire population) is about 10 %.

Specific attention has been paid so far to the audience of these mass-events, without taking into account the rescue people. That's where the Terra 3D Incident Management Platform gets at the forefront.

2 Terra 3D

Terra 3D (see: http://www.fastprotect.net) consists out of three major modules. First, the 3D visualization model control enables display of real-time sensors, objects, tracks and multimedia content in a GIS multi-layer environment. Second, cameras can automatically aim to identified threats or incidents. The operator can send the video or an image snapshot to the (closest) intervention teams together with target coordinates directly evaluated from the video image. The received coordinates can be used as a waypoint for handheld and mobile and airborne navigation to the incident (See Fig. 4). The system supports existing TETRA, 3G/4G networks as well as satellite-based networks. Third, by using a patent pending geo-referenced video calibration procedure, cameras are linked with the digital terrain model (See Fig. 5).

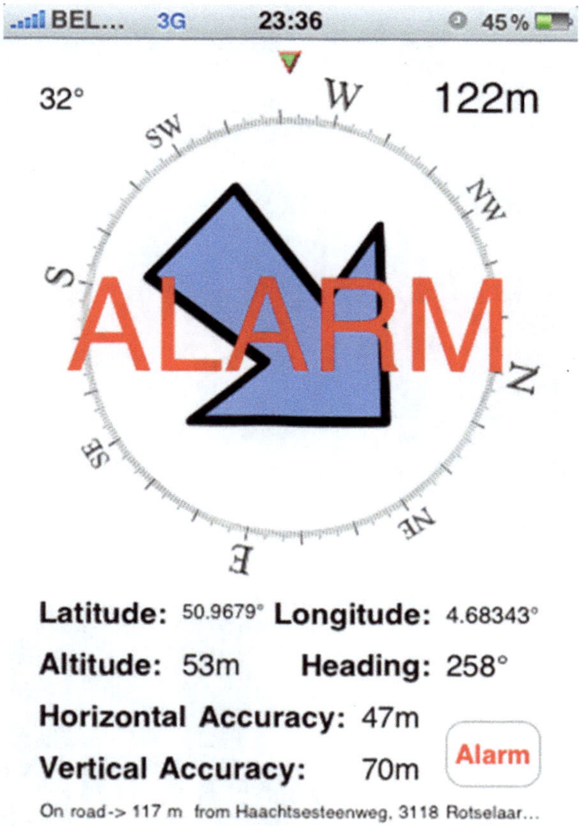

Fig. 4 Terra 3D incident management platform: guidance to waypoint

This allows every pixel of a 2D video image to receive a 3D coordinate. In contrast to traditional video systems, the 3D geo-referenced video provides in real-time object's height/size, position, speed, direction, and location.

Terra3D provides a seamless transition from live video and other sensor data to the 3D virtual environment and vice versa. The structured multi-layer architecture (See Fig. 6) collects information from multiple alarm- and multi-media systems and coordinates the actions that need to be triggered in a unique and transparent way whilst everybody in the operational chain knows what's happening and what actions should be taken. The positions of the emergency services (APL/AVL) can be sent to a central communication server that allows visualizing the teams in the 3D model. Based on the position of the incidents, the operator can dispatch the closest intervention teams to the incident fast and reliable with a very high precision (<2 m accuracy based on geo-referenced video).

During Rock Werchter 2010, the 3D incident management platform was presented as a world premiere to the emergency and security services. Fig. 7 is a 3D visualization + real-time video-feed integration + view position of emergency teams (APL) via L.I.P. protocol EADS/Cassidian THR880i radio's. Fig. 8 is a

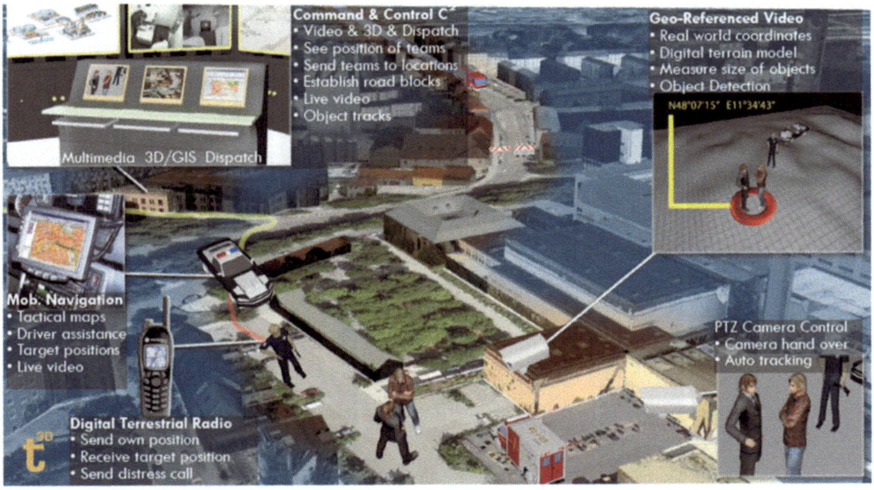

Fig. 5 Terra 3D incident management platform: geo-referenced video calibration procedure

Fig. 6 Terra 3D incident management platform: multi-layer architecture

photo showing public safety/evacuation messages on LED-screens on-stage can be triggered by terra 3D automatically or by manually clicking on the screen in the 3D model.

Fig. 7 Terra 3D incident management platform: 3D visualization + real-time video-feed integration + view position of emergency teams

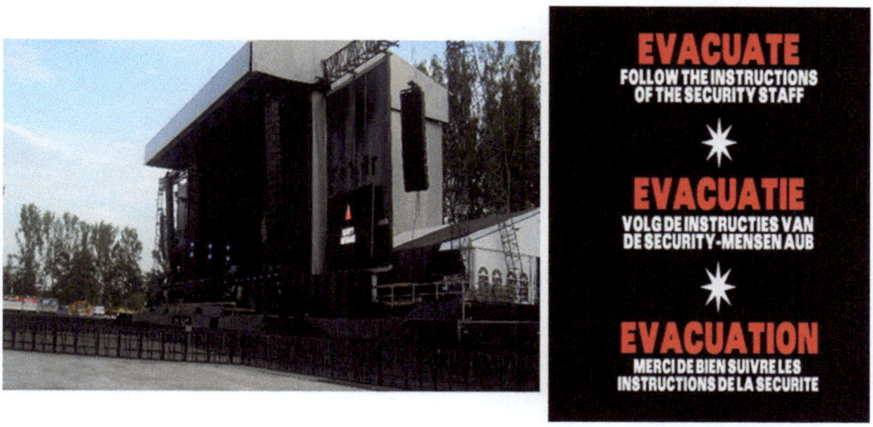

Fig. 8 Terra 3D incident management platform: public safety/evacuation messages on LED-screens on-stage

3 Integrated System

The idea is to integrate both systems (BlueMAP and Terra 3D Incident Management Platform) in order to get an integrated centralized system capable of acquiring, analyzing, and modeling the complex movement interactions occurring at mass-events in order to fully support the organizers in crowd management, security and possible evacuation. Currently, both fundamental research and valorization perspective are taken into account in order to tackle this enormous challenge.

References

1. M. Delafontaine, M. Versichele, T. Neutens, N. Van de Weghe, Analysing spatiotemporal sequences in Bluetooth tracking data. Appl. Geogr. **34**, 507–518 (2012)
2. M. Versichele, T. Neutens, M. Delafontaine, N. Van de Weghe, The use of Bluetooth for analysing spatiotemporal dynamics of human movement at mass events: A case study of the Ghent festivities. Appl. Geogr. **32**(2), 208–220 (2011)

Evaluation of a Support System for Large Area Tourist Evacuation Guidance: Kyoto Simulation Results

Seiki Kinugasa, Tomoko Izumi and Yoshio Nakatani

Abstract Most studies on providing evacuation guidance have targeted residents, with little consideration for evacuation guidance for visitors to the area, such as tourists and businesspeople. Accordingly, this study targets the development of a system that supports the safe and efficient evacuation of tourists from disaster areas to specific safe destinations. The system models the evacuation behavior of tourists and then simulates an evacuation process in which a specific evacuation guidance method is utilized. A major characteristic of tourists in disasters is that they tend to converge on the limited number of railway stations, which may result in severe crowding and panic. The system therefore makes it possible to compare and evaluate the effectiveness of various evacuation guidance methods. The effectiveness of the system was tested by simulation of evacuation processes that utilize a phased evacuation guidance method that is to be introduced in Kyoto, the most popular tourist destination city in Japan.

Keywords Shortest path · Evacuation · Simulation · Visualisation

S. Kinugasa (✉) · T. Izumi · Y. Nakatani
The Graduate School of Science and Engineering, Ritsumeikan University, Kusatsu city, Shiga 525-8577, Japan
e-mail: cc006064@ed.ritsumei.ac.jp

T. Izumi
e-mail: izumi-t@is.ritsumei.ac.jp

Y. Nakatani
e-mail: nakatani@is.ritsumei.ac.jp

S. Zlatanova et al. (eds.), *Intelligent Systems for Crisis Management*, Lecture Notes in Geoinformation and Cartography, DOI: 10.1007/978-3-642-33218-0_22,

1 Introduction

In Japan, which has recently suffered the unparalleled Tohoku earthquake, there is a possibility that further large earthquakes, such as Tonankai and Nankai Earthquakes or a major earthquake in the Tokyo Metropolitan Area, will occur in the near future. Therefore, both the central Japanese and local governments have been taking various countermeasures, and in the meantime the government has positioned tourism as a major economic growth area [1]. The Tourism Nation Promotion Basic Law came into effect in 2007, and the number of tourists has subsequently been increasing. Most countermeasures against disasters, however, target residents and rarely target tourists. These countermeasures are usually obvious because, no matter how famous the region is as a tourist destination, the number of residents is far higher than tourists. However, the countermeasures aimed at tourists are actually extremely important for protecting the livelihood of many residents. Tourists may decrease if there is a risk of suffering severe damage from disasters. Supporting tourists in ways such as protecting them from damage caused by disasters, and providing safe and quick evacuation, emphasizes the importance of tourists to the rest of the world. Thus it is very important that appropriate evacuation guidance and design guidelines are well prepared beforehand.

Devising evacuation guidance for tourists, however, is much more difficult than the usual guidance for residents which is considered only on a theoretical basis. Tourist behavior differs from residents, because tourists need to conduct long-distance evacuation in a large-scale area. Large-scale evacuation experiments are impractical, with computer simulation appearing to be the most effective.

With the aim of solving these problems, the purpose of this study is to provide administrative staff with an environment in which they can plan and verify safety guidance by computer, for quick guidance of tourists to safe destinations in a broad range of areas.

2 Existing Research

Although there are many systems for simulating evacuation guidance via computer, systems that model behavior for guiding tourists are very few. In most cases, the behavior of evacuees is simply implemented as the fluid model, and recently systems that consider interaction between evacuees were developed. Recently, RoboCup Rescue simulation has gained attention as a large-scale multi-agent system [2], but such simulators cannot verify or compare evacuation guidance for tourists in a large-scale area.

Few systems have been developed to support evacuation guidance for tourists. Among them, one example developed by Kyoto University in Japan involves an instructor who conducts evacuations for evacuees via mobile phones [3]. Locations of evacuees are constantly supervised based on location data provided by Global Positioning System (GPS) and are replicated in a virtual city called FreeWalk, which provides a bird's-eye view of the area. The instructor constantly observes the behavior of evacuees in the FreeWalk virtual world while providing them with instructions via mobile phones. Another system, developed by Wakayama University in Japan, enables evacuees to send disaster information and receive evacuation guidance via disaster information stations installed in a city [4]. A number of wireless devices therefore need to be installed throughout the area, and Bluetooth is used for communication between the users' mobile phones and wireless devices.

The evacuation guidance method proposed by Nakatani et al. of Ritsumeikan University, Japan involves evacuees being guided to temporary tourist shelters designated near sightseeing spots in a phased manner, in order to prevent concentration at any one time in the central part of a city [5]. It features "staging posts" as temporary refuges for tourists on the way to their destination (e.g. railway stations). The staging posts function as a buffer zone to prevent tourists from rushing into railway stations in the central parts of cities and thereby enabling fast evacuation of tourists from dangerous areas to safer areas, and reducing the fatigue caused by evacuation on foot. Cooperators in disaster areas, such as souvenir shop staff and tourist agents, are requested to send disaster information to the emergency management center. The emergency management center then determines and changes evacuation instructions based on this information before sending the instructions to the cooperators. The cooperators then guide evacuees to the nearest tourist shelter according to the instructions they receive. The cooperators, during the guidance, send their disaster information to the center at any time. This evacuation method is more practical than other methods because there is no requirement that all evacuees have a cellular phone equipped with GPS.

3 System Construction

The purpose of the proposed system is to compare and verify various evacuation guidance methods suggested by public authorities, and to support the development of effective evacuation guidance. The functions of this system will:

- Visually define evacuee behavior over time on a map.
- Simulate various evacuation methods for tourists and change the evacuation route depending on the situation.
- Show quantitative results and the effects on tourists of various primary factors (width of the road, traffic, etc.)
- Consider factors that keep evacuees in places of refuge on the way to their destination, such as using the staging posts.

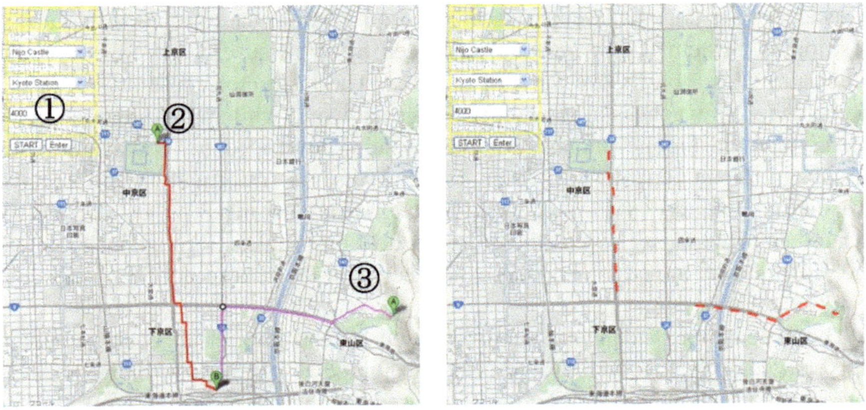

(a) (b)

Fig. 1 System screen. **a** During input of evacuation guidance. **b** During the simulation

3.1 System Screen Configuration

The images shown in Fig. 1 display use of the system during input of evacuation guidance (Fig. 1a). Information about evacuation guidance is input in ①, with checking of the route and other information performed in ②. The details of this information are shown in Table 1. This input determines which shelter evacuees should go to, and via which route, for each destination, and is required for each point of origin. The thick line in Fig. 1a② shows the route during input of guidance (Nijojou to Kyoto Station). The thin line in Fig. 1a③ shows the route with input completed (Kiyomizu Temple to Kyoto Station). Fig. 1a shows the case of Table 1.

Figure 1b shows the system screen after the start of the simulation. The thick line shows the behavior of evacuees following a guide in line. The crowd appears as a dotted line because many evacuees are separated and are individually evacuated in small groups. Each small group has one guide.

Table 1 Contents of input information

Evacuation info.	Input details	Input ex.
Origin	"Address" or "longitude and latitude"	Nijo-jou
Staging post	"Address" or "longitude and latitude" or nothing	(None)
Waiting time	Waiting time at the staging post	0:00
Destination	"Address" or "longitude and latitude"	Kyoto Sta.
Evacuation route	Dragging the route line on the map	Shortest Route
Number of evacuees	Number of evacuees at the origin	4000

Fig. 2 Chain of command

3.2 System Process

This section outlines the process during simulation. This system, after the start of the simulation, designates one "Headquarters", multiple "Guide Leaders" and "Guides" from the viewpoint of the usual chain of command at the time of disaster (Fig. 2). The evacuation guidance inputted before the start of simulation is analyzed at the "Headquarters" and sent to each "Guide Leader" and "Guide". Each origin point has one "Guide Leader". "Guides" receive the orders from the "Guide Leader", and guide evacuees.

Figure 3 is a representation of the system flow after the "Guide" starts evacuation. "Move Evacuees" indicates the transition of evacuees from each step to the next, where a step is the minimum unit from the point of origin to the destination. Each step includes the next step's position, distance, and width of the road. The movement process is calculated based on this information, and transports evacuees in each step from the last step's position to the next step's position. This movement results in a line being drawn from the last step to the next step, thus depicting how the evacuees move in a bold line. The drawing speed of the crowd is proportional to walking speed.

In cases of changes in the width of the road or the merging of two evacuee groups, the density of the crowd and the walking speed change. Fruin [6] discovered that 4 persons per 1 m^2 leads to a sharp decrease in the walking speed. In this system the walking speed is calculated using this information, with equation being based upon it.

$$V(\rho) = 1.1\rho^{-0.7954} \tag{1}$$

ρ: crowd density (person/m^2)
v: walking speed (m/s)

4 System Evaluation

The effectiveness of the system was examined using Japan's most popular tourist destination city, Kyoto. The strong possibility also exists that an earthquake will occur in Kyoto [7]. As the phased evacuation guidance method was proposed by Nakatani for use in the tourist evacuation guidance field in Kyoto, we then examined the method by using it in our system.

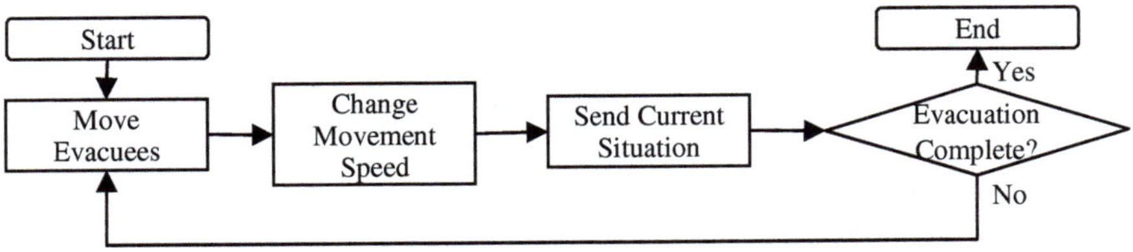

Fig. 3 Flowchart after start of the simulation

Fig. 4 System screen. **a** Screen after 100 min. **b** Screen after 142 min

Figure 4a shows the situation 100 min after the start of the simulation, and Fig. 4b shows the situation after 142 min. Most tourists are likely to head towards Kyoto Station to return to their homes in case the railway service has resumed after an earthquake. The arrow in Fig. 4a shows that each evacuee group merges and crowds together. The arrow in Fig. 4b after 42 min shows a long line of evacuees and dangerous traffic combined with severe vehicle congestion. This system therefore clarifies problems or danger in evacuation guidance methods, including which places will become dangerous at what times and how to improve the situation.

The results were evaluated by two persons from the Kyoto City Fire Department with regard to the effectiveness of the system, with the result being that they considered it effective.

5 Conclusion

This paper has demonstrated the necessity for the proposed system and presented the process and evaluation of the system. In the future we would like to develop a detailed behavioral model that includes merges, splits, stops and slowing of an evacuee line with psychological factors taken into consideration. We then hope to develop and evaluate the system with the help of an expert. We are first considering applying the system in Kyoto City, and the system is applicable to other tourist destination cities,. Finally, our heartfelt appreciation goes to Kyoto City Fire Department officials.

References

1. Japan Tourism agency, Tourism Nation Promotion Basic Law.http://www.mlit.go.jp/kankocho/en/kankorikkoku/kihonhou.html. Accessed 10 July 2011
2. RoboCup Rescue, RoboCup Rescue Agents Simulation Project. http://www.robocuprescue.org/agentsim.html. Accessed 10 July 2011
3. H. Itou, H. Nakanishi, S. Koizumi, R. Ishida, Transcendent Communication: Location-based Navigation for a Large Scale Public Space. *Proceedings of Computer Human Interaction 2004*, CHI2004, 24–29 April, 2004, Vienna, Austria, pp. 655–662
4. K. Nozaki, et al., A Proposal for Dynamic Emergency Navigation System in Disasters. *Proceedings of The Special Interest Group Technical Reports of IPSJ, 2007*, 2007-DBS-141 (29), pp. 185–190.(In Japanese)
5. Y. Nakatani, A verification and research report on collecting and making assistance information available on tourist's disaster prevention, March, 2011, pp. 62–66.(In Japanese)
6. J.J. Fruin, *Pedestrian Planning and Design*, (Elevator World, New York, 1971), pp. 39–43
7. The Headquarters for Earthquake Research Promotion: National Seismic Hazard Maps for Japan, 2005

A Virtual Police Force as Part of an Integrated Community Security Network: The Case of the Dutch VPK Programme

Wim Broer

Abstract The internet revolution, with all of its accompanying positive and negative consequences, has increasingly come to impact on society at large. It also has significant consequences and opportunities for the police. The task is to respond to these challenges with intelligent solutions, and using these solutions to empower police professionals in the field. The desire to meet these challenges has led to a programme called 'the Virtual Police Service (Corps or Service)'. Its ambition is to build a new way of delivering professional policing services, by making optimal use of information cloud technology. This will maximise the rich source of information and communication potentially available on the internet.

Keywords Police · Human factors · User-centric design · Cloud computing

The programme delivers a 'Multi-App' stored in a "Blue App Store" in "App Store" that has its foundations in the current policing theory of a multi-agency approach towards addressing safety problems. It covers both police as well as all possibilities of civilian/community and social media support. Traditionally, the information sharing culture of the Dutch Police force is one of high security, high protectionism and low risk. Where other agencies had adopted: 'share everything, except when it is forbidden', the Police force kept to its old: 'share nothing until you must-' regime. The vision adopted when developing the VPK case was, that instead of building heavily secured 'gateways' to the outside world, it would choose another strategy by working from outside to within. The VPK vision in this case, was also inspired by new architecture concepts like "linked open data", "semantic web" and modern web service based information exchange methods.

W. Broer (✉)
Commissaris van Politie, Programmadirecteur VPK, Virtueel Politie Korps, Innovatieplatform Politie Nederland, Papendorpseweg 68, 3528 BJ Utrecht, The Netherlands
e-mail: Wim.broer@vpkteam.nl

S. Zlatanova et al. (eds.), *Intelligent Systems for Crisis Management*, Lecture Notes in Geoinformation and Cartography, DOI: 10.1007/978-3-642-33218-0_23,

This vision was fuelled by the very real needs of police men out there in the field. In summary, the vision was I.T. system architecture driven and bottom up professionally inspired.

The Multi-App in the case of this paper covers three data elements: Relationship, Location and Role in order to improve the information position of the police professional. This solution is tested in a number of police regions capturing the insights of end users in order to evolve this mobile platform to an enriched solution that can inspire the transition of the wider police service from its legacy systems to a sophisticated architecture of a cloud based platform.

The internet revolution, with all of its positive and negative consequences for society, has impacted on both the community and the policing of that community. Due to the rapid growth of mobile internet and the increasing capabilities of smart phones/tablets etc., information is now ubiquitous [1]. Information is now available in all places at any time. At the same time the police have lost their traditional dominance over (crime related) information that they had during the last century. This was due to the rapid emergence of social networks, new media and the Web 2.0 breakthrough. This internet revolution has enabled all sorts of groups to share information with others. The traditional internet was constricting the police in this area, so it was decided to move towards more up to date and available technological solutions. The police must have access to this new way of working to maximise the communication possibilities, enable for effective operational policing and in essence undertake an innovation revolution with the police involving a wholesale migration to the new internet technologies.

The police service had to maximise the positive opportunities it offered to impact on operational policing and the security and safety of its citizens without been bound by the traditional boundaries of "time" and "place". To translate this vision and theory into reality, they started a new programme called Virtual Police Korps (VPK = Virtual Police Force). The benefits envisioned of such a system would be an increased police visibility, achieved by moving from the physical to a virtual domain of operation. A new 'cloud' enabled police force that could optimise the use of the rich source of information and communication potential available to them.

1 Ambitions of a Virtual Police Force

The following aims underline the start of the VPK

- An Integrated Approach to the Problems of Security

This means reinforcing collaborative working and cooperation both externally and internally. If each component within the security chain has its own independent IT infrastructure, then an efficient flow of information would take a long period of time, obviously. By building a production chain on the internet and using data available within these chains the police could ensure the way cooperation takes place is of a high standard.

The same rules apply for public participation. It goes without saying that the police would benefit greatly from information supplied by the general public, but in order to achieve this, the police must ensure they make every channel available to them through which they can provide us with the information they have. For this to happen, the provision of a website alone would not be sufficient. In addition the Police had to make use of social media networks and provide 'applications' i.e. 'apps' that can tempt people to supply the police men with important information utilising a smart phone or tablet.

1.1 Increasing the Professional Skills of the Police

The professional skills of the police can be increased if the perspective on the tasks of the police is viewed from the outside in rather then from the inside out. In the Netherlands within the context of the programme 'a *Politie in Ontwikkeling*' [=Policing under development], it is argued that a nodal approach should be taken whereby information hubs in society are seen as information anchorages. By making more use of the dynamics in society the police should be able to work more efficiently and effectively. That principle also entails a police presence, not only in the social physical environment but also in the virtual capillaries of society in order to be able to keep an eye on information processes.

- Strengthening Innovation

Now that changes in society are taking place quicker than ever before, the police had to orientate themselves towards, and participate in innovative developments. Given the complexity of the policing world having its own culture and terminology in use, it is important in our view, to forge alliances with science and the world of business to maximise these innovation opportunities. Although institutes, like the establishment of lectureships at the police academy did cover part of the scientific spectrum, there were insufficient actions undertaken to ensure access to innovations applied to the world of business. There was also no Dutch nationwide facility for policing related innovation. In the Netherlands, the arrival of a national police force could provide great opportunities for greater potential and cohesion. The innovation platform established by the National CIO represented the first steps in the Dutch police service innovation revolution. VPK pilots fell under the programme 'remit'. What we were dealing with was more than a mere technological driven activity, however.

In the end the innovation within the police force has to be placed in its wider context. This helps guarantee that the social and business conditions for the success criteria are met, enabling a successful implementation and launch of the new system. One of the relevant criteria is the provision of adequate training for the professionals, for example. Without such a transfer mechanism, the policemen 'out there' would never be able to use any new applications that may be developed.

The use of social media in the context of 'New Work' for example, does provide many learning opportunities for our use case, but these must be counterbalanced will potential threats that the provision/obtaining of information entails in our complex an different Dutch society. In essence this involves a balance between privacy and the process of establishing the 'factual truth'. The police has to be aware, that despite strongly enhanced chances to gather information and to carry out analyses, they need continuously be on guard to ensure respect the boundaries of ethics, privacy and the law.

2 VPK as a Buffer Between the World Within and the World Outside

The initiative for the VPK programme was taken at the end of 2009 for the breakthrough heralding the arrival of a National police force and the current development plan ('Plan of Attack') in terms of the ICT upgrade. In no way did this large scale National Police plan that concentrated many Police organisations in one National Police agency diminish the necessity for gaining small-scale learning experience. This conclusion entailed gaining experience in living and working with modern devices directed towards gathering information that is knowledge-driven. One could imagine the notion of a speedboat heading out in front of a tanker to see which waters are safe, for example. Any such action demands flexibility and agility, which is not the usual strength of the larger bureaucratic system that often characterises Police agencies. The VPK programme started thinking from outside in, by involving members of the public and (information-) chain partners in all of these innovations. In the internet era providing security for society would be impossible without the closest involvement of members of the public, since they have the most at stake. Sharing security facilities with the public helps to meet an internal goal, which is to restore the trust the police has in it's own ICT. This could be done by reducing the administrative burden and increasing user-friendliness, transparency and speed.

Only recently the policemen, were people only allowed to access the internet under severe restrictions. The "big bad outside world" of the World Wide Web apparently constituted too high a risk for the police service. As a consequence the police could only access the mindboggling collection of information the internet has to offer from specially equipped work places or by working from home.

The 'outside-in' perspective offered the advantage that, in principle, there was access to the diverse collections of information available on the outside via the internet. This entailed supporting the police with a strong secure portal where using mobile devices, such as smart phones and tablet PCs, access is provided to selected websites and authorised applications that support working processes leading to greater efficiency and effectiveness.

3 Information Sources to Give the Police Wings

The VPK programme case was based on operational policing theory and practice. Professional police work is directed maintaining safety and security by solving conflicts in society. To do this the police have specific powers ascribed to by law it to gather information, where necessary and permissible, even against the wishes of a person or persons concerned. The solutions that the VPK offers are based on a number of underlying perceptions and assumptions about how to establish the 'factual truth'.

This goal was achieved by offering the possibility of saving the relevant data close to source and then making that data directly available to colleagues. This is made possible by modern smart phones that have the added advantage of reducing mistakes to an absolute minimum. Scanning a car number plate and a driving licence saves time and avoids police personnel having to use pen and paper and making typing/spelling errors, for example.

Apart from data concerning the formal identity, using multimedia could provide benefits such as establishing if initial statements of victims and witnesses compare. Oral statements or a description or a video fragment of the offence can also immediately be circulated, using the user friendly 'app'. The public can access the same application by downloading the application from a public 'appstore'. There is also the VPK's own secured version of the so called 'Blue App Store' with more functionality. The public version encourages and enables the public to act on impulse and make a positive decisive action and clearly contributes towards the '*Heterdaadkracht*' programme recently launched to catch offenders red-handed whilst committing criminal acts.

In essence this use case relates the story of using currently available technology to support processes in the pre-registration phase of police work that often begins on the street due to proactive or reactive actions/incidents. There is no need anymore to return to a police station to start the investigation the recording of facts can take place almost immediately using the 'app'.

4 Three Data Elements

The project has been working on the following three data elements or spearheads based on the policing vision outlined (Fig. 1).

- *Location*; the GIS application under the heading Security Picture to project layers of data onto fine mesh card.
- *Relationship*; which entails the Social Site environment, a professional version of a social network that we can use within a secured environment to maintain relationships, to provide optimal sharing of data and to communicate o the one hand, and on the other hand to provide interactive access to the 'public' social media.

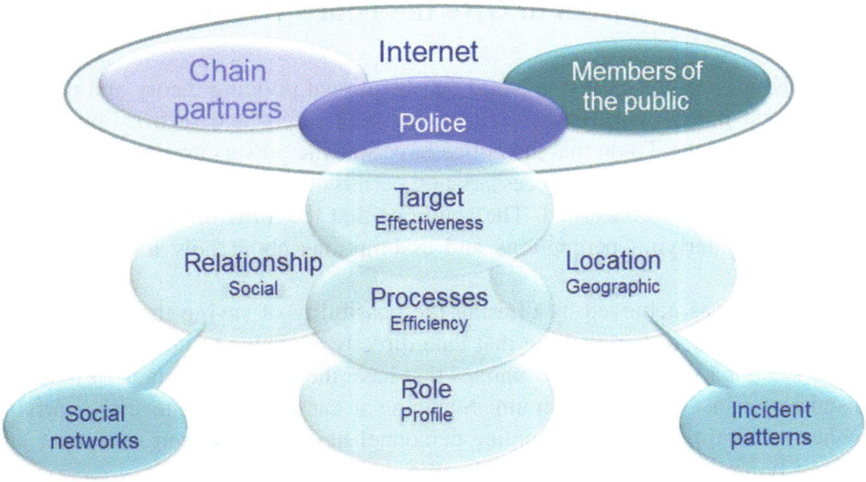

Fig. 1 Diagram showing VPK target groups and spearheads

For every one of these spearheads the police have involved tried and tested applications from the market. They are offered as an integrated 'app' to make them user-friendly. The following "screen grabs" show two windows of this 'Multi-app', the dashboard (Fig. 2) with an accessible overview of the data sources functionality and a work page (Fig. 3) to be able to record any questions and actions taken in a working file.

- *Role*; the setting of a user profile as a basis for the filters that need to be installed to prevent system overload.
- Role

The system has linked a set of roles with data requirements based on the nationwide job classification system, in order to manage profiles in a dynamic way.

Fig. 2 Dashboard

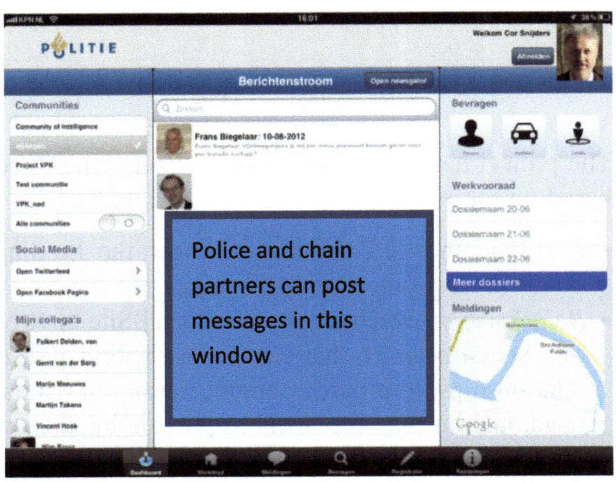

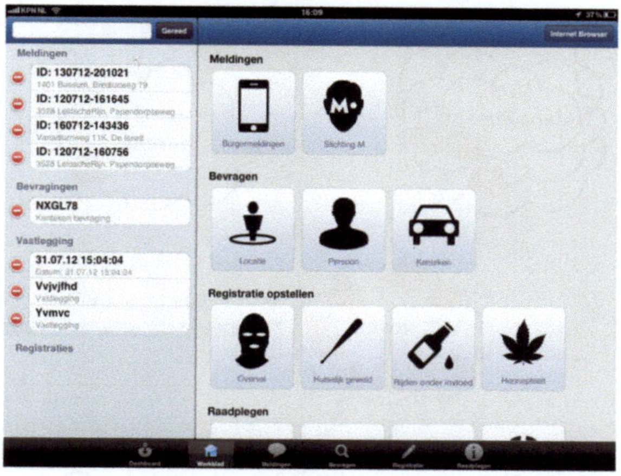

Fig. 3 Working page

In part this has to do with generic needs and in part this has to do with the specific needs of a user. To provide users with simple, but secure access to various services, a secure portal has been developed where users can log in from any internet workplace. The user has access to the Blue 'App Store' where he can reel in 'apps' intended for him, having passed an authorisation and authentication procedure. The 'apps' coupled to his profile form the components of a personal start-up homepage.

- Location

This deals with a geographical system where incidents can be plotted on a geographical map and these numerous fields can be accessed interactivity by both the police and their partners. In addition, the system can access a database by unlocking all of the base registers managed by the municipalities manage in the Netherlands (GBA [=Municipal Personal Records Database, Buildings etc.]) and presenting them by location in geographical order. In this way the system is able to offer an optimal security picture in real time down to neighbourhood level and even down to a specific set of premises in any given neighbourhood.

- Relationship

This relates to a suite of possibilities for professional virtual networking. Everyone is now familiar with the existence of social networks in the public domain. This relates to offering functionalities to police and chain partners within the VPK secure environment to enable finding colleagues and partners inside and outside their part of the organisation with the aim to share data about every subject imaginable. Participation in public social sites is possible directly using the 'Multi-app' that makes it possible to record relevant conversations in the work dossier. An 'app' supports exploration of public social sites using semantic search technology data filtered down to location, emotion and significant relationship.

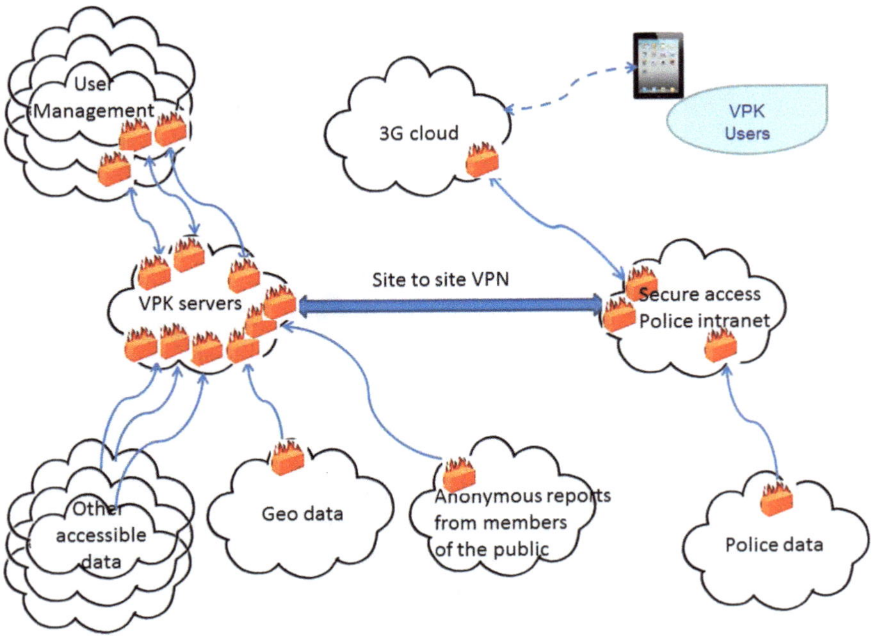

Fig. 4 VPK infrastructure

1. Pilots

To test VPK use case possibilities empirically, a number of pilots will be carried out during the latter half of 2012.

- Cooperation between all Assistant Public Prosecutors and Duty Officers (South Gelderland): working efficiently using mobile access to relevant sources and working with chain partners in an efficient way.
- Supporting everyday police activity in neighbourhood teams (Nijmegen[1] and Woerden): pre-registration phase facilitating the working processes of the police on the street by setting up working dossiers on smart devices and mobile access to relevant sources.

[1] In the run up to the pilot we tested the 'multi-app' on a small scale in Nijmegen, the secure Social Site environment and the 'app' members of the public can use to report back to the police. Our first findings with the 'multi-app' were positive. The technique operated well. Users experienced at first hand the added value of 'multi-app' including moving between live videos that they could switch on and off themselves. The secure Social Site environment that included the possibility of posting a message directly to the site as well as the 'app' for use by the general public that police personnel tested who were present as civilians, were felt to be positive additions to the tool box.

- Professional support for an integrated approach to organised crime and especially combating the organised cultivation of cannabis (Municipality of Tilburg, relevant chain partners): mobile access to relevant sources, setting up work dossiers on smart devices and facilitating mutual file sharing to give effective shape to the combating of crime.

2. Architecture

To set up information and infrastructure architecture in collaboration with a strategic partners, we created two architecture guiding documents namely: Target Architecture and Solution Architecture (Fig. 4). Target Architecture describes the general principles that should be maintained based on general current government architecture guidelines and police enterprise architects. Solution Architecture translates these principles into actual products, services, suppliers, etc.

These architectural documents provide the basis on which to create the environment in which VPK can offer its functionality. We provide an overview of the general principles of the technical safety arrangements in the Fig. 4.

Reference

1. M. Weiser, R. Gold, J.S. Brown, The origins of ubiquitous computing research at PARC in the late 1980s. IBM Syst. J. **38**(4), 693–696 (1999)

Integration of Real-Time UAV Video into the Fire Brigades Crisis Management System

Mark van Persie, Menso C. van Sijl, Edwin Wisse, Janio B. Tjoe-Awie, Arnout J. de Jong and Wim Bakker

Abstract During a fire incident live airborne video offers the fire brigade an additional means of information. Essential for the effective usage of the daylight and infra-red video data from the UAS is that the information is fully integrated into the crisis management system of the fire brigade. This is a GIS based system in which all relevant geospatial information is brought together and automatically distributed to all levels of the organization. In the Dutch Fire-Fly project it was investigated in what way and under which conditions the information obtained with small unmanned aerial systems (UAS) can be integrated into the fire brigades crisis management system in support of their operations. In cooperation with the Dutch fire brigade of the VNOG requirements and a concept of operation were defined, based on which a technical system was developed and demonstrated in

M. van Persie (✉) · M. C. van Sijl
National Aerospace Laboratory (NLR), A. Fokkerweg 2, 1059 CM Amsterdam, The Netherlands
e-mail: Mark.van.Persie@nlr.nl

E. Wisse
Geodan, President Kennedylaan 1, 1079 MB Amsterdam, The Netherlands
e-mail: info@geodan.nl

J. B. Tjoe-Awie
Nieuwland Geo-Informatie, H. van Suchtelenweg 4, 6703 CT Wageningen, The Netherlands
e-mail: janio.tjoe-awie@nieuwland.nl

A. J. de Jong
Delft Dynamics B.V., Molengraaffsingel 12, 2629 JD Delft, The Netherlands
e-mail: info@delftdynamics.nl

W. Bakker
Veiligheidsregio Noord- en Oost-Gelderland (VNOG), Europaweg 79, 7336 AK Apeldoorn, The Netherlands
e-mail: w.bakker@vnog.nl

S. Zlatanova et al. (eds.), *Intelligent Systems for Crisis Management*, Lecture Notes in Geoinformation and Cartography, DOI: 10.1007/978-3-642-33218-0_24,

practice. An existing robot helicopter was equipped with the proper sensors, a geospatial video server was integrated in the local command center and the fire brigades crisis management system was adapted for the display and distribution of real-time geospatial airborne video and derived product to the involved operational levels during a fire incident. This article describes the technical and operational approach and results.

Keywords UAS · Geospatial video · GIS · Fire brigade · Crisis management system

1 Introduction

During a fire incident in an environmental area or complex industrial site it is often difficult for the fire brigade to build up a good overview of the incident. A flexible view from the air provides a better overview of the fire itself, the surroundings and the presence and activities of people. It also can provide details on hot spots or leakage in direct support of the firefighting activities. Small unmanned aerial observation systems have the advantage that they can be used flexibly with a short reaction time and at dangerous or inaccessible locations, see also Fig. 1.

As unmanned aerial systems (UAS) are becoming more mature and easy to operate, these might be useful to fill in this information requirement. Various small UAS systems are on the market and can be equipped with a video camera. Many of them however are not operational systems in the sense that operation of the platform and sensor requires too much skills of the operator, and that the sensor information reaches not further than the control station of the operator. The information cannot be distributed and integrated into the users GIS based crisis information management system.

Fig. 1 Flexible eye in the sky to support the fire brigade

For the involved fire workers and coordination agencies it is important that the system will be easy to use and that the information is well integrated in the crisis management system, so that it can be combined with all other information and distributed to the users in the field, the local command center and the regional coordination center.

In a project Fire-Fly [1] in the framework of the Public Innovation Agenda of the Dutch Ministry of Economic Affairs a project has been carried out with the Dutch Security Region North East Gelderland with the objective to demonstrate the usage of real time UAS sensor information in the command chain. Attention has been paid to the operational and information requirements of the fire brigade and to the geospatial integration of the video information into the GIS based crisis management system. Extensions to an existing crisis management system were developed to incorporate the video and aerial photo functionality into the crisis management system and thus into the working practices of the crisis response teams.

The project ended with a demonstration in which the system was used and evaluated in two different scenarios: a forest fire approaching an industrial complex and a large traffic accident.

2 Fire Brigade Information System Concept

Based on discussions with people from the fire brigade and a demonstration of technical capabilities at the start of the project a concept of operation and requirements to the UAS system could be specified. The work focused on the usage of mini and small type of UAV's in support of complex fires at relatively local scale. This was worked out in a system architecture as shown in Fig. 2.

The system is based on a UAS with visible and infra-red video cameras. Also attention has been paid to chemical measurements, but this is not part of this article.

In the architecture the central point for the UAS coordination and information handling is the local command center that is placed at the incident location. In this shelter a video server is defined that automatically processes, archives and distributes the live video and products to client systems. Multiple UAS systems can be supported.

In the shelter a Video Editor Station is included that is operated by an information analyst, who inspects the live video, generates products, and is a liaison for information requests to the UAS ground control stations. In the shelter also the plotter station of the crisis management system is present, with which the video products can be retrieved from the video server and combined with all other relevant information.

Based on the crisis management system, local coordination of the activities can take place. All information however is also distributed to the other regional coordination levels and operational persons in the field. In discussion with the fire

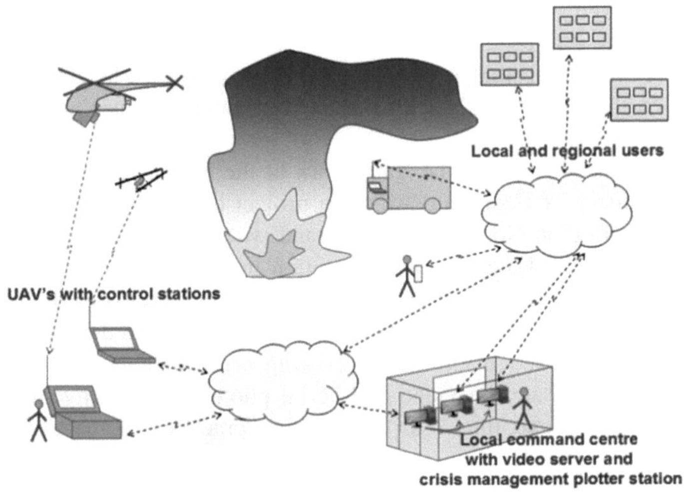

Fig. 2 System architecture

brigade a number of relevant information products have been distinguished: the live video stream, video clips, geo-referenced or geo-corrected snapshots, video mosaics. The information manager also does have a capability to send request for the imaging of specific areas of interest to the ground control station of the UAS operator.

Interesting is that comparable UAS usage and requirements occur in the defense area. Also here real-time geospatial video is required to determine and analyze locations and objects of interest for intelligence, and to combine the sensor information with other information sources to support command and control. Different data standards have been developed in this context of which some also could be applied for this civil application. An example where the same ground station technology was used for this type of usage is the Dutch military exercise Purple Nectar [2].

3 System Description

The system as developed for support of the fire brigades activities is described in this chapter. Attention is paid to the UAS systems, the video handling and the crisis information management system.

3.1 UAS Systems

The FireFly project was based on the usage of the Delft Dynamics Robot Helicopter [3]. The RH2 Stern robot helicopter system consists of a helicopter with camera or other sensor(s) and a ground station from which the operator controls and monitors the helicopter, see Fig. 3. The helicopter is equipped with sensors and a computer system that provide stabilization and control. The helicopter weights about 15 kg and has an endurance of more than an hour. The control system developed by Delft Dynamics allows the operator to safely control the helicopter after a minimum of training. Thanks to its low weight the robot helicopter is very quiet and environmental friendly compared to its manned counterpart in terms of pollution and disturbance.

For usage in the context of fire incidents the system was adapted and extended at a number of points. The video data was formatted conform the STANAG 4609 standard, so that geospatial video could be made available. An infrared camera was integrated. Display of the video footprint on the control station was realized in order support the sensor operator with a better overview. The grabbing of still frames was made possible, including storage of the metadata in Exif format.

Further discussions took place with the VNOG fire workers on the required skills and training and the operational circumstances in which the UAV can be used.

For the final demonstration also the Aerovironment PUMA fixed wing platform was coupled successfully to the system. Due to the used STANAG 4609 the video stream could easily be integrated into the video server and Eagle and CCS-M crisis management systems.

Apart from these systems NLR paid attention to the development of a user friendly UAS ground station. Based on discussions with the fire workers a tablet based simple and intuitive graphical user interface was developed and demonstrated for operating the NLR Pelican UAS from Ascending Technology, see Fig. 4.

Fig. 3 Delft dynamics RH2 Stern robot helicopter system

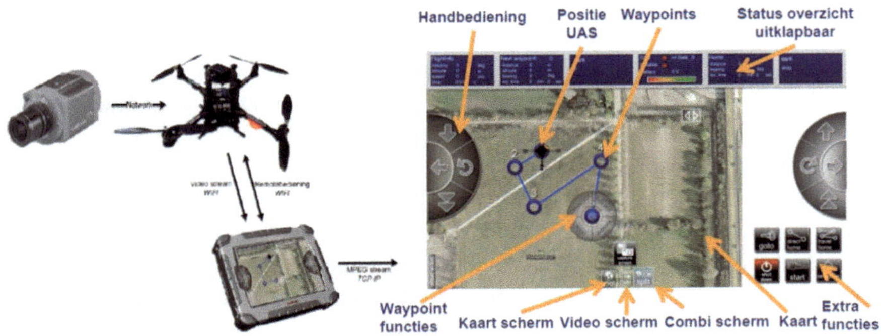

Fig. 4 Tablet based operation of the NLR Pelican UAS

Special point of attention was an inventory of the current and foreseen Dutch regulations for the operation of small unmanned platforms. Based on this operating limitations and operational procedures were discussed with VNOG.

3.2 Video Handling

For the handling and translation of the UAS video streams to relevant crisis information products a video chain was set up. The different components are discussed.

Video processor: The raw video as received in the local command center in a first step is formatted. In this step also some enhancement of geometric quality of the meta-information is done. For the geospatial handling of the video it was decided to apply the military NATO standard STANAG-4609 format [4, 5] in which meta-information on the geospatial parameters of the platform and sensor are stored in the MPEG-2 Transport Stream. A Video Multiplexing System was developed with which flexibly several formats of video- and meta-data can be read-in and converted.

Video server: For the handling and distribution of the geospatial video and the generation of products the Keystone Image Management System is used as a basis. This is a server-based system that archives, catalogues, processes and delivers digital images and video from satellites and airborne digital sensors. The system is developed by Spacemetric [6]. In co-operation with NLR adaptations and extensions are made for clients and research. Keystone is Java/Eclipse based and has a client–server approach in which the OGC standards are applied. Focus of the Keystone system is the accurate geospatial handling of the imagery. For video the standardized video and meta-data is ingested, enhanced and stored in the server database and catalogue. With the Keystone Client the live video stream can be viewed or catalogue queries on the raw video or derived imagery products can be made to the server, after which the stored video or products can be viewed.

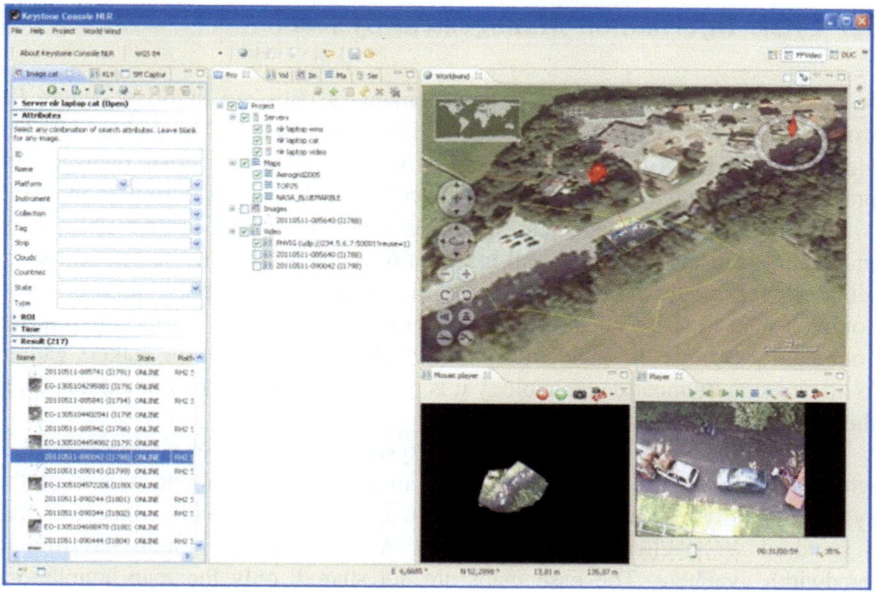

Fig. 5 Keystone based video editor with database part, project overview and display of video, perspective map and mosaic

Video editor: As a basis for the Video Editor station the Keystone client has been used, see also Fig. 5. Display of the live video is done in the client, both in a raw video window and a map window in which the video is displayed perceptively together with the aircraft position and in combination with other raster maps and imagery and vector overlays. A mosaicking function has been added in which a dynamic mosaic of the video is built up for a better overview and also static mosaic products can be generated and stored in the database.

Based on the video, products can be made, stored and distributed: mpeg clips and projected frames or mosaics as GeoTiffs. Distribution is done by an ftp server. Special attention was given to situations where the video is partly pointed over the horizon. In this case relative mosaics can be generated, based on relative image stitching. For these the geo-location information is stored in the Exif fields.

CoT function: For commands to the UAS ground station controllers a Cursor on Target (CoT) protocol is used [7] that makes it possible to directly do requests from the crisis management system clients to the ground control station. A server guiding the requests was added to the Video server station.

3.3 Crisis Information Management System

In the Fire-Fly project two crisis management systems are involved, Eagle [8] from Geodan based on ArcGIS and CCS-M developed by Nieuwland [9].The video and imagery functionality had to be integrated into the crisis management practice in such a way that it does not affect the work routines during an incident.

The Eagle crisis management system is a system for net-centric crisis communication. During an incident all information has a geographic component since all messages relate to a location near the incident. The crisis workers use a common map to share all information. Traditionally this was a paper map in the command post near the incident around which the local decision makers gather. With new communication technology it is possible to share this common map over the network.

Eagle: In the Eagle crisis management system a peer to peer system is used in which the map updates are shared between all actors in the incident. As a result the system is highly robust and does not rely on a continuous network connection like a standard server-client application does. As a result the system has limited bandwidth available. The base maps are not shared, only the map annotations related to the current incident.

The Eagle crisis management system consists of a number of different user interfaces. There is a version intended for use in crisis response vehicles, this versions is intended to show the map and to allow users to add simple annotations to the map like locations of road blocks, the command post of the local commander etc. The Eagle desktop version allows the user to perform geospatial analyses on the available data, for instance to estimate the number of people affected by an incident. Finally there is the Eagle surface table, this is a multi-touch, multi-user table computer intended as a collaboration tool by groups of decision makers.

Because of the bandwidth restrictions on location it was decided to implement aerial video functionality into the Eagle desktop and Eagle surface components.

In aerial video an aircraft sends a video stream from a bird's eye perspective. In practice it can be difficult to see what the video is actually displaying. When flying over more or less uniform terrain like forests or even urban areas the video image itself will not give the operator clues about where the camera is looking. The video stream used in this project was enhanced with location data about the corner points. This means the video stream can be displayed on a map. In order to keep the crisis map clear and understandable it was decided only to display the frame of the video stream on the map. This means only the lines between corner-points will be displayed (Fig. 6). The actual video stream is displayed on a separate screen. The desktop versions shows a similar map image, again with the current location of the video stream indicated as a polygon.

The surface and desktop can also display still images on the map (single or mosaicked video frames) on a map (Fig. 7). These still images are selected by the operator/screener. In Eagle these still images can be shared between the

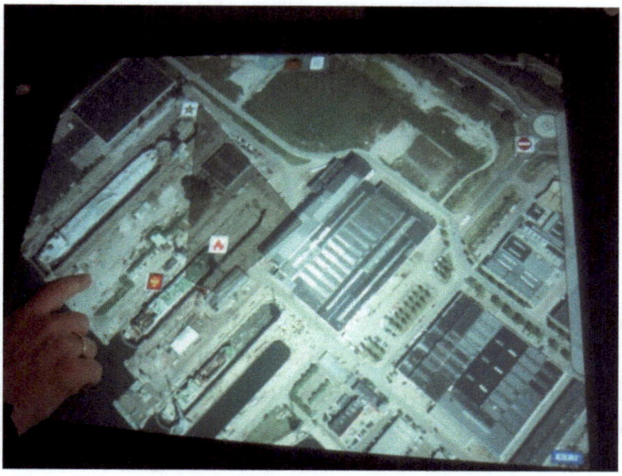

Fig. 6 Eagle surface showing the frame of the video stream as a footprint

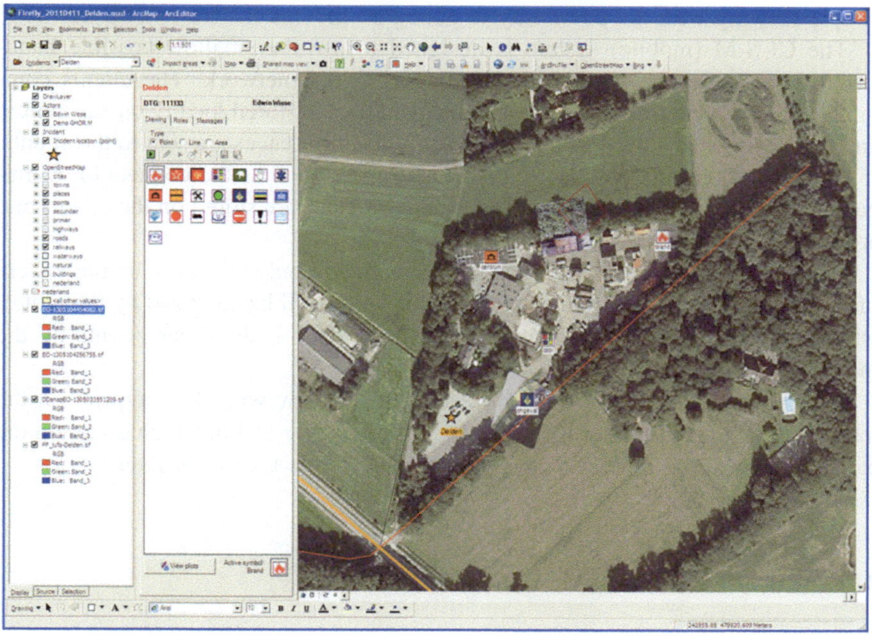

Fig. 7 Eagle desktop showing an exercise incident, the *red frame* indicates the area the video camera is pointed at. Also, there are two still images displayed in this map

participants using the peer to peer network. Within the peer to peer network the images are shared between all users using the aerial video module. The images are not shared with the mobile terminals in fire trucks because of the bandwidth limitations.

By using the peer to peer system the operator/screener can share aerial images with the other participants in the incident. The Fire-Fly modules for the Eagle surface and desktop automatically display new images which are shared through the peer to peer workspace. In practice the shared map gets updated continuously with new aerial images. The choice to import automatically was based on the principle that the participants should see the same shared map image on their screen. In practice this shared map, the common operational picture, gets updated continuously with the messages and map annotations from the other involved parties in the incident. By importing images automatically this principle is extended to the aerial imagery as well. In practice this ensured that the modules developed for this project behaved in a manner familiar to the users of Eagle.

CCS: The CCS crisis management system is the same main idea as the Eagle system. The system is based on the Tatuk framework and is built up around one type of client and a server. The clients can be configured in different ways for different use. For example, the version in the command room can be configured to do analyses on the spreading of the smoke and indicate the area to be evacuated. On the other hand the client in the crisis response vehicles can be configured for simple viewing with the possibility to configure the buttons for touchscreen use.

The CCS-M (mobile) system is designed to use the smallest possible bandwidth. It can even be used over the (Dutch) C2000 network. The clients communicate over a central server. All the incident data is stored locally on each client and on the server. To use less bandwidth only the differences are sent to the clients. If the connection fails the data will be queued. As soon as the connection comes back up the data will be synchronized. The base maps are stored on the clients and are not synchronized with the other clients.

Because of the bandwidth of video, the stream will only be shown on the system of the command unit. The information in this video will be analyzed by the plotter who will decide which information to be passed on to the other clients by the means of incident information.

The plotter will see the video in a separate window with the footprint on the map. This will give him/her the possibility to see where and in which direction the camera is looking. Figure 8 shows on overview of the CCS-M screen.

4 Demonstration and Conclusions

A final demonstration and evaluation of the Fire-Fly system was held in May at the G4S training site in Delden. Two scenarios were exercised, one of a car accident with multiple cars including a truck with chemicals, another with a fire in an industrial complex next to a railway. In an earlier demonstration in ‘t Harde also scenarios for detection of leaking liquids, a smoke plume and missed people walking in forested area were executed, see also Fig. 9.

During the demonstration live geospatial video and GIS products based on it were made available in the local command center, while the video products were

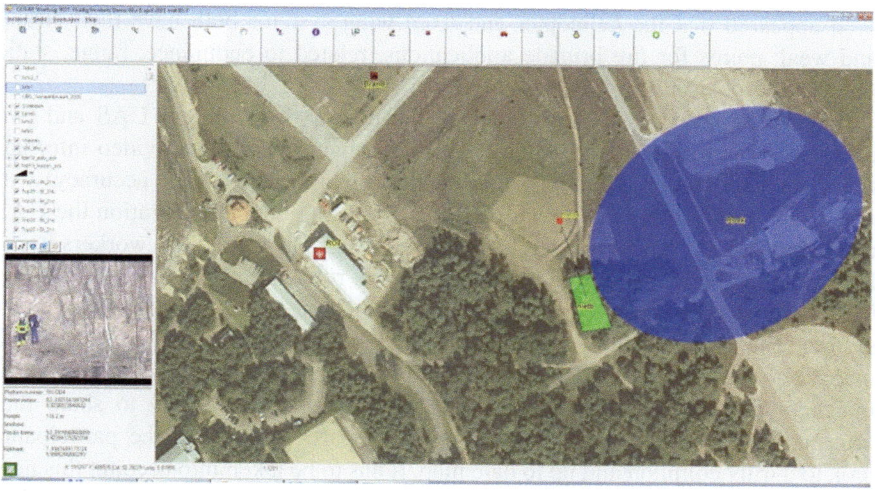

Fig. 8 CCS-M crisis management system with integrated video imagery and footprint

Fig. 9 UAS observing traffic accident at demonstration in Delden

distributed to the regional coordination centers and local fieldworkers by the crisis management system. The relevance of the data was judged positive by the involved participants. The live geospatial video from the air helped in the detection, localization and support of actions.

The concept of operation with the UAS positioned as observation and reconnaissance means and the tasking by the local information manager was experienced as correct. Important point for operational practice is the usability of the system: quickly deployable, minimal constraints in operation (beyond vision, above people, permission to use airspace, usage of frequencies), and no interference with other activities.

It appeared that the helicopter and fixed wing systems both have their strong and weak points for fire brigade applications, related to endurance, range, static observation possibility.

Interoperability was demonstrated by the use of two different UAS and two different crisis management systems. Further enhancement of the video information (like stabilization, contrast, super-resolution, higher geometric accuracy) will contribute to the easier interpretation and use. During the demonstration the UAS and video edit system were operated by technicians and not the fire workers itself. Further attention is required to ease of operation, robustness and operator training.

The aerial video and imagery can be well integrated into a crisis management system. The choice to limit visualization of the video stream to a footprint on the map (with an external video display) worked well in practice. Sharing aerial imagery adds useful information during an incident. The ability to share that information (even between a limited group of participants) lets the participants look to a truly common and up to date map. It has to be taken into account this new information stream requires a new role within the organization of crisis relief workers.

In a next step further operationalization of the concept needs to be worked out. The idea is to make one or more pilot systems available to the fire workers in order to gain operational experience based on which the system and procedures can be optimized. Points of attention during this phase will be robustness of the system, enhancement of the video and product quality, the communication infrastructure, establishment of simple procedures for arrangement of flight permission and for deployment and operation.

Meanwhile a separate project Explorer has been started up with the objective to extend the operational usage of small UAS by applying detect and avoid technologies and discussing procedures and legislation with the Dutch authorities.

Acknowledgments Thanks to the Dutch Ministry of Economic Affairs who made the Fire-Fly project possible.

References

1. M. van Persie et al., FireFly Eindrapportage (in Dutch), NLR Report NLR-CR-2011-421, Amsterdam (2011)
2. R.E. Bransen et.al., NLR Support 'Purple NECtar 09'. NLR Report NLR-CR-2009-356, Amsterdam (2009)
3. Delft Dynamics Robot Helicopter, http://www.delftdynamics.nl
4. NATO Standardization Agency, Digital Motion Imagery Standard. STANAG 4609 JAIS (Edition 3) (2009)
5. Motion Imagery Standards Board, MISB, Engineering Guideline 0601.2 UAS DataLink Local Metadata (2008)
6. Spacemetric Keystone Product Description, http://www.spacemetric.com/products
7. JEITA, Exchangeable image file format for digital still cameras: Exif Version 2.3. CP-3451B (or CIPA DC-008-201) (2010)

8. J.M.M. Neuvel et al., From spatial data to synchronized actions: the network-centric organization of spatial decision support for risk and emergency management. Appl. Spatial Anal. Policy. 5, 51–72 (2012)
9. Nieuwland GeoSafeAssist | CCS product description (in Dutch), http://geo.nieuwland.nl/index.cfm?subid=5497&pid=5495&p=5357

Agent-Enabled Information Provisioning While Retaining Control: A Demonstration

Peter de Bruijn and Niek Wijngaards

Abstract The project 'SlimVerbinden' addresses the challenge of retaining autonomy while sharing information among multiple parties. Based on a web of trust, information providers can grant and deny access to information, while information consumers can delegate access to specific members within their 'organization' (which can be defined within and/or across existing organizations). The policy- and PKI-based realization enables an agent-based secure shared distributed dataspace where no single party knows 'everything' and the barriers to information sharing are lowered. The use-case involves public–private cooperation during the mitigation of an incident and drives the development of an operational pilot.

Keywords Information sharing · Control systems · Agent-based

1 Introduction

Information provisioning forms the basis of the ability to coordinate actions, assets and other resources when addressing a complex, dynamic situation. The involved parties have different needs and responsibilities and in most cases only have to

P. de Bruijn (✉)
Stichting Studio Veiligheid, Den Haag, The Netherlands
e-mail: peter.debruijn@studioveiligheid.net

N. Wijngaards
Research & Technology, Thales Nederland BV, Delft, The Netherlands
e-mail: niek.wijngaards@d-cis.nl

N. Wijngaards
D-CIS Lab, Delft, The Netherlands

S. Zlatanova et al. (eds.), *Intelligent Systems for Crisis Management*, Lecture Notes in Geoinformation and Cartography, DOI: 10.1007/978-3-642-33218-0_25,

cooperate during a crisis situation. As these parties rely on each other, they need to take each other into account when planning their activities. Furthermore, also other parties are involved that provide relevant information on e.g. the availability of resources, incident specific information, weather forecast and danger estimations. As many incidents differ from previous incidents, workers are frequently facing novel situations: we cannot predict in advance what information will become relevant. In addition, time pressure is high since human lives are often at stake and decisions also have environmental and economic consequences: information provisioning must be efficient to be effective. Coordination is particularly complex at the beginning of an incident, when a large number of interdependent activities is initiated in parallel by many parties: sharing information enables cooperation.

There are several ways in which information supply can fail. First, information may not reach the right person because information providers do not realize that their information is needed. Second, information providers do not desire to share information because they are afraid it will be misused, even though they (can) know that their information is important. Third, people may receive more information than they can process, and thereby miss relevant information.

In sum, the challenge that we focus on in the SlimVerbinden project is: Can we enable secure information sharing while ensuring relevant information provisioning? Our approach to this challenge is based on the concept of autonomy (or control) of organizations: information providers are in control of their information and trust some information consumers, while information consumers control who in their organization can access information. To further assure information providers, there is no single controlling entity with full information access.

In this short paper we present the demonstration of an agent-based support system for information provisioning for daily practice and during crisis management.

2 Societal Relevance

The SlimVerbinden project works with a scenario-driven method to showcase the research results [1]. By consultation with various experts from industry, municipality, safety region, police, etc., a plausible and realistic scenario is defined based on an actual threat analysis of a steel factory in a densely populated area and the known issues in information sharing among multiple parties. This method has proven to bridge the gap between science and practice and paves the way via operational pilot(s) towards implementation in products.

The steel factory is a complex contraption spanning a large area with many installations, pipelines, valves, storage units, etc. A number of the chemicals and products in this factory can be of hazard to the environment, health and safety of the population. If an incident occurs on the terrain of the factory (e.g., a valve releasing a gaseous substance), an effective and efficient cooperation must start among the factory (private company) and police, fire brigade, municipality, etc.

During the incident, there is a growing number of parties that become part of the (ad-hoc) incident management community, including parties such as industries, media (i.e., journalists), etc. Near the end of incident lifecycle, the number of parties reduces until the incident is finished and the logs are archived.

Currently there is a barrier to sharing information: once information is shared, control is given up: anyone can get access. Solutions are proposed to mitigate this problem by either always connecting everybody with full access to their data sources, and/or ensuring that a data-warehouse is set up with access control: apart from introducing information updating challenges, the information sharing and access control is beyond the control of the information supplying party. How to enable information sharing while retaining control? And, while having access to a lot of potentially relevant information: how to realize context-aware information provisioning?

3 Information Provisioning Approach

Our analysis on information provisioning in crisis management is substantiated by the steel factory use-case, in which two roles are distinguished: information consumers and information providers, combined with a general view on teamwork and access control. People are considered to be grouped in communities (where people may come from different organizations such as companies, public safety organizations, municipality, etc.). The basis for information sharing between communities is trust: an information provider trusts a community of information consumers. This trust is further grounded by audit logs and ultimately by the legal system to rectify violations of trust and/or misbehavior. Our analysis results:

Information consumers (1) need permission to access information in their context; (2) want information that is relevant in their context and understand why this information is relevant for them; (3) need to actually receive the information; and (4) should be able to estimate the reliability of the information.

Information providers (1) grant access to information; (2) understand which information is needed for what reason; (3) monitor access, (4) understand why their policies for sharing are too strict cq. conflict with other information providers' policies; and (5) need to actually provide the information.

When multiple information providers and information consumers work together (in sharing and accessing information, from possibly multiple communities), the web of trust needs to be ascertained. In *teamwork*: (1) (in)direct information access is checked against information sharing policies; and (2) information access conflicts are escalated to information providers and information consumers, respectively. This ensures that valid requests for information can be allowed.

More information is not necessarily better: context-aware information filtering allows information consumers to be alerted to relevant information.

4 Agent-Based Approach

The multi-agent metaphor provides a useful modeling perspective when analyzing multiple, autonomous, entities that engage in unforeseen yet coherent interaction patterns—a loosely coupled, open, distributed system allowing for join and leave of participating entities. Part of the design of any multi-agent system is the emphasis placed on orchestrating emergent behavior. In particular in SlimVerbinden: orchestrating the access to the information in the secure shared distributed dataspace (S^2D^2S) [2]. Our architecture distinguishes:

- The secure, shared, distributed dataspace, in which information sharing is allowed according to policies (named: 'dataspace' hereafter) including information provisioning, consumption, and explanation services. Any agent can be both information provider and information consumer.
- Within our architecture, many entities are represented by software agents:

  - Humans (e.g., policemen, fire fighters, medical doctors, civilians, mechanics, etc.) may have their own personal assistants (software agents on e.g. smartphones) that manage information based on the human's context.
  - Each human (and its personal assistant) is represented by an agent in the dataspace, to arrange authorized access to information, and controlled sharing of information. These 'representation agents' are dynamically created for new users of the dataspace.
  - Sensors, services, and other processes (including legacy systems) that need to provide or consume information are also represented by agents.
  - Context-aware information filtering software is encapsulated by agents with appropriate access to relevant information.
  - Information providers' information sources are guarded by agents that stipulate sharing policies and oversee changes to these policies (including escalating policy change requests to an authorized human [3]).
  - The dataspace needs additional agents for its own management, as well as the security infrastructure, including the issuing of certificates, parsing of sharing policies for individual information consumers and teamwork.

5 Demonstrator

The demonstrator is based on the steel factory scenario and features a hybrid system-of-systems including a secure shared distributed dataspace (S^2D^2S). A number of information providers and consumers are distinguished in the scenario, with appropriate information sharing policies and web of trust. Each of the entities is represented by an agent with access to the dataspace.

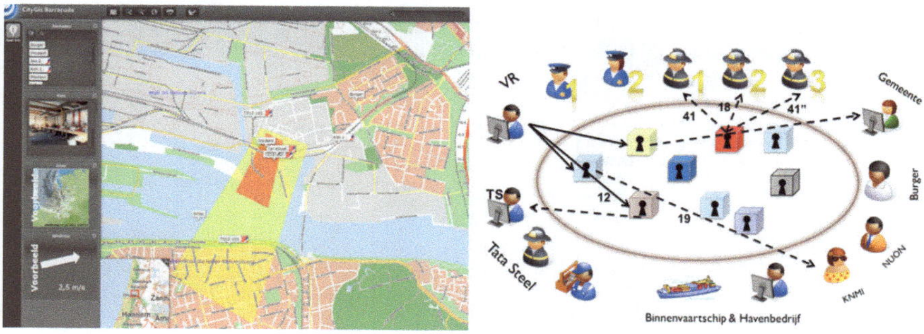

Fig. 1 Screenshot from the CityGIS viewer, and interaction pattern from the scenario

The demonstrator shows that authorized agents (as information consumers) can connect with data from different sources. The demonstrator also shows that agents can have both roles of information provider and information consumer. Furthermore, access to information can be denied and an 'escalation' procedure is available to ask for access at the information provider who can (temporarily) set a more lenient policy for a specific information consumer's community. Additional processing of information is shown, where intelligent software agents monitor information in the dataspace, and pro-actively inform humans about relevant information, including brief explanations *why* information is relevant.

The demonstrator integrates different techniques, including agent-based technology (the AgentScape agent operating system [4]), Paige agents and smartphones [5], and the CityGIS geo-information viewer [6]. The demonstrator is self-contained: information interoperability is beyond the scope of this project. The current SlimVerbinden demonstrator is the second version, runs on multiple computers and smartphones, and has been lightly evaluated by a group of domain experts (from industry, municipalities, police, fire brigade, and other public safety and security organizations). The scenario is considered plausible and realistic, and the underlying technologies are sufficiently portrayed.

Figure 1 shows on the left the CityGIS viewer of the (simulated) GPS positions, and one of the effect zones (as well as a demo-picture live-fed to the dataspace). On the right is an interaction pattern derived from the scenario BPMN specification, in which the safety region updates the secured plume shape information (shown in a box with a keyhole) to which specific information consumers have access (agents are not shown).

Figure 2 shows screenshots of the smartphones of two police officers. The officer on the left is standing in the projected plume shape, and automatically receives information with an explanation: the position is safe for about an hour. The officer on the right likewise receives information: this position is safe.

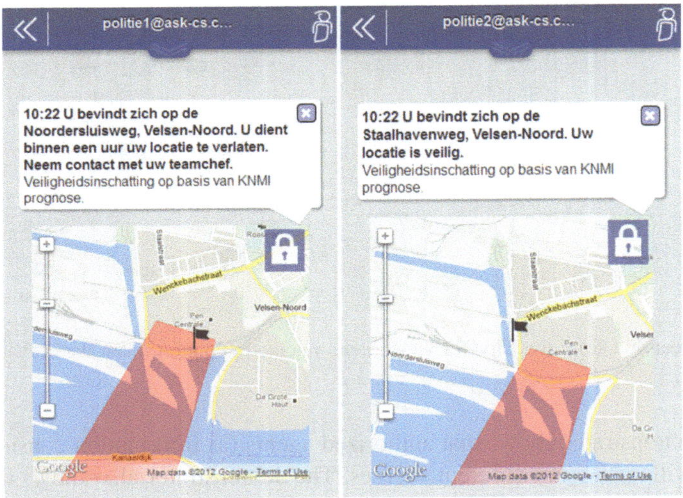

Fig. 2 Screenshots from smartphones of two police officers with context-specific information

6 Conclusion

The SlimVerbinden project aims to lower the barriers for information sharing. The progress in our research and with the demonstrator shows that it is possible to share information without losing control, while respecting the autonomy of both information providers and information consumers. Teamwork is supported by various mechanisms, including access permissions and escalation mechanisms.

The ongoing work in the SlimVerbinden project is focused on further detailing the security infrastructure within the demonstrator, coupled with the teamwork workflow authority. The project, supported by business partners, brings the first working components as products and best practices to the market.

Acknowledgments

The work described in this paper provides an overview of the "Slim Verbinden" project, that is sponsored by "Programma Innovatie voor Maatschappelijke Veiligheid" by Agentschap.nl, project no. IMV1100038. The authors thank their colleagues in the project: Fons Panneman and Kees Zeeman (from Studio Veiligheid), David Mobach, Reinier Timmer, Thomas Quillinan, Sorin Iacob, Gerard Toonstra (from Thales Research & Technology), Frances Brazier, Michel Oey, Martijn Warnier, Sander van Splunter, Zülküf Genç, and Faridah Heidari (from Delft University of Technology, TBM), Catholijn Jonker, Maaike Harbers and Joost Broekens (from Delft University of Technology, MMI), Duco Ferro, Dennis Dortland, Tymon de Jonge, Sven Stam, Ludo Stellingwerf and Xiaoyu Mao (from Ask Community Systems), Kees Jan Buter, Renato Bianchessi, Erik van der Berg and Adrie de Kooter (from Cap Gemini), and Robert Kieboom, Stephan Miegies and Ming Chan (from CityGIS). The authors thank the experts from public safety & security organizations and private companies for their contributions to the use-case and constructive feedback. The authors are grateful for the constructive comments of the reviewers.

References

1. A.M.J. Panneman, E. de Vries, P. de Bruijn, *Network Information Management for Collaboration in Disaster Management: Concepts and Case Study*. In this volume (2012)
2. Z. Genc, F. Heidari, M.A. Oey, S. van Splunter, F.M.T. Brazier, *Agent-Based Information Infrastructure for Disaster Management*. In this volume (2012)
3. H. Harbers, J. Broekens, Q. Quillinan, M.B. Van Riemsdijk, N. Wijngaards, Goal-Based Explainable Security Certificate Requests. in proceedings of the Gi4DM (2012 in press)
4. www.agentscape.org (e.g., Oey, M. A., van Splunter, S., Ogston, E., Warnier, M. and Brazier, F. M. T. (2010), A Framework for Developing Agent-Based Distributed Applications, in *Proceedings of the 2010 IEEE/WIC/ACM International Conference on Intelligent Agent Technology (IAT-10)*, IEEE/WIC/ACM, Toronto, Canada.) Date of visit: 5 Oct 2012
5. www.ask-cs.nl/nl/paige Date of visit: 5 Oct 2012
6. www.citygis.nl, Date of visit: 5 Oct 2012

Agent-Based Information Infrastructure for Disaster Management

Zulkuf Genc, Farideh Heidari, Michel A. Oey, Sander van Splunter and Frances M. T. Brazier

Abstract The success of a disaster management process depends on effective, secure and efficient information flow. This paper proposes an agent-based distributed information infrastructure to enable the realization of such secure information flows in disasters. The proposed infrastructure uses software agents in the exchange and processing of information, secure and dynamic information sharing and automated information flow generation and configuration.

Keywords Crisis management · Disaster management · Information flow · Information management · Multi-agent systems · Secure distributed information sharing

Z. Genc (✉) · F. Heidari · M. A. Oey · S. van Splunter · F. M. T. Brazier
Delft University of Technology, Delft, The Netherlands
e-mail: Z.Genc@tudelft.nl

F. Heidari
e-mail: F.Heidari@tudelft.nl

M. A. Oey
e-mail: M.A.Oey@tudelft.nl

S. van Splunter
e-mail: S.vanSplunter@tudelft.nl

F. M. T. Brazier
e-mail: F.M.Brazier@tudelft.nl

S. Zlatanova et al. (eds.), *Intelligent Systems for Crisis Management*, Lecture Notes in Geoinformation and Cartography, DOI: 10.1007/978-3-642-33218-0_26,

1 Introduction

Disasters, from man-made hazards to natural catastrophes, are complex situations requiring rapid decision-making and coordination of autonomous participants. Provision of efficient and secure flow of information, which includes geo- and contextual information, between the participants is critical [1].

Information exchange between multiple participants refers to (1) acquiring information from participants, and (2) providing information to participants. A large number of participants are involved in disaster management including the fire brigade, the police force, the government, other experts, and owners of sensor networks. Participants have their own preferences and policies on when, how, and with whom information should be shared as information providers and/or as information consumers.

Designing effective information flow is a challenge: (1) ensuring interoperability and timeliness [8, 12], and (2) ensuring secure information exchange between multiple participants [1], and (3) ensuring flexibility and adaptability between a dynamic set of participants. Access policies should be reconfigurable in real-time depending on the level of the disaster, the context of the information and the participants.

This paper proposes an agent-based distributed information infrastructure to enable secure realization of effective information flows in disaster situations. In the following sections of the paper, three challenges are addressed separately. The first part, Sect. 2, outlines the use of *agents* in interoperable exchange of information and automated data processing. The second part, Sect. 3, describes the secure and distributed information infrastructure, the Secure Shared Distributed Data Space (S^2D^2S). The S^2D^2S allows providers to dynamically make information available to consumers and keep control on sharing conditions via dynamic access policies. The third part, Sect. 4, describes a configuration approach for dynamic management of the information infrastructure.

2 Data Exchange and Processing with Agents

The complexity of processing data, which comes from diverse sources in many different formats, and time sensitivity of exchanged information in disasters require automation of the information management. Software agents can be utilized to provide this automation in different stages of the information management process that includes information extraction, information retrieval, information filtering, data mining and decision support [5].

The term agent is used in this paper to describe software that is capable of autonomous actions designed to meet its design objectives [6]. The agent-paradigm provides a means to model the way in which participants in disaster management work together, as autonomous entities in control of their own data and actions. In this paper each participant in a disaster situation is represented by a corresponding agent. Each agent is an information provider and/or consumer with information related tasks such as processing and sharing. Information exchange is loosely coupled to individual agents: it is defined across a system.

Participants deploy different types of devices that interact with each other in different formats. Their agents automatically convert the data these participants exchange into pre-defined types and formats such as EDXL [9], requesting guidance when needed, filtering out irrelevant information. Additional disaster specific agents are included in the system to actively monitor distributed information from information providers. They combine pieces of information and create more meaningful or valuable information, that is then distributed to other agents. For example, in case of a forest fire, the weather forecast can be combined with a street map in order to assess which buildings should be evacuated. If the agents have data inference capabilities then people living in those building can be *pro-actively* warned by the agents deployed in their mobile devices.

3 Secure Shared Distributed Data Space

In disaster management data comes from different sources. The gathered data is generally unrelated and heterogeneous, which makes it challenging to fit this data into conventional data base models and systems [3]. The infrastructure proposed in this paper uses a secure and distributed data space solution for sharing the information. S^2D^2S, depicted in Fig. 1, creates a virtual structured shared data space with which information providers can dynamically make their information accessible to specific classes of authenticated information consumers via agents. Information sharing is based on a publish-subscribe mechanism. Information providers publish their information on a specific topic for specific types of consumers who can subscribe to this information specifically. Information providers control who can access which data items and under what conditions and restrictions, and specify this in access policies. S^2D^2S restricts access to information according to these policies.

Note that an information provider can delegate application and configuration of these policies to an agent that is under a provider's control. This agent then instructs S^2D^2S to restrict access to the data items accordingly. The agent can change specific access rules during runtime as long as they are in accordance with the more high-level access policies of the information provider. This flexibility is necessary during disasters. In unclear cases, an agent can contact an information provider if his/her attention is required to grant or deny access to a particular information request.

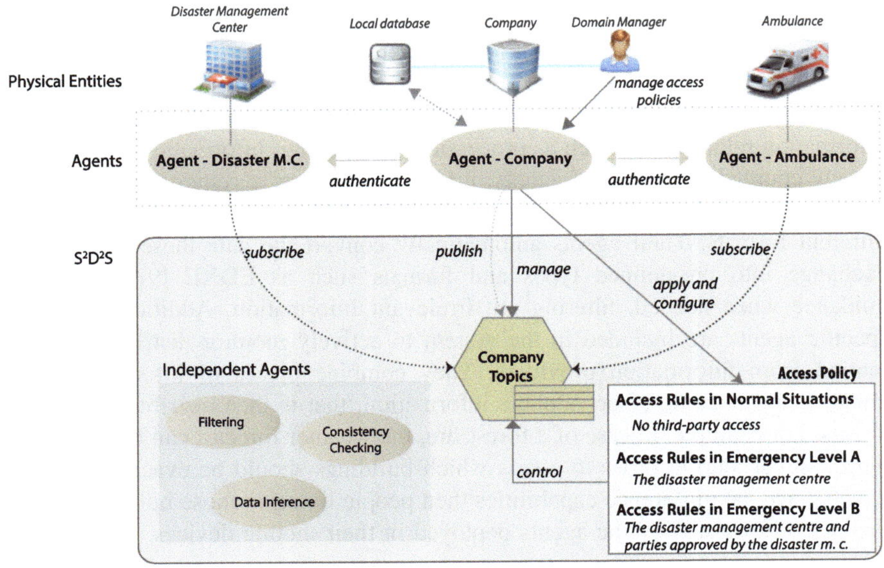

Fig. 1 An overview of S^2D^2S

In organizations, a domain manager role can be defined to control access to an organization's information. Access control can be fine-grained, defining access rights of individual users, but it can also be coarse-grained, defining access rights to users based on their roles, for example. Furthermore, access policies can be adaptable. They can be modified over time or they can change in line with the type and the level of a disaster. The actual enforcement of these access restrictions is acquired through data encryption. Data published on topics is encrypted and only the consumers who are granted access with the necessary certificates will have received the corresponding key to decrypt the data. Security certificates are automatically generated by agents in line with their directives (a goal-based explainable approach [4]).

4 Automated Information Flow Generation

Manual management of a secure information infrastructure is challenging: multiple access policies are defined when an information consumer requires information from various providers. Management complexity increases with cross dependencies between information providers. To support such management, this section describes a template-based configuration approach used to help automated management of information sharing.

It is common in disaster management to have pre-defined management protocols and plans for different types of disasters. These protocols and plans contain both disaster-specific and generic patterns for information sharing. Such patterns are either based on recurring generic information needs between organizations, or based on common practice indicating how specific data from different participants can be combined to fulfil a specific information need. These abstract participants represent information providers and consumers not by referring to specific agents, but by specifying required properties only, such as roles, organizations, capabilities, or kind of information provided. Once identified, such patterns can, in theory, be automatically linked [10].

Automated Workflow Generation as proposed in this paper, enables compositional refinement of an information need, based on pre-defined templates. To facilitate workflow generation, a graphical editor has been developed. In an emergency incident, crisis managers select a workflow pattern that fits a current situation and customise the pattern where necessary. The resulting pattern-based workflow is instantiated by a Workflow Authority agent. This agent negotiates with other agents representing information providers and consumers to determine which participants play which role in a template.

The forest fire example mentioned above, is an example. To inform the general public of the fire, a human communication manager gathers as much information as possible, and then creates and approves a message that is placed on a public website. If the fire continues to grow and its smoke plume is endangering parts of a nearby city, the inhabitants of this city also need to be warned. To this purpose another message is created advising the citizens to close their doors and windows and to remain inside. An existing template structures this process. This template contains an information pattern that first combines information from the fire department on the type of fire, its location, its size and weather information from a weather station to predict how the smoke plume will develop. Second, the pattern combines the predicted development of the plume and a map of the city from a GIS to identify the affected areas within the city (e.g., streets or neighbourhoods). A public announcement including this information is then generated and sent to the human communication manager to be assessed for approval.

The Workflow Authority supports flexible and adaptive information exchange between provider and consumer agents, changing data access conditions and restrictions according to given access and security policies. If during a disaster, some information becomes unavailable, the Workflow Authority can reconfigure a specific flow of information by finding another information provider able to provide the same type of information. For example, if the weather forecast service becomes unavailable, another information provider for weather forecasting can be searched for by the Workflow Authority based on the required properties specified in the pattern-based workflow. If there is no information provider available, then the Automated Workflow Generation checks whether a template is available to construct the requested information using different sources, and have the Workflow Authority integrate this new template into the information flow affected by the unavailability of the weather forecast service.

5 Conclusion

This paper addresses the key challenges facing the realization of secure and efficient information flows in disasters and proposes an agent-based secure and distributed information infrastructure to cope with these challenges.

The use of agents in data exchange and processing is explored. S^2D^2S enables secure sharing of information between participants in a dynamic and distributed way by addressing the concerns of information providers, using agent technology. Automation of information flow generation and configuration by agents is designed to reduce the complexity in the management of information infrastructure.

This agent-based disaster management approach is currently being studied within the Slim Verbinden project [2, 7]. The S^2D^2S data space has been implemented on top of the AgentScape, a large-scale, distributed, multi-agent platform [11].

Acknowledgments The work described in this paper is part of the Slim Verbinden project, sponsored by 'Programma Innovatie voor Maatschappelijke Veiligheid' by Agentschap.nl. The project partners are: Studio Veiligheid, Cap Gemini, CityGIS, Ask Community Systems, TU Delft and Thales Research & Technology. The authors thank Niek Wijngaards and Thomas Quillinan for their input and feedback.

References

1. N. Bharosa, J. Lee, M. Janssen, Challenges and obstacles in sharing and coordinating information during multi-agency disaster response: propositions from field exercises. Inf. Syst. Frontiers **12**(1), 49–65 (2010)
2. P. de Bruin, N.J.E. Wijngaards, Agent-enabled information provisioning while retaining control: a demonstration, in *Proceedings of the Gi4DM* (2012) (in press)
3. M. Franklin, A. Halevy, D. Maier, From databases to dataspaces: a new abstraction for information management. ACM Sigmod Rec. **34**(4), 27–33 (2005)
4. M. Harbers, J. Broekens, T. Quillinan, B. van Riemsdijk, N.J.E. Wijngaards, Goal-based explainable security certificate requests, in *Proceedings of the Gi4DM* (2012) (in press)
5. V. Hristidis, S.C. Chen, T. Li, S. Luis, Y. Deng, Survey of data management and analysis in disaster situations. J. Syst. Softw. **83**(10), 1701–1714 (2010)
6. N.R. Jennings, M.J. Wooldridge, *Applications of Intelligent Agents* (Springer, Berlin, 1998)
7. Agentschap NL. Slim verbinden (2012), http://www.agentschapnl.nl/onderwerp/slim-verbinden
8. Y. Peng, Y. Zhang, Y. Tang, S. Li, An incident information management framework based on data integration, data mining, and multi-criteria decision making. Decis. Support Syst. **51**(2), 316–327 (2011)
9. M. Raymond, S. Webb, P.I. Aymond, Emergency data exchange language (edxl) distribution element, v. 1.0. *OASIS Standard EDXL-DE v1*. 0, 1 (2006)

10. J.B. van Veelen, S. van Splunter, N.J.E. Wijngaards, F.M.T. Brazier, Reconfiguration management of crisis management services, in *The 15th conference of the International Emergency Management Society (TIEMS 2008)*, 2008
11. N.J.E. Wijngaards, B.J. Overeinder, M. van Steen, F.M.T. Brazier, Supporting internet-scale multi-agent systems. Data Knowl. Eng. **41**(2–3), 229–245 (2002)
12. S.W. Yoon, J.D. Velasquez, B.K. Partridge, S.Y. Nof, Transportation security decision support system for emergency response: a training prototype. Decis. Support Syst. **46**(1), 139–148 (2008)

Using a Base Registry Key in Disaster Information Management: A Dutch Case Study on Linked Data

Jan-Willem van Aalst, Bart van Leeuwen and Rob Peters

Abstract This paper describes the process of using the Dutch Base Registration Identification number (BAG) as a primary key to link various sorts of information about a particular built-up object. This greatly increases the quality and effectiveness of the common operational picture. Consequently, the paper investigates the usefulness and affordability of Linked Data (a method of publishing structured data so that it can be interlinked). Several practical applications are considered and evaluated. The primary conclusion is that the BAG Base address registration greatly enhances the possibilities for safety workers to gain a common insight about particular built-up objects. This pertains to the "cold" risk-analyses and preparatory work phases, as well as in the "warm" incident and disaster management phases. It is in the combination of these two phases where the profit is maximized: when attempting to assess the consequences of the disaster management processes during and after the incident, relevant data about the particular object is now much easier than before, and with much better quality. This is already turning out to be very profitable for the fire brigade workers, the police officers, the ambulance drivers, as well as the municipal officers.

Keywords Linked data · Physical safety · Geographical information

J.-W. van Aalst (✉)
Imergis, Utrecht Area, The Netherlands
e-mail: jw@imergis.nl
URL: www.imergis.nl

B. van Leeuwen
Netage, Almere, The Netherlands
e-mail: bart_van_leeuwen@netage.nl
URL: www.netage.nl

R. Peters
Public Safety & Health Region Kennemerland, Haarlem, The Netherlands
e-mail: rpeters@vrk.nl

S. Zlatanova et al. (eds.), *Intelligent Systems for Crisis Management*, Lecture Notes in Geoinformation and Cartography, DOI: 10.1007/978-3-642-33218-0_27,

1 Introduction: On the Significance of Addresses

For physical safety workers first responders such as police, fire brigade and ambulance staff, finding the correct address can literally be a matter of life or death. While most safety workers are familiar with the local situation, there have been many reported cases of ambulances not being able to find the correct address, for example in a newly built-up area that has not yet been mapped completely, in cases of crossings over waters, and in cases of large properties with private infrastructures to provide water. In other cases, mismatches in the way a particular street name is spelled has led to delays in arriving at the correct address. To make matters worse, inconsistencies in address spelling abound in the various address registrations of police, fire brigade, ambulance and municipal administrations. One particular fire brigade department decided to analyze their data registrations and found that they had no less than 34 address registrations with many inconsistencies, outdated entries, and no consistent address maintenance. This leads to a myriad of errors, not only in actual incident management, but also in the preparatory work, for example in juridical advisory tasks and regional risk analyses.

The Dutch government has acknowledged the need for national standards on a variety of data types and addresses are amongst these. About a decade ago, the Dutch government initiated a National Infrastructure of Core Base Registrations (1) (Dutch: Stelsel van Basisregistraties). The implementation of Base Registries is defined as one of the key-actions of the European eGovernment interoperability programme ISA. This infrastructure constellation started with personal data of civilians, and soon expanded to include topography, work and income, car license plates, and more. As at 2012, the Constellation of Data Base Registrations infrastructure holds no less than 13 Base Registrations. Each of these has been carefully designed and modeled according to national and European standards. For example, the Topography registration is largely compliant with Inspire guidelines and derived related modeling such as the IM-GEO. For addresses, this has resulted in the Core Base Registry of Addresses and Built-up objects (Dutch: Basisregistratie Adressen en Gebouwen, BAG). In Fig. 1, we present a (simplified) scheme of the structure of this Registration.

Not only has the data been uniformly designed, but the process of maintenance and assurance of the quality of the actual data has been organized. Various public instances have now been assigned the task of keeping these national registrations up-to-date, and the obligatory usage of these data (for public institutes) has been secured by national Law. In the case of addresses, this means that every public organization is now required to make use of the national Address and Built-up Object Base Registration, and although there is still discussion about what "making use of" means exactly, the BAG Base Registration is now an existing and mandatory data set for the entire Dutch public service community.

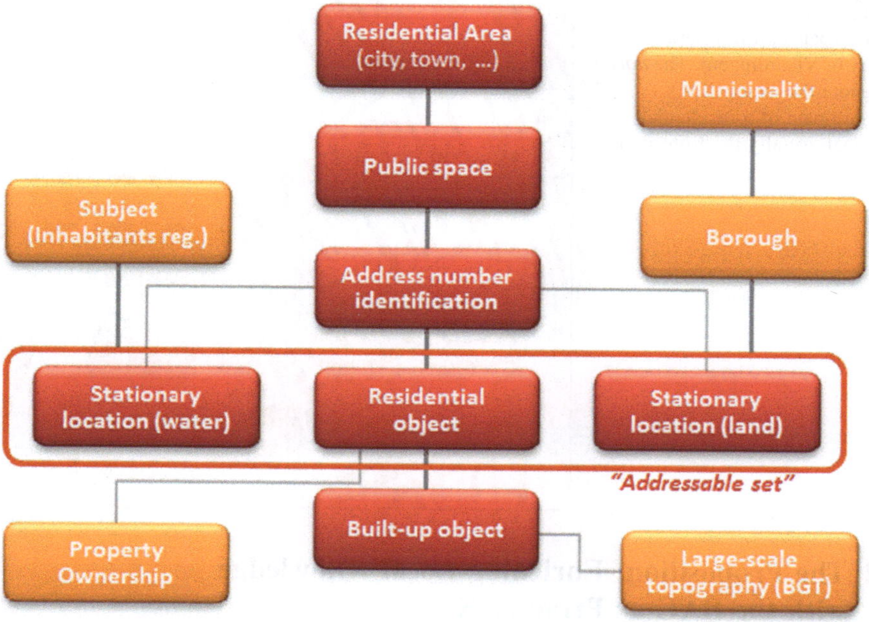

Fig. 1 Data model schema of the Dutch core base registration of addresses and built-up objects ("BAG"). The *red boxes* are integral part of the BAG; the *orange boxes* represent adjacent base registrations

As the name implies, the "Address and Built-up Object Base Registration" actually consists of two main parts:

1. A comprehensive listing of all address in the Netherlands, including a dictionary of how each name is to be spelt correctly;
2. Geographical data about built-up objects. This includes any building or construction that is (a) somehow durably anchored to the ground; (b) lockable; and (c) human-usable as shelter (i.e., it can be entered). The Netherlands contains some 9 million objects that meet these criteria. The BAG not only contains the exact geometry of these objects, but also the exact (X,Y) coordinate. In effect, this means that each address in The Netherlands now has (a) a single uniform way of spelling the name, and (b) an exact location on a map, and these make it relevant to geographical information systems. Figure 2 shows a simple map containing BAG building objects—again, the BAG registry holds both the exact geometrical shape and the point coordinate of the address.

One interesting consequence of having such a uniform nationwide registration, which is the main topic of this paper, is that any data that is meta-tagged with at least an address can now be positioned on a map in a reliable manner.

Fig. 2 Example of a BAG map (part of a Borough in the city of Vlaardingen). The *dots* denote the exact (X,Y) coordinate of a residential object, usually the front door

2 The Proposition: Enriching Local Knowledge with the BAG as Primary Key

With the BAG Base Registration, each built-up object that meets the criteria described above is assigned a unique number. Since the BAG identification number is now available to any application that records information about a particular building, useful analyses become possible to aid in disaster management. We illustrate these benefits using a fictitious incident scenario.

In the scenario, a fire is accidentally started in an underground parking garage in the vicinity of a shopping center. Within minutes, smoke spreads across the parking decks. Since the garage complex is ventilated according to legal regulations, the smoke—and the flames!—soon find their way to the open air. The Parking garage has an installation that automatically signals the regional incident control room—again due to legal regulations (for the control and dispatch room, incidentally, such a signal is no guarantee for a real incident since over 70 % of all such signals turn out to be false). Additional calls validate the genuineness of the event, and safety services are dispatched to the location.

Until recently, both the control room and safety workers had to rely solely on their 'private' knowledge of the local situation. This became increasingly problematic when larger regional authorities were established because personnel was distributed all over the District. Using the digitalized maps based on the BAG, all these people can access a wealth of extra information about the object in particular, and adjacent objects in the surrounding area.

In the "cold", pre-incident safety work (i.e., administration, risk analysis and preparation), a variety of spatial analyses become possible, including:

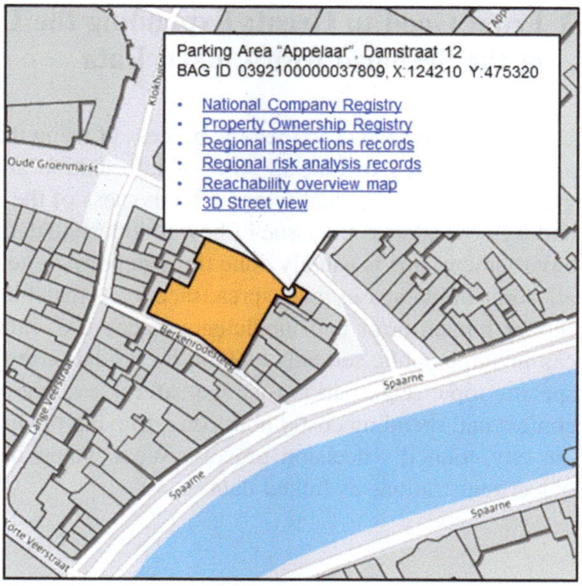

Fig. 3 Example of a BAG object containing links to other information sources about the same object, all linked through the unique BAG identification key (The hyperlinks have been translated into English for the purpose of this paper)

1. The national registry of companies, usually for insurance purposes;
2. The national Property Ownership Registry, to inform the legal owner of the property;
3. Regional inspection records, to assess whether or not this object meets legal safety regulations;
4. Approachability overview map, a standard map product for the fire brigade.

The approachability map (in Dutch; "bereikbaarheidskaart") is an important tool for Fire brigades to have digital information in the Vehicle Garage, or in the vehicles while driving to that object during an incident or a fire. A National 2 million Euro programme has fed the standardization of those maps in order to ensure maintainability and interoperability. The Approachability map is standardized in open GML, WFS and OWL/Skos, and these standards became part of the Safety Region reference architecture (VERA) when the National Commander Council and the Dutch National Security Council adopted its content in 2012 (Fig. 3).

The prerequisite, of course, is that the administrative organization of the safety workers must have mapped all existing documents (inspections, warrants, incident evaluation reports, etc.) about a particular location to the specific BAG identification number. This may seem like a time-consuming task, but it pays off in times of disaster management.

3 From Good to Great: Expanding the Usefulness of the Scenario with Linked Data

Up to now, we have described the process of using the BAG identification number as a primary key to link all sorts of information about a particular object. This greatly increases the quality and effectiveness of the common operational picture, and we now address the issue of how to link such data. In an average public policy environment, this is usually done through copy-paste of specific numbers between all sorts of databases and spreadsheets. A matching process like that induces further human error and inefficiency because of information retrieval time. One can imagine public servants looking for a relationship between budgets spent on a specific topic (e.g., parking garage safety), evaluations, decisions, policies, legal context and spending costly hours on retrieving the information for the next step in the city council's decision process. What would happen if we organized these object registrations as linked data?

3.1 Linked Data

The term 'linked data' describes a method of publishing structured data so that it can be interlinked and become more useful. It builds upon standard Web technologies such as HTTP and URIs, but rather than using them to serve web pages for human readers, it extends them to share information in a way that can be read automatically by computers. This enables data from different sources to be connected and queried [1].

3.2 Matching Local Registrations to the BAG Object Using Linked Data

The procedure described above, where a government worker uses cut & paste to search with BAG identifiers reveals an interesting problem. A BAG object identifier lacks some important aspects:

- The number itself does not reveal what it is about since it could be an address or building;
- There is no hint of how and where to retrieve authoritative, validated information about the object that the number represents;
- There is no guarantee of the uniqueness of the number.

If we would use linked data principles, we could solve these problems in one go. A URI for a BAG object solves the uniqueness and access issue, and the data retrieved at the URI will tell us in great detail about the nature of the object.

We can then use this URI as a linking point to other applications: instead of inserting just the number in there, we use the full URI.

While the BAG dataset in essence is geo data, it contains much more information and so is of interest to those not involved with geo applications. The BAG also contains information on how various objects are administratively connected, and all of these connections can be perfectly expressed using linked data principles. In this way, and without having to actually plot objects on a map, we can browse the relations they have solely using linked data. Linked data itself is heavily based on vocabularies describing the connection between the subjects and objects in the data, and to describe administrative connections the EU actually has some vocabularies in place which will assist us in exposing the BAG as linked data.

If we have other datasets using the same vocabularies and BAG URIs we can then let smart software do the "mix and matching" based on this semantic similarity instead of a visual queue based on a map image. One further interesting aspect is that matching linked data URI's is far less processor (CPU) intensive then Geospatial queries, making linked data a highly attractive format for light weight 'in the field' devices.

4 Linking Local Administration to the BAG Registration: A Practical Field Experiment

The Public Safety Authority Kennemerland governs the crisis management information concerning 'risk objects' related to Schiphol Airport and Tata steel, and has recently given the assignment to suppliers to integrate the entire document management system, the web portal content management system, and the Geo warehouse using the BAG as key.

The aim is to create a permanent link between 'cold' and 'warm' information sharing. Documents related to an object, such as evaluations, exercise results, inspections and factual information about objects like elevators or ventilator piping's can be retrieved in a manner of seconds. A second reason for this design approach is the integration of knowledge between Police, Fire brigade and Ambulance personnel who have a tendency to work within their own crisis management protocols and ways of working. Figure 4 shows the safety documents relating to various BAG objects on a map.

The Dutch Police innovation programme aimed at street level information provision to police personnel (The so-called Virtual Police Corps) has agreed to use the same semantic standards and the same BAG based search key to share information among the different 'emergency colors'.

A third reason to use the BAG (or any other such Base Registry on addresses) as an information key is to enable map based retrieval of information during a large incident or crisis. The European GMES initiative is maturing and the Galileo based satellite imagery is also intended to support crisis management. This facility is

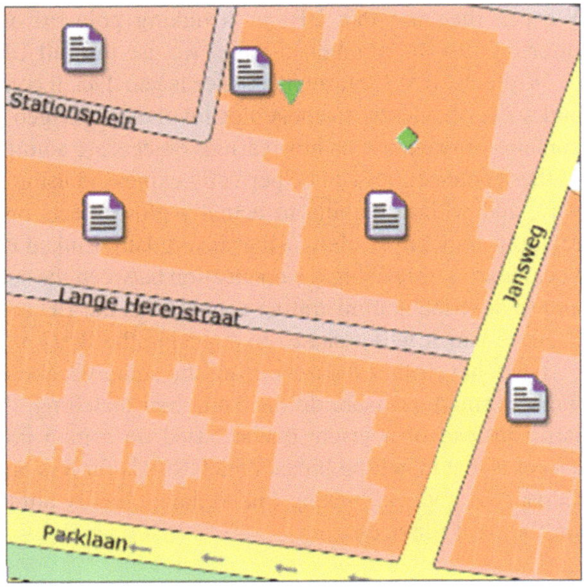

Fig. 4 Linking local administration (usually Office documents) to the BAG registration, and consequently on a map, as each document is meta-tagged with a BAG ID. Clicking on an icon opens the relevant document(s)

supposedly backed by 'localized data' from National information providers, and in the Netherlands this role is allocated to the Dutch National Environmental planning agency [Dutch: Planbureau voor de leefomgeving [2] (PBL 2)].

The Dutch National Emergency management systems (LCMS) and the approachability map standards are object oriented and much more detailed.

Using the BAG as key, the related X,Y coordinates could be used to combine GMES satellite images with new positioning systems and the existing local emergency management system.

One of the latest developments in the application of Base registries in The Netherlands has been the introduction of GEOSTUF. This new standard combines the messaging characteristics of workflow management (permit procedures, mutations, demolition, increase of some attributes like Tunnel height) with GIS based objects. This is a "marriage" between two radically different paradigms. It was necessary to introduce this combination, because most city processes are supported by caseload software, rather than object oriented GIS software. The solution chosen by the system architects was to create an XML messaging envelope carrying the process identifier and 'catching' the identifier again when the object mutation (with base registry identification) had been processed in the GIS database.

This construction will enable a number of maintenance processes of risk-objects and availability figures, including hospital capacities, bridge loads, number of trailers in a camping site, and other safety related pieces of information.

5 Conclusion and Further Research

The BAG Base address registration greatly enhances the possibilities for safety workers to gain a common insight about particular built-up objects. This pertains to the "cold" risk-analyses and preparatory work phases, as well as in the "warm" incident and disaster management phases. In fact, it is the combination of these two phases where the profit is maximized: when attempting to assess the consequences of the disaster management processes during and after the incident, relevant data about the particular object now becomes available much more easily than before, and with much better quality. This is already turning out to be very profitable for the fire brigade workers, the police officers, the ambulance drivers, and the municipal officers.

The use of base address registrations and its application via Linked data is in scope of other research projects, notably the EU Disaster FP 7 Project [3].

5.1 The Use of Linked Data in Disaster Management

The following observations have been made regarding improvement of the usage of this data in disaster management:

1. The BAG registration is not yet "up-to-standard", in that it still contains some faulty information about a limited number of objects.
2. The BAG registration is not yet available as linked data. As a result, experiments with linked data remain fragmentary.
3. Connecting the BAG unique identification number to local addresses and objects is still a cumbersome process in many cases. Automated tools are only beginning to emerge.
4. Although Middle and Top level management need not be convinced about the usefulness of maps in general, they are oftentimes hard to convince about the usefulness of the BAG registration because the investment is largely invisible ("under the hood"), and the results are not always visibly related to that investment.

References

1. ISA access to Base registries, Typology of D1.0.2. Typology of base registries. ISA report, European Commission, http://ec.europa.eu/isa/actions/documents/isa_1.2_access_to_base_registers_workprogramme.pdf
2. Dutch Public Government site, http://www.pbl.nl/dossiers/leefomgeving/modellen
3. C. Bizer, T. Heath, T. Berners-Lee, Linked data—the story so far. Int. J. Semant. Web Inf. Syst. **5**(3), 1–22 (2009). doi:10.4018/jswis.2009081901. ISSN 15526283. Accessed 18 Dec 2010

Using Icons as a Means for Semantic Interoperability in Emergency Management: The Case of Cross-Border Moor Fires and Schiphol Airport

Rob Peters, Jan-Willem van Aalst, Frank Wilson and Til Hofmann

Abstract The understanding of terms and maps is considered a vital part of emergency management by its practitioners. This paper describes a solution chosen by the emergency management information officers of the Twente and Kennemerland Dutch regional authorities to bridge the gap of understanding between first responder officers of different agencies. After having implemented a National emergency management system with a strong tendency to standardize protocols and tooling the designers involved realized that one cannot superimpose a single symbology for everyone involved. A need for a more flexible harmonization with partners like colleagues from cross the border and the IATA governed Schiphol Airport was necessary. To achieve flexibility and understanding at the same time a method for on-the-fly translation of icons was developed. The technology that powers the translation is the application of SKOS semantic web trippels. The paper describes some of the challenges that occurred when different icon sets based on different standardization regimes were analyzed.

Keywords Iconography · Symbols · GIS · Semantics

R. Peters
Public Safety & Health Region Kennemerland, Haarlem, The Netherlands
e-mail: rpeters@vrk.nl

J.-W. van Aalst (✉)
Imergis, Utrecht, The Netherlands
e-mail: jw@imergis.nl

F. Wilson
Interaction Design Ltd., London, UK
e-mail: frankwilson@btconnect.com

T. Hofmann
Cologne University of Applied Sciences (CUAS), Köln, Germany

S. Zlatanova et al. (eds.), *Intelligent Systems for Crisis Management*, Lecture Notes in Geoinformation and Cartography, DOI: 10.1007/978-3-642-33218-0_28,

1 Introduction: The Map a Main Instrument for Incident: and Disaster Management

Few people will debate the significance of maps as a key instrument to handle a crisis. It is also clear that (digital) maps play a vital role in day to day actions of Police, Fire brigade and Ambulances. The Dutch forces are all equipped with some form of digital map support, ranging from navigational aids to full- scale AI-supported, on-board emergency management information systems.

The higher management levels of crisis management do not require convincing either. The Dutch National emergency system has a strong GIS support section. The European GMES initiative, however criticized it may be, is basically a satellite image based map support system, thereby illustrating the consensus about the crucial role of maps.

The problem with these maps is twofold.

First of all there is the usability problem of interfaces being able to handle the reduction of information to avoid overload. GIS technicians have a tendency to put too much information in those systems. The second problem is addressed in this paper, and in some ways is connected to the first one (Fig. 1).

Icons and symbols are used in cartography from the beginning of paper maps. They seem to 'belong' to maps as an inseparable instrument to 'point' at locations with a specific significance. This is the 'affordance' (Gibson 1960) of icons that represents a small simulation of the intended significance, usually a small version of the physical object (cars, people, and factories) (ref Gibson). Icons and symbols are used because they convey the meaning efficiently and quickly without using much 'paper space'. This feature has not changed when the paper was replaced by digital means in GIS systems. Perhaps the challenge has become even greater, since GIS systems can combine many 'map layers' all having a range of icons and symbols (Fig. 2).

There are a number of issues with these icons in digital Geographic information systems. One of the issues involved is the problem of conversion from one system (buildings used to bed designed in Computer aided Design (Cad) software and GIS

Fig. 1 The risk of information overload in a fire truck cabin

Fig. 2 Two widely-used map symbols for the same physical object: an (*underground*) water hydrant in a public area

layers work differently, so conversion is costly). Another issue covers concerns about representation in colors and proportions and other lay-out issues. Some may think that everything is a database and the rest is a matter of representation. Cartographers and GIS specialists see this differently; the icon or representation layer is or at least should be a well-defined object with attributes in itself.

What we address here is the problem of standardization and representation. We propose a SKOS[1] based semantic web solution to allow for variation and harmonization where standardization is not likely to happen. We applied this semantic solution to the case of mutual understanding among emergency management officers at the border between Germany and The Netherlands. We collected the available domain specific sets of icons and symbols used in emergency management with team of specialists in medical care during Crisis (Dutch: "GHOR") and the Fire brigade units. We combined those listings with the Dutch standards for icons and symbolism used in the construction regulations and ISO icons. These we further combined with the sets for emergency management systems in The Netherlands and eventually those of the European GMES. The resulting set has been put forward as a discussion set among professionals from a range of countries in the context of the European FP7 research project 'Disaster'.[2]

The goal of this operation is mutual understanding. For that purpose it was necessary to define a common language to describe incident scenarios. We have used three of those:

1. The Tata steel toxic cloud scenario
2. The Cross border Moor Fire scenario
3. The Schiphol airplane crash scenario.

All three scenarios have actually occurred in recent years and have resulted in a number of very specific information needs. These needs have been systematically collected, investigated and expressed in listings with the aid of many professionals. The resulting lists could be seen as the starting point or version 0.1 of a controlled vocabulary. In this case the vocabulary is inseparably linked with a well defined icon, or set of icons.

The RDF SKOS[3] based technology enables the retrieval of the icon, the related object and the meaning as understood by the user in his context. As a consequence,

[1] SKOS—Simple Knowledge Organization System.

[2] http://disaster-fp7.eu/

[3] Resource description framework (RDF), Simple knowledge organisation model (SKOS and Ontology web language (OWL) are W3C semantic standards http://www.w3.org/2004/02/skos/.

in cross-border scenarios, relevant disaster aid parties on both sides of the border can act effectively, since they recognize their own context instantly. The same information is shared cross-border, but in a local context with local, recognizable representation using their own icons. This does require an international based harmonization structure 'underneath' the visual screen.

The claim of this paper is that such a harmonized icon structure would be an improvement of the GMES system. It could also be the instrument to combine GMES with local data resources in the same GIS environment. One could even combine these layers of objects with INSPIRE geographic data infrastructure and with open linked data resources.

The key idea is that the structure of the *identification* of icons is standardized, and that the *identification structure of the icon* allows for local representation on the screen. Here we consider a classic example in disaster management: the connection points for the fire hoses of the fire brigade workers. Figure 3 shows the variance in the official icon representation:

Although the visual representation is different, the underlying meaning is the same for all parties involved. Instead of a cumbersome conversion of digital data between countries, we propose to use a standardized structure for representing the same object. The SKOS methodology offers excellent possibilities to this end.

In the case of the Moor Fire Scenario on the border between The Netherlands and Germany (see map), fire brigade professionals from both countries want to share a *common operational picture*, which in theory is possible (Fig. 4).

All professionals involved rely on having their map presented in their own local context. In the case of this scenario, the same data should be presented in two local contexts. This is possible only if the underlying structure is harmonized in some way, preferably built on existing international standards.

The German federal state of Nordhein-Westfalen has set a number of rules regarding the usability of icons in emergency management. The basic mapping symbology used in the German emergency management is explained in the "Feuerwehrdienstvorschrift 102—Taktische Zeichen (FwDV 102)"(Fire Service Regulation 102—Tactical Signs). Historically, the current EMS symbology is based on an outdated symbology used by the German Federal Armed Forces, creating a very high resemblance to a military symbology.

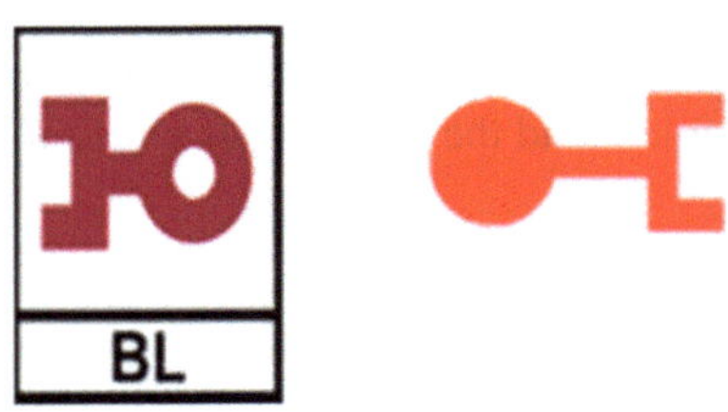

Fig. 3 Different icons for connection points for the fire hose. *Left*: Dutch; *right*: Germany

Fig. 4 A moor fire on both sides of the national border between Germany and The Netherlands (schematically; map image courtesy Google). Fire brigades of both countries are involved. The common operational picture can be presented to all parties in their own local context

This symbology is supposed to be oriented to the following basic principles: The symbols must be:

- Consequential and clear.
- As easy and as self-explanatory as possible.
- Portrayable with simple means.
- Applicable across organisations, the country and internationally.
- Designed in a way that they are adaptable to the respective management structure.
- Designed in a way that they are adaptable to the respective legal and administrative structures.
- As compatible as possible within the emergency management system as a whole (Police, Nato, Federal Armed Forces)
- Suitable as a basis for European and international standardization.

The symbology works with generic basic icons, filling icons, leadership items and colours. Additionally, the strength of a tactical formation can be added with numbers. Basic Icons 1.1–1.4 represent actors and forces within an operation. The basic icon (Table 1) shows the basic type of the item. The filling icon (Table 2), placed in the basic item, marks the function of the object (i.e. fire fighting vehicle, ambulance, command and support facilities). The colour of the item indicates to which branch of the EMS the show item belongs. Leadership items do not indicate the size of a unit, but the rank of a person or facility if it is a HQ).

The Dutch standard for symbology is based on the permit structure and regulations involved with building and safety; the NEN 1414.[4] The NEN prescribe very specific rules related to icons, colours, descriptions and rules with regard to ISO.

[4] http://www.nen.nl/web/Normshop/Norm/NEN-14142007-nl.htm

Table 1 Basic icon outline in Germany

Number	Symbol	Meaning	Number	Symbol	Meaning
1.1		Tactical formation (unit/group)	1.6		Measure
1.2		Command post (acting)	1.7		Incidence
1.3		Location/facility	1.8	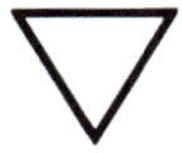	Hazard
1.4		Person	1.9		Stationary/place-bound
1.5		Area/expanse	1.10	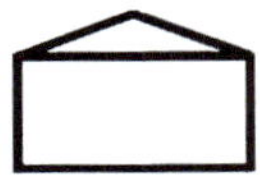	Building

Table 2 Translation of effect area icons

Dutch	English	Icon	Dutch	English	Icon
Logistiek punt	Emergency distribution point		Ambulance	Ambulance vehicle	
C2000 mast	Digital comms support mast		Aanwezigheid Brandweer	Presence fire brigade forces	
WAS paal	Public alarm horn		Brandweer voertuig	Fire brigade vehicle	
Zendmast	Broadcasting mast/ antenna		Commando plaats incident	Location command center (bronze)	

Over the last 2 years a working group of specialists in emergency management, health care and fire brigade preparative information management has suggested a new set in addition to the exiting 2012 NEN icons.

1.1 Codification of Icons in a SOA Based GIS Environment

This working group also proposed a new codification schema to enable the use of icons in a RDF semantic environment. By providing a number and a uniform resource locater (URI) the icon sets in different domains would become maintainable by the domain experts themselves, but the system would recognise the variations and translate them into the required icon in the other sector.

The working Group suggested a set of preferred icons related to the codes, while realizing that 10 % of those icons would be different. The RDF structure would allow for such variation, thereby allowing two important aspects in harmonization:

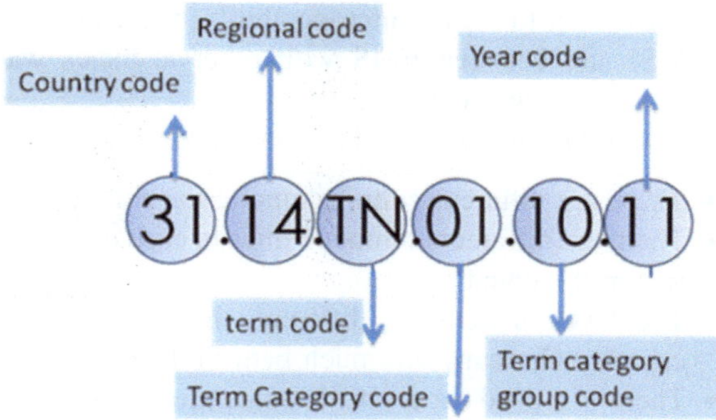

Fig. 5 Proposed codification schema emergency management icons

- A degree of freedom creates acceptance and much easier uptake
- Some organizations, like the international airport umbrella organisation IATA, have their own icons (Fig. 5).

Below you will find a small example set combining symbols with the new codification.

Icon	Code	New	Terms
	Tb1.007a	31.15.TB.01.08.11	Droge blusleiding afnamepunt
	Tb1.007	31.00.TB.01.07.11	Droge blusleiding voedingspunt
	Tb.1.003	31.00.TB.01.03.11	Sleutelkluis/Sleutelbuis
	Tb.1.004	31.00.TB.01.04.11	Brandmeldpaneel
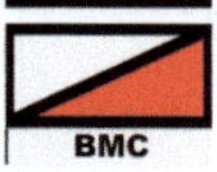	Tb.1.004a	31.00.TB.01.09.11	Brandmeldcentrale
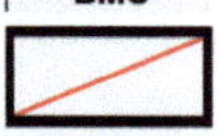	Tb1.005	31.00.TB.01.05.11	Nevenpaneel
	Tb2.001	31.00.TB.02.01.11	Schakelaar neon
	Tb2.002	31.00.TB.02.02.11	Schakelaar cv
	Tb2.003	31.00.TB.02.03.11	Schakelaar electriteit
	Tb2.004	31.00.TB.02.04.11	Schakelaar luchtbehandeling

Next step to create interoperability between Emergency management systems (EMS) and between EMS with other agencies was to translate all terms into several languages (Fig. 6).

The translation emphasised several findings:

- There are different cultures among the different emergency agencies;
- There are different command lines in each country and sometimes in each region within the country.
- The level of detail is different, the levels of hierarchy in vehicles size and capacity Germany are much better relayed in German icons, for example.
- There is a need for a set of source area icons and a need for a set of effect area icons.
- There is a temporal factor; one needs a first emergency 20 min set and a more elaborate 40 min plus set, due to emergency communication requirements.

Both listings cover many aspects of Micro incident management at the micro level (earthquakes with collapsed buildings, Tunnels, etc. can be ionized in a GIS

UK	GER	NE	Description	tactical value (UK)	tactical value (GER)
fire fighter	Feuerwehrmann	Brandwacht	Responsible for day-to-day firefighting and fire safety work.	-	-
squad	Trupp	1 Ploeg	small (the smallest) tactical unit used for fire fighting		2-4 firefighters
squad leader	Truppführer	-	Leader of one squad		1 firefighter
squadron	Staffel	2 Ploegen	A tactical unit that comprises more fire fighters than a squad but less than a pump crew.		5 firefighters
squadron leader	Staffelführer	Bevelvoerder	Leader of one squadron		1 firefighter
pump crew	Löschgruppe	Bezetting Tankautospuit	A tactical unit that comprises more fire fighters than a squad / squadron.	4 - 5 firefighters	9 firefighters
crew manager	Gruppenführer	Bevelvoerder	In charge of the watch at smaller fire stations or the crew of a fire appliance. Carries out day-to-day fire fighting and fire safety work. Will attend incidents as officer in charge of an appliance and will also take command of **small-scale**		1 firefighter

Fig. 6 Example table with translations about Firebrigade 'capacity' standards

system with those symbols to indicate new problem area's for emergency management personnel) (Fig. 7).

This last aspect, concerning the temporal factor, has lead to a discussion about the required core-ontology for emergency management. The core ontology would

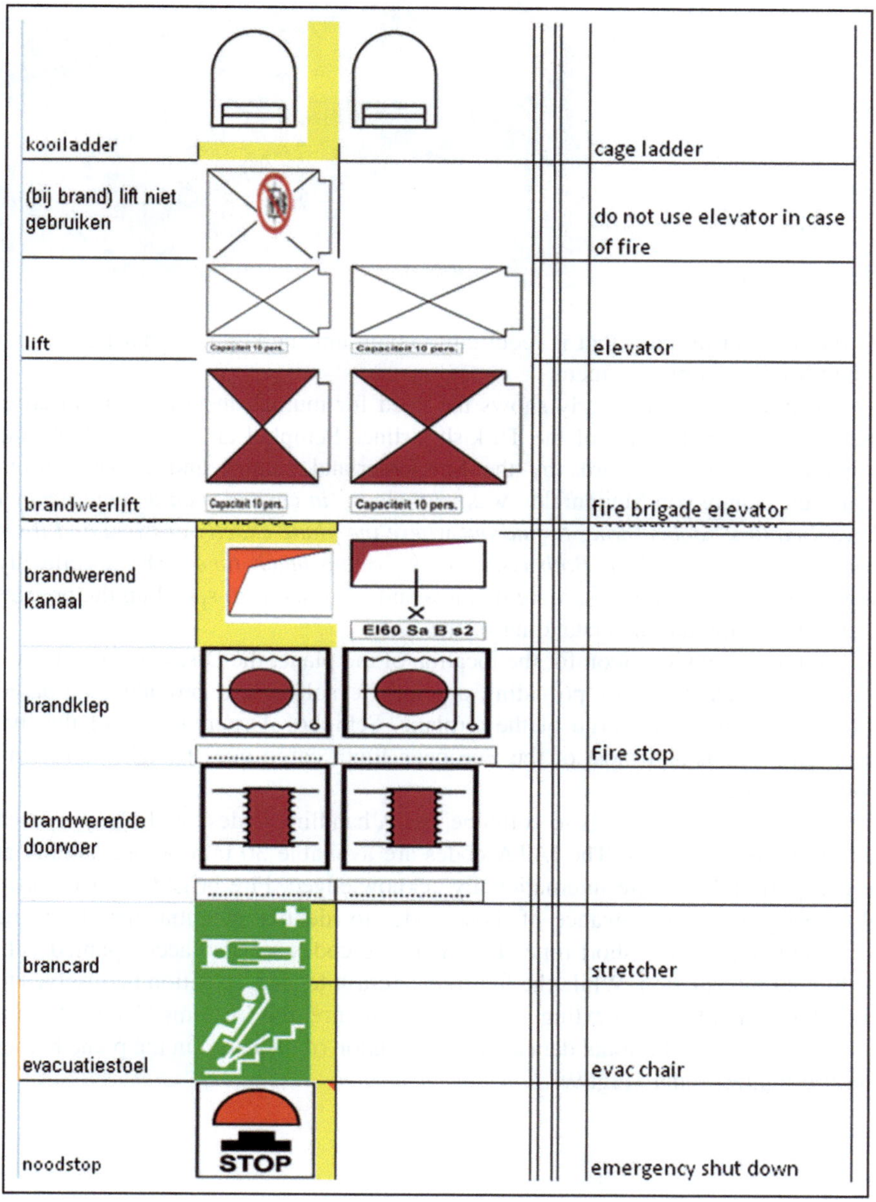

Fig. 7 example of source area icons set

Fig. 8 Schiphol Firebrigade team practicing with the 'Firefly' practicing airplane

be the base set of icons that is required to communicate key facts during the first minutes of crisis management.

The moor fire case clearly shows the need for mutual understanding in cross border cases. In the case of the Turkish airlines Schiphol air crash of 2009, the statements made afterwards by the Kennemerland control and dispatch room manager, Gonny von Meijenfeldt, was very clear: '*in case of such an emergency I have 4 control rooms trying to find out where the plane exactly crashed and if we could safely approach it. Both issues took far too much time'*. She specifically suggested using more maps, more pictures and more icons to speed up the process and to reach mutual understanding of the facts.

Next to the obvious icon for the location of the plane, the Disaster project team took on the challenge of providing a semantic solution to provide map-based information about the cargo of the airplane. Schiphol Airport is one of the five main cargo hubs and most of the European flight cargo goes trough these main ports (Fig. 8).

The proposed solution is to combine IATA handling codes[5] with the proposed set of emergency icons. The IATA codes are available 30 min before landing at any airport and they are internationally acknowledged. Fire brigade experts have also recognised the relevance of those codes to identify potential threats in the airplane within a very short time. The Handling codes are also accompanied with additional information. While the fire trucks (crash tenders) position themselves to cool down the plane and extinguish the kerosene fire, the Customs Control Centre is able to retrieve additional data about the location of the cargo in the plane before the fire fighters enter (Fig. 9).

[5] http://www.iata.org/ps/publications/pages/code-search.aspx.

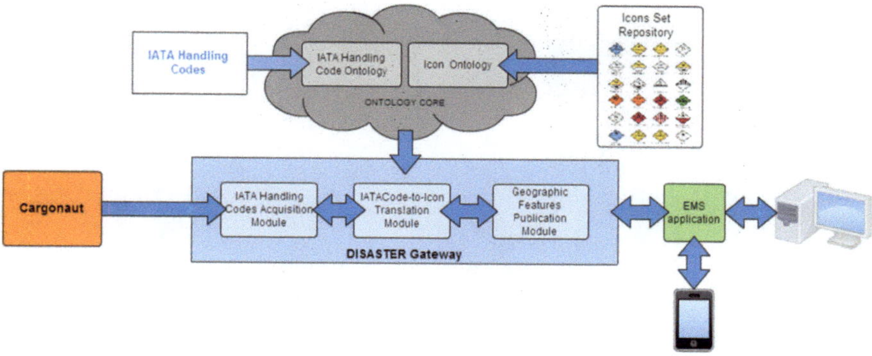

Fig. 9 IATA handling codes on-the fly translation into different icons for different roles in the emergency management process

2 Future Tests and Research

We suggest the European standard set of icons would not only help to create more understanding between different countries, but also between different agencies involved in a major incident. Further research should explore the opportunities of combining GMES with regional sources. The test case with handling codes would substantiate some of the ideas of semantic solutions to improve emergency management. The proposed core ontology of icons will be one of the public emergency management deliverables of the EU-funded project Disaster. It will be tested in a major exercise in 2013.

Network Information Management for Collaboration in Disaster Management: Concepts and Case Study

Fons Panneman, Erik de Vries and Peter de Bruijn

Abstract Interorganizational information exchange has become crucial to increase resilience and agility in disaster management. New collaborations between organizations to exchange information through innovative technologies create challenges for information management. It implies different stakeholders, diversified policies, multiple interpretations of innovative technologies, restricted knowledge of each other's organizations and shared governance of processes of innovation, cooperation, design and decision making. We offer an approach for Network Information Management that allows for and takes advantage of these differences between stakeholders and their organizations. An approach in which innovation, cooperation, design and decision-making is shared among stakeholders. The approach is based on three basic components: process management, scenario based development and organizing vision. These components reinforce each other resulting in convergence of stakeholder's positions throughout the process. The approach has been applied and tested in an innovation project for disaster management in the Netherlands, which serves as a case study in this paper.

Keywords Scenario based development · Network information management · Innovation management

F. Panneman (✉)
Stichting Studio Veiligheid, Saturnusstraat 60, 2516 AH Den Haag, The Netherlands
e-mail: Fons.panneman@studioveiligheid.net

P. de Bruijn
e-mail: Peter.debruijn@studioveiligheid.net

E. de Vries
Haagse Hogeschool, Johanna Westerdijkplein, 2521 EN Den haag, The Netherlands
e-mail: E.j.devries@hhs.nl

S. Zlatanova et al. (eds.), *Intelligent Systems for Crisis Management*, Lecture Notes in Geoinformation and Cartography, DOI: 10.1007/978-3-642-33218-0_29,

1 Introduction

Hazard prediction, risk mitigation and crisis response increasingly ask for collaboration between different organizations, agencies or stakeholders and have become an issue of interorganizational information exchange. The recent fire disaster at Moerdijk, nearby Europe's largest harbor Rotterdam (The Netherlands) showed over again the need for information exchange between public and private organizations (Nationale Raad voor Veiligheid) to improve crisis response and to minimize economic damage [1]. Increasingly organizations create alliances to share information. This accounts for organizations involved in disaster management (the demand side) as well as for organizations providing the technology for disaster management (technology partners). On the demand side, private and public institutions need to cooperate to exchange information by applying new technologies. These technologies can support cooperation in existing business chains, but also enable new collaborations that were previously considered impossible. Technology partners on their side need to collaborate to develop an ecosystem of innovative technologies, simply because it is rarely one technology that provides the innovation. It is the 'new combination' of organizations and technologies that lead to innovation [2].

Information management is the primary discipline in organizations concerned with relating patterns of information exchange between organizations to innovative technological innovations supporting this exchange. Information management is defined as the integrative, balanced management of different domains related to information usage in and between organizations. It concerns strategic, structural and operational information-related issues, which relate information and communication processes in the organization to supporting ICT [3]. This view on information management is congruent with that of others in the field who put the usage of information in organizations central and not the technology [4–7].

Innovative collaborations between organizations to exchange information through innovative technologies create challenges for information management. Organizations wanting to innovate to increase collaboration and to reach common goals, face challenges of alignment, conflicting interests and trust [8]. Collaboration requires coordination on processes of information exchange, requirements determination, design and decision-making. Interorganizational collaboration implies different stakeholders, diversified policies, multiple interpretations of innovative technologies, restricted knowledge of each other's organizations and shared governance of processes of innovation and cooperation. Innovative interorganizational collaboration implies ambiguity. Nor the collaboration itself, nor the abilities to enable collaboration with technologies and how the ecosystem of technologies needs to be designed are known in advance. These need to be determined throughout the innovation process. Moreover innovation processes are not straightforward. They tend to be iterative and multi interpretable. Thus, interorganizational innovation breaks with organizational routines and is ambiguous, which ask for processes of sense making [9]. Therefore, information management faces the challenge of networking and sense making, a process of getting to a collective vision on information exchange between organizations [10]. We call this Network Information Management (NIM).

In this paper we describe an approach for Network Information Management based on three basic components: process management, scenario based development and organizing vision. This approach is being applied and tested in an innovation project for disaster management in the Netherlands, which serves as a case study in this paper. The paper ends with discussion and conclusions. With our NIM approach we contribute to knowledge and practice in the field of information management for disaster management specifically and on knowledge on information management in networks of collaborating organizations in general.

2 A Network Information Management Approach

Network Information Management (NIM) is an approach in which process management, scenario based development and the development of an organizing vision is central. The three reinforce each other.

2.1 Process Management

The application of innovative technology to achieve new collaborations between organizations (eventually even between organizations that did not collaborate before) is not unambiguous. Involved stakeholders will have different views on how they might benefit from the collaboration and/or the technology. Furthermore, how the technology might enable collaborations needs to be proven. Besides this ambiguity, stakeholders' interest might differ or even conflict and involved parties might work together on other projects, which contradict with the current project. Therefore they interact in a network of mutual dependencies. Project management, although inextricably linked to the application of ICT, has its limitations in these situations. Such situations do not offer the opportunity to define an unambiguous project definition. Unambiguous commissioning is rarely the case and financial business cases are hard to determine. The application of innovative technologies for new collaborations requires freedom for experimentation and sense making. Moreover, interests in this project might contradict with interests of the stakeholders in other collaborations. Any new form of collaboration, eventually with new partners, requires discovery of each other's interests, positions and expectations. Process management is a more sensible approach in such situations [11].

In process management, the focus is on the quality of the decision making *process*, not on the content of decisions, or on reaching predefined goals with predefined means (cf. project management). Process management accommodates stakeholders to bring their interests, problems and visions to the front. The initiator recognizes its interdependence from the other stakeholders fully and encourages them to take initiative themselves. This creates commitment and support. Problem definitions benefit from this approach as stakeholders bring in their perceptions, values and beliefs. Crucial in process management is the design of an open and socially safe environment in which the stakeholders are invited to co-create the agenda and decision-making process while they can protect their core values. Furthermore, it is important to stimulate progress in the process. In our NIM approach this is ensured by working with scenario based development.

2.2 Scenario Based Development

It is important to involve potential recipients of innovative technology as early as possible in the design and development process as their requirements are hard to articulate upfront. From the beginning on the different parties need to be involved in a process in which they discover the innovation potential of the technology for new practices of collaboration. Scenario-Based Development (SBD) is a method to bring technical and customer parties together around a specific problem taken from work practice. A scenario is a description of the dynamics in a work situation, in which actors, backgrounds and assumptions are included in more or less detail. Scenarios allow for interaction, for articulation of meaning. With SBD, commitment and imagination of users from different (public/private) parties, with their specific backgrounds, are encouraged. Mutual understanding and a common vocabulary are co-created [12]. The (future) information technology application and its innovative potential are being created in an interactive process. The outcome of SBD is a demonstrator in which the dynamics of the scenario can be shown, and possibly interacted with. By means of a demonstrator technical researchers can trigger potential customers to understand the added value of new technology for their working practice. This creates a cyclical process in with supply and demand parties interact and articulate user requirements and specifications. The goal of SBD is to show innovative potential of technology and to stimulate articulation of new ways of collaboration in the first place, not to prove technological feasibility or to develop a first version of a production system: a common goal of traditional prototyping.

2.3 Organizing Vision

During the different phases of the process that is being management by process management, different demonstrators are build based on mutually defined scenarios. Based on these scenarios stakeholders develop an organizational vision. An organizational vision (OV) is a focal community idea for the application of ICT in organizations. Organizational visions are collective sense making of the innovative application of ICT for organizations or networks of organizations [13]. People recognize what the technology can mean for their organization and the collaboration with other organizations. As such an OV serves as a *memory of the future.* Developments in technology and changes in the organization are not only interpreted through this memory but also created according to this memory. The OV is formed through experiences with creating and interacting with the scenarios and demonstrators. As such the demonstrators serve as a *memory of the present* regarding what is currently possible. As several demonstrators are being developed and stakeholders have build trust in each other and the demonstrated technology, the OV tends to become more visionary, more futuristic. The OV goes beyond today's possibilities of demonstrators and becomes more directive for the articulation of (new) scenarios. The memory of the future begins to direct the creation of memories of the present. OV's leave room for interpretation so that sense making is not being blocked. OV's are plausible stories of the future of the organization in which agendas, policies and core values can be recognized and continued [14]. But at the same time OV's can be used to redirect agendas and policies in the direction of the OV. Demonstrators can be used to test the feasibility of these directions, thus decreasing ambiguity of the situation.

2.4 Reinforcement of the Three Basic Building Blocks in the NIM Approach

Process management, SBD and OVs reinforce each other by incrementally decreasing the ambiguity of the situation whilst undergoing the innovation process. Process management contributes by providing clarity about the decision making process and social safety throughout this process. Scenarios and resulting demonstrators provide stakeholders with the opportunity to bring in what they find important and to decrease ambiguity by showing the abilities of the technology. This provides a memory of the present and secures progress in the process, which is an essential requisite in process management to keep the process going and preventing stakeholders from leaving the process [11]. The OV that develops throughout the process further reduces ambiguity because it tends to become shared amongst stakeholders adding to trust and giving directions to scenario's and demonstrators. Every time demonstrators show the feasibility of the OV, trust in the collaboration and the technology is reinforced and ambiguity is decreased. As ambiguity decreases the OV becomes increasingly what it is meant for: a device for interpretation, mobilization (of people within organizations) and legitimization (of people's actions) [13]. The process management that started quite open closes step by step, eventually resulting in a situation where traditional project management can do the rest.

3 Case Study Slim Verbinden

Slim Verbinden (SV) (Connecting Smartly) is an innovative 2-year research program (2011–2013), which has been established by the Dutch Ministry of Economic Affairs. The program focuses on the use of innovative technology for achieving greater safety for workers and citizens during disasters and crises by providing relevant information from multiple sources. A great number of reputable technology parties are involved (from industry and academia), as well as a number of private and public organizations in different security regions in the Netherlands. Within the program, an ecosystem of various innovative technologies is developed and tested, including multi-agent technology for distributed sensing and multi-sensor data collection, multi-agent middleware software for secure data sharing enabling information providers to control who can access their information, crowdsourcing and self organization, context awareness technology, geographic information visualization and artificial intelligence explanation [15]. Studio Veiligheid, a Dutch foundation stimulating innovation in the field of safety, transport, traffic management and law enforcement, manages the SV program. Our approach for Network Information Management is being applied and tested in this program.

Process Management

At the start of the program it was unclear for the involved parties what the common end result would be. Each participating organization had its own interests. On the demand side, several private and public parties have their own role and responsibility in crisis management. On the technology supply side various parties offer specific innovative technologies and have their own market interest. Types of future collaborations and how technology could enable it were ambiguous and so was the innovation process to arrive at it.

We created an open and secure process leaving room for participative agenda setting, discovery of each other's interests and protection of core values. At the same time we offered enough structure to get to decision-making.

We encouraged the technology partners to bring to the front the innovative potential of an ecosystem of their technologies for disaster management, helping them to discover their mutual strengths and benefits of cooperation. We involved demand side parties in the process to bring in real life disaster management scenarios to stimulate these parties to envision future collaboration opportunities based on new technologies. The technology parties developed demonstrators based on these scenarios to show potential. These demonstrators made it easier for the demand side parties to imagine potential applications. Throughout the process, the interests and viewpoints of the participants converged and demand and technology supply parties began to understand their mutual dependence. Doing so we close the gap between science and practice and pave the way towards working solutions.

Scenario Based Development

One of the scenarios that have been developed during the Slim Verbinden program is the Tata Steel scenario. In collaboration with Tata Steel in IJmuiden, the safety region Kennemerland municipality Velsen and several international logistic partners a scenario is developed around an imaginary 'toxic cloud' incident. This scenario encourages parties to think over their future cooperation and information exchange in crisis management in relation to the innovative Slim Verbinden technology. In this way these parties intensified cooperation in their disaster management, including official communications with citizens.

Another scenario has been developed in cooperation with the Municipality of The Hague, a private security organization and the Police Haaglanden (The Hague and surroundings). At Parkpop, a major pop music event in the city, innovative technology was used to locate and position emergency workers and security guards to route them quickly and effectively to a diverse set of incidents. New ideas emerged on interorganizational cooperation at festivals and created enthusiasm amongst different administrative levels at the participating organizations. Furthermore the users, offering technology partners advantages to bring their technology to market, playfully brought in rather detailed specifications for the technology.

Organizing Vision

At another Slim Verbinden project 'Raebell' a feasibility study of the application possibilities of low airspace radar technology was conducted. Based on scenarios and demonstrators an 'organizing vision' emerged. Initially, scenarios focused on "Prinsjesdag", a major yearly event in which the Dutch Queen rides in her nineteenth century golden carriage through The Hague to the Houses of Parliament to present the Government's plan to Parliament. At first scenarios were focused to protect the Queen, Parliament and the public against low-flying objects. However the scenario and the demonstrations provided parties the opportunity to think further and to express views on future scenarios, which are currently hard to imagine, such as increase in mobile air traffic equipment in lower airspace (like drones or even flying cars). Such a future situation requires new and different ways of coordination and information exchange as well as a different usage of radar technology, eventually in combination with innovative imaging and sensor technology. Throughout the process a memory of the future emerged from the demonstrators, the memories of the present.

4 Conclusion

The case study demonstrated that the combination of process management, scenario-based development and organizing vision is a powerful approach to improve the quality and effectiveness of an innovation process in networks of organizations involved with disaster management. Our approach for Network Information Management is also suitable for other situations where new or changing forms of cooperation need to become envisioned by existing or new participants and in which new technologies are being developed to support this new collaboration. Essentially, NIM designs a process in which cooperation (and accompanying vision) develops *in* a process where parties test their cooperation through concrete scenarios in which innovative technology is being applied. We are convinced that using our approach as a base structure in the innovation process will not only lead to a better fit between developed solution and business process but also lead to a better return on innovation funding.

References

1. Nationale Raad voor Veiligheid, Brand bij Chemie-pack te Moerdijk (Den Haag, 2012)
2. L. Oerlemans, P. Kenis, Netwerken en innovatieve prestaties, in Ondernemen in allianties en netwerken, een multidisciplinair perspectief, J. Boonstra (red.) (Kluwer, Deventer, 2007)
3. R. Maes, An integrative perspective on information management, in *Information Management: Setting the Scene*, ed. by A. Huizing, E. de Vries (Elsevier Science, Oxford, 2007), pp. 15–24
4. C.W. Choo, *The Knowing Organization; How Organizations Use Information to Construct Meaning, Create Knowledge, and Make Decisions* (University England, Oxford, 2006)
5. L.D. Introna, *Management Information and Power* (MacMillan, Hampshire, 1997)
6. D.A. Marchand, W.J. Kettinger, J.D. Rollins, Information orientation: People, technology and the bottom line. Sloan Manag. Rev. **41**, 69–80 (2000)
7. C. Ciborra, *The Labyrinths of Information. Challenging the Wisdom of Systems* (Oxford University Press, Oxford, 2002)
8. J. Grijpink (red.), Geboeid door ketens; keteninformatisering (Lemma, Den Haag, 2007)
9. K.E. Weick, *Sense Making in Organizations* (Sage Publications, Thousand Oaks, 1995)
10. J. Boonstra, Samenwerken in allianties en netwerken; spelen met paradoxen, in Ondernemen in allianties en netwerken, een multidisciplinair perspectief J. Boonstra (red.) (Kluwer, Deventer, 2007)
11. H. de Bruijn, E. ten Heuvelhof, R. in 't Veld, Procesmanagement; over proces- ontwerp en besluitvorming (SDU, Den Haag, 2008)
12. K. Go, J.M. Caroll, The blind men and the elephant: Views of scenario-based system design. Interactions, 45–53 (2004)
13. E.B Swanson, The organizing vision in information systems innovation. Organ. Sci. **8**(5), 458–474 (1997)
14. K.E. Weick, K.M. Sutcliffe, D. Obstfeld, Organizing and the process of sense making. Organ. Sci. **16**(4), 409–421 (2005)
15. P. de Bruin, N. Wijngaards N, *Agent-Enabled Information Provisioning while Re taining Control: A Demonstration*. In this volume (2012)

MIX
Papier aus verantwortungsvollen Quellen
Paper from responsible sources
FSC® C105338

If you have any concerns about our products,
you can contact us on
ProductSafety@springernature.com

In case Publisher is established outside the EU,
the EU authorized representative is:
Springer Nature Customer Service Center GmbH
Europaplatz 3, 69115 Heidelberg, Germany

Printed by Libri Plureos GmbH
in Hamburg, Germany